PLASMA AND FUSION SCIENCE

From Fundamental Research to Technological Applications

PLASMA AND FUSION SCIENCE

From Fundamental Research to Technological Applications

Edited by
B. Raneesh, PhD
Nandakumar Kalarikkal, PhD
Jemy James, MSc
Anju K. Nair, MPhil

Apple Academic Press Inc.
3333 Mistwell Crescent
Oakville, ON L6L 0A2 Canada

Apple Academic Press Inc.
9 Spinnaker Way
Waretown, NJ 08758 USA

©2018 by Apple Academic Press, Inc.

First issued in paperback 2021

Exclusive worldwide distribution by CRC Press, a member of Taylor & Francis Group
No claim to original U.S. Government works

ISBN 13: 978-1-77-463043-3 (pbk)
ISBN 13: 978-1-77-188453-2 (hbk)

Library and Archives Canada Cataloguing in Publication

Plasma and fusion science : from fundamental research to technological applications / edited by B. Raneesh, PhD, Nandakumar Kalarikkal, PhD, Jemy James and Anju K. Nair.
Includes bibliographical references and index.
Issued in print and electronic formats.
ISBN 978-1-77188-453-2 (hardcover).--ISBN 978-1-315-36594-7 (PDF) 1. Plasma (Ionized gases).
2. Nuclear fusion. 3. Controlled fusion. I. Raneesh, B., editor II. Kalarikkal, Nandakumar, editor III.
James, Jemy, editor, IV. Nair, Anju K., editor
QC718.P63 2017   530.4'4   C2017-904218-1   C2017-904219-X

Library of Congress Cataloging-in-Publication Data

Names: Raneesh, B., 1986- editor. | Kalarikkal, Nandakumar, editor. | James, Jemy, editor| Nair, Anju K., editor.
Title: Plasma and fusion science : from fundamental research to technological applications / editors, B. Raneesh, PhD, Nandakumar Kalarikkal, PhD, Jemy James and Anju K. Nair
Description: Oakville, ON ; Waretown, NJ : Apple Academic Press, [2017] |
Includes bibliographical references and index.
Identifiers: LCCN 2017027546 (print) | LCCN 2017030738 (ebook) | ISBN 781315365947 (ebook)
| ISBN 9781771884532 (hardcover : alk. paper) | ISBN 1771884533 (hardcover : alk. paper) | ISBN 1315365944 (eBook)
Subjects: LCSH: Plasma (Ionized gases) | Nuclear fusion.
Classification: LCC QC718.5.P5 (ebook) | LCC QC718.5.P5 P53 2017 (print) | DDC 530.4/4--dc23
LC record available at https://lccn.loc.gov/2017027546

Apple Academic Press also publishes its books in a variety of electronic formats. Some content that appears in print may not be available in electronic format. For information about Apple Academic Press products, visit our website at **www.appleacademicpress.com** and the CRC Press website at **www.crcpress.com**

CONTENTS

ABOUT THE EDITORS

B. Raneesh, PhD
Assistant Professor, Department of Physics, Catholicate College, Pathanamthitta, Kerala, India

Dr. B. Raneesh, PhD, is currently an Assistant Professor in the Department of Physics, Catholicate College, Pathanamthitta, Kerala, India. His current research interests include nanomultiferroics, metal oxide thin films, plasma science, and electron microscopy. He has published many research articles in peer-reviewed journals and co-edited one book. Dr. Raneesh has received his MSc and MPhil degrees from Bharathiyar University, Coimbatore, India, and PhD in physics from Mahatma Gandhi University, Kerala, India.

Nandakumar Kalarikkal, PhD
Director, International and Inter University Centre for Nanoscience and Nanotechnology, Mahatma Gandhi University, Kerala, India

Dr. Nandakumar Kalarikkal, PhD, is the Director of the International and Inter University Centre for Nanoscience and Nanotechnology as well as an Associate Professor in the School of Pure and Applied Physics at Mahatma Gandhi University, Kerala, India. His current research interests include synthesis, characterization, and applications of various nanostructured materials, laser plasma, and phase transitions. He has published more than 80 research articles in peer-reviewed journals and has co-edited seven books. Dr. Kalarikkal obtained his Master's degree in physics with a specialization in industrial physics and his PhD in Semiconductor Physics from Cochin University of Science and Technology, Kerala, India. He was a postdoctoral fellow at NIIST, Trivandrum, and later joined Mahatma Gandhi University, Kerala, India.

Jemy James, MSc
*International and Inter University Centre for Nanoscience and
Nanotechnology, Mahatma Gandhi University, Kottayam, Kerala, India*

Mr. Jemy James obtained his Integrated MSc degree in photonics from
Cochin University of Science and Technology, Kochi, Kerala, India. He is
currently pursuing his PhD program in International and Inter University
Centre for Nanoscience and Nanotechnology at Mahatma Gandhi
University, Kottayam, Kerala, India. His current research interests include
laser plasma and polymer nanocomposites.

Anju K. Nair, MPhil
*Research Scholar, International and Inter University Centre for
Nanoscience and Nanotechnology, Mahatma Gandhi University,
Kottayam, Kerala, India*

Ms. Anju K. Nair is currently a Research Scholar in the International and
Inter University Centre for Nanoscience and Nanotechnology at Mahatma
Gandhi University, Kottayam, Kerala, India. Her current research interests
include graphene-based hybrid structures for sensing applications. She
received her MSc and MPhil degrees in physics from Mahatma Gandhi
University, Kerala, India.

LIST OF CONTRIBUTORS

M. K. Ahmed
Department of Physics, Birjhora Mahavidyalaya, Bongaigaon – 783380, Assam, India,
E-mail: mnzur_27@rediffmail.com

S. Amal
Institute for Plasma Research (IPR), Bhat, Gandhinagar, Gujarat – 382428, India,
E-mail: amal@ipr.res.in

C. P. Anilkumar
Equatorial Geophysical Research Laboratory, Indian Institute of Geomagnetism, Krishnapuram,
Tirunelveli, Tamil Nadu – 627011, India, E-mail: cvgmgphys@yahoo.co.in

Parveen Bala
Department of Mathematics, Statistics & Physics, Punjab Agriculture University, Ludhiana –
141004, India, E-mail: pravi2506@gmail.com

M. Bandyopadhyay
ITER-India, Institute for Plasma Research, A-29, GIDC, Electronic Estate, Sector-25, Gandhinagar –
382025, Gujarat, India, E-mail: mbandyo@yahoo.com

Ujjwal K. Baruah
Institute for Plasma Research (IPR), Bhat, Gandhinagar, Gujarat – 382428, India

S. Belsare
Divertor and First Wall Technology Development Division, Institute for Plasma Research, Bhat,
Gandhinagar – 382428, India

Anil Bhardwaj
Space Physics Laboratory (SPL), Vikram Sarabhai Space Centre (VSSC), Thiruvananthapuram –
695022, Kerala, India

Kedar Bhope
Divertor and First Wall Technology Development Division, Institute for Plasma Research, Bhat,
Gandhinagar – 382428, India

S. Borthakur
Centre of Plasma Physics, Institute for Plasma Research, Sonapur, Kamrup, Assam – 782402, India

T. K. Borthakur
Centre of Plasma Physics, Institute for Plasma Research, Sonapur, Kamrup, Assam – 782402, India,
E-mail: tkborthakur@yahoo.co.uk

Christian Brandt
Center for Energy Research, University of California at San Diego, San Diego, CA 92093, USA

A. Chakraborty
ITER-INDIA, Institute for Plasma Research, Bhat, Gandhinagar 382428, Gujrat, India,
E-mail: arun.chakraborty@iter-india.org

M. Chakraborty
Centre of Plasma Physics-Institute for Plasma Research, Nazirakhat, Sonapur – 782402, Kamrup, Assam, India, E-mail: monojitc@yahoo.com

Vishnu Chaudhary
Institute for Plasma Research, Gandhinagar – 382428, India

Prashant Chauhan
Department of Physics and Material Science and Engineering, Jaypee Institute of Information Technology, Noida – 201307, UP, India, E-mail: prashant.chauhan@jiit.ac.in

Ankur Chowdhury
Applied Physics Division, Bhabha Atomic Research Centre, Trombay, Mumbai – 400085, India, E-mail: ankurc@barc.gov.in

B. K. Das
Centre of Plasma Physics-Institute for Plasma Research, Nazirakhat, Sonapur – 782402, Kamrup, Assam, India, E-mail: bdyt.ds@rediffmail.com

P. Das
Ravenshaw University, Cuttack, Odisha – 753003, India

Pankaj Deb
Energetics and Electromagnetics Division, Bhabha Atomic Research Centre, IDA Block B, 4th Cross Road, Autonagar, Visakhapatnam – 530012, India, E-mail: pankajdeb24@gmail.com

Apul Narayan Dev
Centre for Applied Mathematics and Computing, Siksha 'O' Anusandhan University, Khandagiri, Bhubaneswar – 751003, India, E-mail: apulnarayan@gmail.com

E. Savithri Devi
School of Pure and Applied Physics, Mahatma Gandhi University, Kottayam – 686560, Kerala, India

J. Ghosh
Institute for Plasma Research, Bhat, Gandhinagar, Gujarat – 382428, India

Tarsem Singh Gill
Department of Physics, Guru Nanak Dev University, Amritsar – 143005, India, E-mail: tarsemgill50@gmail.com

Deepika Goel
Department of Physics and Material Science and Engineering, Jaypee Institute of Information Technology, Noida – 201307, UP, India, E-mail: deepika7nov@yahoo.co.in

Alexey Goncharov
Institute of Physics NAS of Ukraine, Kiev, 03650, Ukraine, E-mail: gonchar@iop.kiev.ua

K. G. Gopchandran
Associate Professor, Department of Optoelectronics, University of Kerala, Kariavattom, Thiruvananthapuram, India, E-mail: satheeshr83@gmail.com

RF-ICRH Group
Institute for Plasma Research, Bhat, Gandhinagar – 382428, India, E-mail: parihar@ipr.res.in

Satish C. Gupta
Applied Physics Division, Bhabha Atomic Research Centre, Trombay, Mumbai – 400085, India, E-mail: satish@barc.gov.in

J. P. Gurung
Department of Natural Science, Kathmandu University, Dhulikhel, Nepal

P. Hazarika
Centre of Plasma Physics-Institute for Plasma Research, Nazirakhat, Sonapur – 782402, Kamrup, Assam, India, E-mail: hazarikaparismita@rediffmail.com

H. M. Jadav
Institute for Plasma Research, Bhat, Gandhinagar – 382428, India

Ashutosh Jaiswar
Applied Physics Division, Bhabha Atomic Research Centre, Trombay, Mumbai – 400085, India, E-mail: ashuj@barc.gov.in

Zdeňka Jeníková
Czech Technical University of Prague, Faculty of Mechanical Engineering, Department of Materials Engineering, Karlovo náměstí 13, CZ-120 00 Prague, Czech Republic, E-mail: Zdenka.Jenikova@fs.cvut.cz

Ramesh Joshi
Institute for Plasma Research, Bhat, Gandhinagar – 382428, India

M. D. Kale
Applied Physics Division, Bhabha Atomic Research Centre, Trombay, Mumbai – 400085, India, E-mail: kalemd@barc.gov.in

Latika Kalita
Department of Mathematics, Kamrup Polytechnic, Baihata Chariali, Kamrup, Assam, India, E-mail: latika84k@rediffmail.com

T. C. Kaushik
Applied Physics Division, Bhabha Atomic Research Centre, Trombay, Mumbai – 400085, India, E-mail: tckk@barc.gov.in

S. S. Khirwadkar
Divertor and First Wall Technology Development Division, Institute for Plasma Research, Bhat, Gandhinagar – 382428, India

S. V. Kulkarni
Institute for Plasma Research, Bhat, Gandhinagar – 382428, India

Ajai Kumar
Institute for Plasma Research, Gandhinagar – 382428, India

Manoj Kumar
Institute for Plasma Research, Gandhinagar – 382428, India, E-mail: mkg.ipr@gmail.com

Punit Kumar
Department of Physics, University of Lucknow, Lucknow – 226007, India, E-mail: punitkumar@hotmail.com

Sunil Kumar
Institute for Plasma Research, Bhat, Gandhinagar – 382428, India

Hitendra K. Malik
Department of Physics and Electronics, Rajdhani College, University of Delhi, New Delhi – 110015, India

Manesh Michael
School of Pure and Applied Physics, Mahatma Gandhi University, Kottayam – 686560, Kerala, India

Prakash Mokaria
Divertor and First Wall Technology Development Division, Institute for Plasma Research, Bhat, Gandhinagar – 382428, India

S. P. Nayak
Applied Physics Division, Bhabha Atomic Research Centre, Trombay, Mumbai – 400085, India, E-mail: spnayak@barc.gov.in

N. K. Neog
Centre of Plasma Physics, Institute for Plasma Research, Sonapur, Kamrup, Assam – 782402, India

R. Paikaray
Christ College, Cuttack, Odisha – 753001, India

Arun Pandey
ITER-INDIA, Institute for Plasma Research, Bhat, Gandhinagar – 382428, Gujarat, India, E-mail: arun.pandey@ipr.res.in

Chesta Parmar
Institute for Plasma Research, Gandhinagar – 382428, India

Alpesh Patel
Divertor and First Wall Technology Development Division, Institute for Plasma Research, Bhat, Gandhinagar – 382428, India

Nikunj Patel
Divertor and First Wall Technology Development Division, Institute for Plasma Research, Bhat, Gandhinagar – 382428, India

R. K. Pensia
Department of Physics, Govt. Girls College, Neemuch (M.P.), 458441, India

S. Pradhan
Institute for Plasma Research, Bhat, Gandhinagar, India

Upendra Prasad
Institute for Plasma Research, Bhat, Gandhinagar, India, E-mail: upendra@ipr.res.in

O. P. Sah
Department of Physics, Birjhora Mahavidyalaya, Bongaigaon – 783380, Assam, India, E-mail: opbngn@gmail.com

G. Sahoo
Stewart Science College, Cuttack, Odisha – 753001, India, E-mail: gsahoo@iopb.res.in

Vivek Sajal
Department of Physics and Material Science and Engineering, Jaypee Institute of Information Technology, Noida – 201307, UP, India, E-mail: vsajal@rediffmail.com

S. Samantaray
Ravenshaw University, Cuttack, Odisha – 753003, India

S. Sankar
Assistant Professor, Department of Physics, S. N. College, Chathannur, Karamkode (P.O.), Kollam, India

A. Sanyasi
Institute for Plasma Research, Bhat, Gandhinagar Gujarat, 382428, India

Jnanjyoti Sarma
Department of Mathematics, R. G. Baruah College, Guwahati – 781025, Assam, India,
E-mail: jsarma_2001@yahoo.com

R. Satheesh
Assistant Professor, Department of Physics, S. V. R. NSS College, Vazhoor, T. P. Puram (P.O.),
Kottayam, India

Kumar Saurabh
Institute for Plasma Research (IPR), Bhat, Gandhinagar, Gujarat – 382428, India

Sijo Sebastian
School of Pure and Applied Physics, Mahatma Gandhi University, Kottayam – 686560, Kerala, India

A. Shrestha
Department of Natural Science, Kathmandu University, Dhulikhel, Nepal

R. Shrestha
Department of Physics, Basu H.S.S./Basu College, Kalighat, Byasi, Bhaktapur, Nepal,
E-mail: rajendra.ts2002@gmail.com

Anurag Shyam
Energetics and Electromagnetics Division, Bhabha Atomic Research Centre, IDA Block B,
4th Cross Road, Autonagar, Visakhapatnam – 530012, India

Abhisek Kumar Singh
Department of Physics, University of Lucknow, Lucknow – 226007, India

D. B. Singh
Laser Science and Technology Center, Metcalfe House, New Delhi – 110054, India,
E-mail: dbsingh2@rediffmail.com

Divya Singh
Department of Physics, PWAPA Laboratory, Indian Institute of Technology, New Delhi – 110016,
India, E-mail: divyasingh1984@gmail.com

K. P. Singh
Divertor and First Wall Technology Development Division, Institute for Plasma Research, Bhat,
Gandhinagar – 382428, India, Tel.: +91-79-2396-2107; E-mail: kpsingh@ipr.res.in

Manoj Singh
Institute for Plasma Research, Bhat, Gandhinagar – 382428, India

Sanjay Singh
Applied Physics Division, Bhabha Atomic Research Centre, Trombay, Mumbai – 400085, India,
E-mail: sanjay@barc.gov.in

Shiv Singh
Department of Physics, University of Lucknow, Lucknow – 226007, India

Petr Špatenka
Czech Technical University of Prague, Faculty of Mechanical Engineering, Department of Materials
Engineering, Karlovo náměstí 13, CZ-120 00 Prague, Czech Republic,
E-mail: Petr.Spatenka@fs.cvut.cz

G. Sreekala
School of Pure and Applied Physics, Mahatma Gandhi University, Kottayam – 686560, Kerala, India

Y. S. S. Srinivas
Institute for Plasma Research, Bhat, Gandhinagar – 382428, India

D. P. Subedi
Department of Natural Science, Kathmandu University, Dhulikhel, Nepal

Rajesh Subedi
Department of Science and Humanities, College of Science and Technology, Rinchending, Phuentsholing – 450, Bhutan, E-mail: rajesh@cst.edu.bt

Dass Sudhir
ITER-INDIA, Institute for Plasma Research, Bhat, Gandhinagar 382428, Gujarat, India, E-mail: dass.sudhir@iter-india.org

D. L. Sutar
ResearchScholar Mewar University Gangrar, Chittorgarh, 312901, Rajasthan, India, E-mail: Devilalsutar833@gmail.com

N. Talukdar
Centre of Plasma Physics, Institute for Plasma Research, Sonapur, Kamrup, Assam – 782402, India

V. L. Tanna
Institute for Plasma Research, Bhat, Gandhinagar – 382428, India

SST-1 Team
Institute for Plasma Research, Bhat, Gandhinagar – 382428, India

Urmil M. Thaker
Institute for Plasma Research (IPR), Bhat, Gandhinagar, Gujarat – 382428, India

Saikat Chakraborty Thakur
458, EBU-II, CER – UCSD, 9500 Gilman Drive, Mail Code 0417, La Jolla, CA 92093, USA, E-mail: saikat@ucsd.edu

George R. Tynan
Center for Energy Research, University of California at San Diego, San Diego, CA 92093, USA

Taťána Vacková
Czech Technical University of Prague, Faculty of Mechanical Engineering, Department of Materials Engineering, Karlovo náměstí 13, CZ-120 00 Prague, Czech Republic, E-mail: Tatana.Vackova@fs.cvut.cz

Gautam R. Vadolia
Institute for Plasma Research, Near Indira Bridge, Bhat, Gandhinagar – 382428, Gujarat, India, E-mail: gautamv@ipr.res.in

Anshu Varshney
Department of Physics and Material Science and Engineering, Jaypee Institute of Information Technology, Noida – 201307, UP, India, E-mail: anshu.varshney@jiit.ac.in

Chandu Venugopal
School of Pure and Applied Physics, Mahatma Gandhi University, Kottayam – 686560, Kerala, India

Vipin K. Yadav
Space Physics Laboratory (SPL), Vikram Sarabhai Space Centre (VSSC), Thiruvananthapuram – 695022, Kerala, India, E-mail: vipin_ky@vssc.gov.in

LIST OF ABBREVIATIONS

AO	analog output
APPJ	atmospheric pressure plasma jet
ASME	American Society of Mechanical Engineers
BHEL	Bharat Heavy Electricals
BRFST	Board of Research for Fusion Science and Technology
CCD	charge couple device
CCP	capacitive coupled plasma
CCU	camera control unit
CMERI	Central Mechanical Engineering Research Institute
CMTI	Central Manufacturing Technology Institute
CNM	cold neutral medium
CPS	compact plasma system
CSDX	Controlled Shear De-correlation eXperiment
DA	dust acoustic
DAQ	data acquisition
DC	direct current
DIA	dust ion acoustic
DM	de-mineralized
DO	digital output
DPD	double plasma device
EAWs	electron-acoustic waves
EEC	electrically exploded conductor
EEF	electrically exploded foil
ESCA	electron spectroscopy for chemical analyses
ESL	equivalent series inductance
FLR	finite Larmor radius
FTL	flat top laser
FWHM	full width at half maximum
GL	Gaussian laser
GUI	graphical user interface
HCG	hot coronal gas

HDPE	high density polyethylene
HEAT	helicon electrodeless advanced thruster
HIM	hot ionized medium
HIP	hot isostatic press
HV	hardness value
HV	high voltage
IASs	ion-acoustic solitons
IASWs	ion acoustic solitary waves
IAW	ion-acoustic wave
ICE	International Cometary Explorer
ICP	inductively coupled plasma
IMD	internal Mg diffusion
IPM	interplanetary medium
IPR	Institute for Plasma Research
ISM	interstellar medium
ITER	International Thermonuclear Experimental Reactor
KASWs	kinetic Alfvén solitary waves
KdV	Schamel Korteweg-de Vries
KH	Kelvin-Helmholtz
LHe	liquid helium
LIF	laser-induced fluorescence
LPA	low power amplifier
LTD	linear transformer drivers
MAW	magnetoacoustic waves
MITL	magnetically insulated transmission lines
mK-P	modified Kadomstev-Petviashivili
MW	microwave
NBI	neutral beam injector
NDT	non-destructive testing
OD	outer diameter
PC	personal computer
PC	polycarbonate
PE	polyethylene
PEEW	pulsed electrical exploding wire
PFC	plasma facing component
PFN	pulse forming network

PU	polyurethane
QHD	quantum hydrodynamic
R&D	research and development
RDW	resistive drift wave
RF	radio-frequency
RPT	reductive perturbation technique
RRR	residual resistance ratio
RT	Rayleigh Taylor
SC	superconductor
SEM	scanning electron micrographs
SOL	scrape-off-layer
SP	Sagdeev potential
SPS	spark plasma sintering
SPW	surface plasma waves
SS	stainless steel
SWs	solitary waves
TCT	thermal cyclic test
TDE	time delay estimation
TMF	transverse magnetic field
UT	ultrasonic testing
VIM	vacuum induction melting
VPS	vacuum plasma spraying
WIM	warm ionized medium
WKB	Wentzel-Kramers-Brillouin
WNM	warm neutral medium
XRD	x-ray diffractometer

PREFACE

Recently, plasma science and nanotechnology have attracted the attention of researchers because of the enormous applications of these techniques in the modern world to create novel power sources and devices that have unique applications. Plasma is often called as "the fourth state of matter," along with solid, liquid, and gas. Plasma science is the study of the ionized states of matter. Most of the observable matter in the universe is in the plasma state. Because plasmas are conductive and respond to electric and magnetic fields, and can be efficient sources of radiation, they are usable in numerous applications where such control is needed or when special sources of energy or radiation are required. Much of the understanding of plasmas has come from the pursuit of controlled nuclear fusion and fusion power, for which plasma physics provides the scientific basis.

This book is the collection of peer-reviewed papers presented at the 29th National Symposium Plasma Science & Technology and International Conference on Plasma and Nanotechnology (PLASMA-2014) held in Kottayam, India. In fact, this edited book showcases the research of an international roster of scientists and students. The conference was organized by International and Interuniversity Centre for Nanoscience & Nanotechnology at Mahatma Gandhi University, Kottayam, Kerala. The conference had over 500 delegates from all over the world—good representation from France, Slovenia, Australia, USA, Netherlands, Germany, Iran, Sweden, Spain, and Italy; and as well as a substantial number of eminent Indian scientists. The conference was designed so as to generate excellent opportunities for international researchers interested in plasma science and technology to meet and discuss issues related to current developments in the field of power generation. The goal of the conference emphasizes interdisciplinary research on plasma science and nanotechnology.

This volume, *Plasma and Fusion Science: From Fundamental Research to Technological Applications* collects together a selection of 29 papers presented during the conference. The papers include a wide variety of high attractive topics in the plasma science with emphasis on basic plasma physics, computer modeling for plasma, exotic plasma (including dusty plasma), industrial plasma applications, laser plasma, nuclear fusion technology, plasma diagnostics, plasma processing, pulsed power, space astrophysical plasma, and plasma and nanotechnology. In this important work, an excellent big team of international experts provide an exploration of the emerging plasma science that are poised to make the plasma technology to become a reality in the manufacturing sector. We trust that this special issue will stimulate new ideas, methods, and applications in the field of plasma science and technology.

We would like to thank all who kindly contributed their papers for this issue and the editors of Apple Academic Press for their kind help and co-operation. We are also indebted to the Apple Academic Press editorial office and the production team for their assistance in the preparation and publication of this issue.

PART I

BASIC PLASMA

CHAPTER 1

PLASMA TREATMENT OF POWDER AND GRANULATES

PETR ŠPATENKA, TAŤANA VACKOVÁ, and ZDEŇKA JENÍKOVÁ

Czech Technical University of Prague, Faculty of Mechanical Engineering, Department of Materials Engineering, Karlovo Náměstí 13, CZ-120 00 Prague, Czech Republic, E-mail: Petr.Spatenka@fs.cvut.cz, Tatana.Vackova@fs.cvut.cz, Zdenka.Jenikova@fs.cvut.cz

CONTENTS

1.1 INTRODUCTION

The cold plasma surface modification is a low-cost and highly effective technology for influencing surface properties of polymers without altering the bulk material. Prevalent industrial applications for surface modification of bulk polymers—for example, packaging foils, headlights, or automobile bumpers preceding metallization—are frequently used. But for many other different applications, such as basic construction materials, synthetic polymer fillers, agriculture, pharmaceutical, cosmetic, paintings, and food industry polymer the materials are required in the form of granules or powder. The worldwide annual production of all kinds of granules and powders reaches approximately 10 billion metric tons [1]; nevertheless, the plasma modification of polymer powders has not found plenty of applications as modification of flat bulk polymers. This is due to the problems connected with (i) three-dimensional geometry of powders, (ii) their large surface area, and (iii) the aggregation phenomenon; and therefore, there is a necessity of solid mixing of treated powder.

The reactors designed for plasma powder treatment can be divided into three main groups. The first one is based on an application of mechanical mixture of the powder. As an example, a radio-frequency (RF) plasma reactor consisting of a large rotating and rocking Pyrex glass vacuum chamber can be mentioned. The RF power is coupled with a coil or clamshell electrode arrangement. Batches of several thousands of rubber parts—for example, intended for the pharmaceutical industry—are loaded into the reactor in this process. The uniformity of the surface modification is assured by the motion of the reactor while the batch is treated in a propylene-helium mixture [2, 3]. Another type of reactors used for the mechanical mixing is described elsewhere [4]. The RF or microwave (MW) antennas are placed on the top of the plasma reactor and the system employs continuous mixing during the treatment process. This is accomplished by a mixer arranged inside a tubular stainless steel reactor. One batch contains up to 70 kg of polyethylene powder.

The second group is based on a vertical tubular plasma reactors. The polymer powder is gravity-fed in a continuous stream through the RF plasma zone from a hopper at the top to a collecting bin at the bottom [3, 5]. Such a system was proposed to improve properties of thermoplastic polymer matrix composites. The treatment conditions are fixed and therefore they could be applied for only one type of material. One of the treatment objectives is to incorporate a few atomic percent of chemically bonded oxygen into particles surface. The efficiency of the plasma modification in this type of reactor is approximately 50 kg/h for the particles with diameter of 60 μm on average.

The third group, plasma fluidized bed techniques, is one of the most frequently reported laboratory-scale method [6]. The principle of the reactor is porous plate placed on the reactor bottom on which the powder is spread and the working gas is injected through the porous reactor bottom. Subsequently, the gas passes through a bed of the powder. When the gas flow increases above the critical flow rate the drag on the individual powder particle increases. As a result, the powder starts to move and becomes suspended into the fluid. This state is called fluidization.

Application of low-temperature atmospheric plasma was also demonstrated by several authors. Hladík et al. [7] and Píchal et al. [8] employed dielectric barrier discharge for treatment of polyethylene (PE) particles. The particles passed an approx. 25 cm long channel for several times. They demonstrated hydrophility saturation after 20 transitions of the powder through the plasma channel. Based on electron spectroscopy for chemical analyses (ESCA) they proved incorporation of hydroxyl polar groups on the PE polymer chain. Gilliam et al. [9] treated poly(methylmethacrylate) and polypropylene in the downstream of a low-temperature plasma torch. They proved significant hydrophility enhancement of the both polymers. Addition of 10 g/min of water flow into the torch resulted in subsequent decrease of the measured contact angle from 110° to 100°. Spillmann et al. [10] presented a sophisticated device based on inductively coupled discharge excited in fluidized bed reactor. This type of reactor was used by Roth et al. [11] for powder encapsulation by SiO_2. The reactor operates with a capacity of 2 kg/hour.

Most of the published papers describe laboratory-scale experiments, where the portion of treated material is of several tens of grams. For real applications, the industrial/scaled-up devices with a capacity at least of some hundreds kilograms per day are necessary. The objective of this chapter is to analyze the conditions for the powder treatment in industrial-scale application, to present an industrial scale reactor, and to demonstrate some examples of plasma treated powder applications.

1.2 THEORETICAL BACKGROUND

A powder represent a body with a very large total surface area in comparison with bulk materials or foils. To ensure treatment of the whole surface of each particle, the whole particle should be in contact with plasma (or more precisely with radicals produced by plasma). Two methods are usually applied—mechanical mixing and fluidized bed reactor. A mechanical mixing is convenient for the larger particles; with lowering particle diameter fluidization is necessary. A detailed calculation can clarify the limits of the mechanical mixing.

Assuming free packing of the powder particles in a container, the particles do not fill the whole container volume. A quantity of the powder can be characterized by a filling factor. The filling factor is defined as a ratio of volume summation of all powder particles in the container to the container volume. Provided the density of the particles is constant, the filling factor F can be expressed as:

$$F = V_{powder}/V_{box} = m_{powder}/m_{bulk}$$

where V_{powder} and V_{box} are the volumes of powder in container, respectively; m_{powder} and m_{bulk} are the masses of powder of a bulk body occupying the whole container volume, respectively. Typically, the filling factor varies between 0.45–0.64 [12]. The total number of all powder particles N in the container can be determined as:

$$N = (V_{box} F)/(4/3\ \pi R^3)$$

where R is the particle diameter. The total surface area S_{total} of particles present in the container can be then expressed as a multiple of the total number of powder particles N and the surface area of a powder particle $S_{particle}$:

$$S_{total} = N.S_{particle} = N.4\pi R^2 = \frac{V_{box}F}{\frac{4}{3}\pi R^3}.4\pi R^2 = \frac{3V_{box}F}{R}$$

It follows that the total surface area increases with the particle diameter decrease. Thus, reduction of the particle diameter results in increase of the treated total surface area. Consequently, it leads to a prolongation of the process time. Application of the fluidized bed reduces the filling factor. This leads to the reduction of the total powder amount in the reactor and thus further reduction of the process efficiency. For example, Bretagnol et al. [6] describes plasma treatment of 3 g of a low density PE in the fluidized bed reactor and the process reaches the saturation after 200 s.

Not surprisingly, the random filled powder in a plasma reactor does not form a compact body, but resembles a porous material. Although the direct contact of the treated bulk material with the plasma is assumed as a necessary condition of successful plasma treatment; in several references, the penetration of active species into the porous material is mentioned [13–15]. This indicates that the penetration of the active species under the upper layer of the powder can treat particles placed in deeper layers even without any fluidization. As it is shown in Spatenka et al. [16], certain degree of the powder hydrophilization was observed even at 10 mm deep under the surface layer (Figure 1.1).

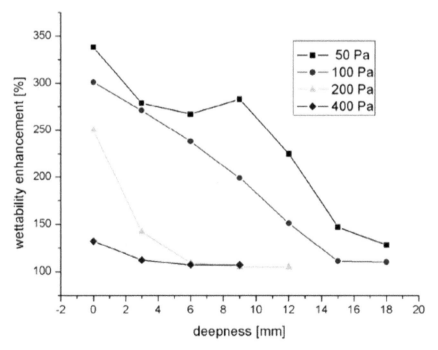

FIGURE 1.1 Dependence of the wettability enhancement on the deepness for different pressure. MW discharge, working gas O_2 treatment time 2 min., particle diameter 200 μm.

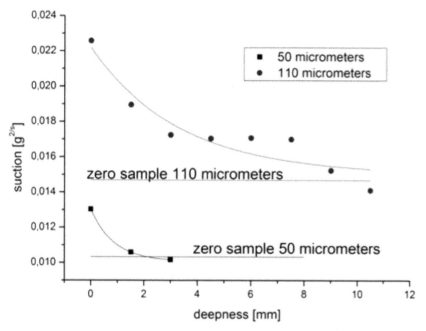

FIGURE 1.2 Dependence of the suction on the deepness for two different powder particle sizes.

According to our estimations, in case of 250 μm powder, the penetration effect resulted in modification of more than 40 top powdered layers. Such penetration significantly enhances the effectiveness of treatment and potentially reduces the process time. On the contrary, the penetration ability decreases with the working pressure. This could probably happen due to higher quenching of radicals in the plasma bulk. Another fact is that the particle diameter strongly influences the penetration ability. With decreasing particle diameter, the capillary between particular particles became smaller and strongly reduces the diffusion of the active radicals. As it can be seen in Figure 1.2, the active radicals can penetrate to ~8 mm for particles of 110 μm in diameter, whereas only negligible penetration was found for 50 μm particles.

Model calculations of hydroxyl radical diffusion depth performed for different filling factors (Figure 1.3) give similar results. The radical penetration reaches few millimeters for particles of 50 μm for the

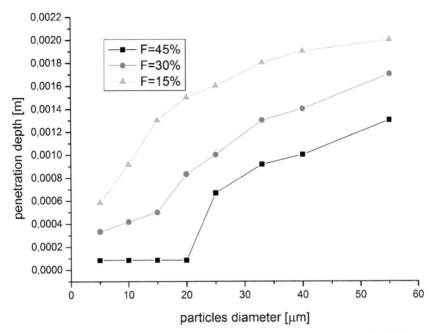

FIGURE 1.3 Dependence of the penetration depth on the particles diameter for different filling factor F.

filling factor F = 45%, which corresponds to the random filling. Further lowering the particle size, it reduces the diffusion depth significantly. No penetration occurs for particles smaller than 20 μm. The penetration increases with lower filling factor. Thus, whereas mechanical mixing is sufficient for plasma treatment of particles larger than 100 μm in diameter, the effective treatment of lower particles in a fluidization is necessary.

1.3 CONCEPT OF PILOT PLANT FOR POWDER TREATMENT

Based on the above experimental and theoretical considerations, a concept for the industrial-scale unit was designed and a pilot plant was constructed. The pilot plant LA650 (*see* Figure 1.4) is based on vacuum process. It is equipped with mechanical stirrer. The plasma is excited by two microwave power sources of the total power up to 2 kW. The

FIGURE 1.4   The pilot plant LA650 (property of the SurfaceTreat a.s.).

sources operate in pulsed regime. Depending on the particle diameter, up to 15 kg of powder can be treated in one batch. The mixing technique enables effective treatment of granulates of several millimeters up to micro-powder with particles smaller than 30 μm in diameter. As an example the high density PE copolymer (Liten ZB29 by Chemapol Litvinov, Czech Republic) was treated for less than 30 min. This means that the total capacity assuming day production time of 12 hours and 200 working days in a year is possible to produce 48 tons of powder per year. In case of higher demands, a 25-kg batch plant is constructed. According to the theoretical calculations, the year production of this plant is more than 100 tons.

1.4 APPLICATIONS

The plasma treatment significantly increases powder wettability, which was quantitatively determined by dynamic capillarity rising measurements—Washburn method. A glass tube with open bottom is filled with powder and immersed into the testing liquid. The resulting wettability is obtained by weighting an enhancement yielded by the penetration of the

FIGURE 1.5 Polymer powder dispersion in water: left test tube (plasma treated), right test tube (original powder).

liquid. The suction of the treated powder was found to be more than 4 times higher than those of the untreated one [17]. Polyolefin powder is usually hydrophobic which restricts its application, for example, for production in composite materials or water-based paintings. The untreated polymer powder mixed in water forms large agglomerates on the water surface. The plasma treated powder is characterized with incorporation of polar groups on its surface resulting in significant enhancement of its hydrophility. Such materials can easily dissolve in water and creates homogeneous suspension as demonstrated in Figure 1.5.

Abrasion and adhesion is of vital importance in the lifetime of a varnish surfaces. Various fillers including silicon or titanium dioxides are tested to enhance the wear resistance of polyurethane paints. Enhancement of powder wettability results in effective dispersion of powder in liquid environment enabling to apply the powder as a filler in water-based paints. The influence of addition of an ultra-high molecular weight PE into water-based paint on the abrasion resistance was tested. A typical result of an abrasion test was performed on the Elcometer 1720 abrasion, scrubbing, and washability tester. Resulting surfaces of PE foils after abrasion tests, according to the norm EN 60730, are demonstrated in Figure 1.6. Paints without the filler evidences a significant distortion after 5,300 cycles, whereas addition of 20% of modified filler resulted in standing more than 135,000 cycles.

Elcometer 1720 Abrasion, Scrubbing & Washability Tester
according to the norm EN 60730

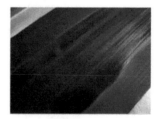

Surface of pure varnish after 30 000 cycles using a boar hair brush tool.

Surface of varnish mixed with 10 percent by volume modified PE powder after 135 000 cycles using a boar hair brush tool.

FIGURE 1.6   Resulting surfaces of water-based varnishes PE foils after abrasion tests.

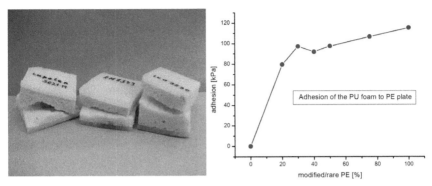

FIGURE 1.7 Illustrative samples of PU isolating foam applied on PE plates. The graph depicts the dependence of the adhesion on the modification percentage of PE plate.

When the plasma treated powder is molded into a solid body by proper method (e.g., rotomoulding), the surface of the final body keeps the higher surface tension of the treated raw powders. Such product can be directly painted; moreover, it exhibits higher adhesion to the glues. Figure 1.7 depicts adhesion of treated powder to polyurethane (PU) isolating foam. The adhesion increases even when only 25% of the plasma treated powder is mixed into the raw material.

1.5 CONCLUSION

The chapter summarized appropriate conditions for powder plasma treatment in industrial-scale applications. Industrial-scale plasma plant producing hydrophilic polymer powders, for example, different types of PE was presented and examples of applications were demonstrated. Some characterization techniques of resulting plasma treatments are also mentioned.

ACKNOWLEDGMENTS

This work was supported by the Ministry of Education, Youth, and Sport of the Czech Republic, program NPU1, project no. LO1207 and project no. CZ.02.1.01./0.0./0.0./16_019/0000826.

KEYWORDS

- cold plasma
- granulate
- hydrophility
- industrial-scale application
- plasma treatment
- polymer
- powder

REFERENCES

1. Duran, J. (2000). Interactions in Granular Media. In: *Sands, Powders, and Grains. An Introduction to the Physics of Granular Materials*. Springer, New York, pp. 19–52.
2. Spatenka, P., Hladík, J., Peciar, M., & Zítko, M. (2009). Plasma treatment of micropowder – from laboratory experiments to production plant. *52nd Annual Technical Conference Proceedings of the Society of Vacuum Coaters, 52*, 537–540.
3. Wertheimer, M. R., Thomas, H. R., Perri, M. J., Klemberg-Sapieha, J. E., & Martinu, L. (1996). Plasmas and polymers – from laboratory to large-scale commercialization. *Pure and Applied Chemistry, 68*(5), 1047–1053.
4. Berger, S., Depner, H. R., Marek, H., Messelhaeuser, J., Welnitz, R., & Wingen, H. (1998). Installation for Low-Pressure Plasma Processing. Patent WO 1998028117 A1.
5. Babacz, R. J. (1993). Continuous Plasma Activated Species Treatment Process for Particulate. US Patent No. US 5,234,723 A.
6. Bretagnol, F., Tatoulian, M., Arefi-Khonsari, F., Lorang, G., & Amouroux, J. (2004). Surface modification of polyethylene powder by nitrogen and ammonia low pressure plasma in a fluidized bed reactor. *Reactive and Functional Polymers, 61*(2), 221–232.
7. Hladík, J., Špatenka, P., Aubrecht, L., & Píchal, J. (2006). New method of microwave plasma treatment of HDPE powders. *Czechoslovak Journal of Physics, 56*(Supplement B), B1120–B1125.
8. Píchal, J., Hladík, J., & Špatenka, P. (2009). Atmospheric-air plasma surface modification of polyethylene powder. *Plasma Processing and Polymers, 6*(2), 148–153.
9. Gilliam, M., Farhat, S., Zand, A., Stubbs, B., Magyar, M., & Garner, G. (2014). Atmospheric plasma surface modification of PMMA and PP micro-particles. *Plasma Processing and Polymers, 11*(11), 1037–1043.
10. Spillmann, A., Sonnenfeld, A., & von Rohr, P. R. (2007). Flowability Modification of Lactose Powder by Plasma Enhanced Chemical Vapor Deposition. *Plasma Processes and Polymers, 4*(S1), S16–S20.

11. Roth, Ch., Künsch, Z., Sonnenfeld, A., & von Rohr, P. R. (2011) Plasma surface modification of powders for pharmaceutical applications. *Surface & Coatings Technology, 205,* S597–S600.

12. Smith, L. N., & Midha, P. S. (1997). A computer model for relating powder density to composition, employing simulations of dense random packings of monosized and bimodal spherical particles. *Journal of Materials Processing Technology, 72*(2), 277–282.

13. Bartoš, P., Volfová, L., & Špatenka, P. (2009). Thin film deposition in limited volume of geometrically complicated substrates. *The European Physical Journal, D 54*(2), 173–177.

14. Friedrich, J. F., Unger, W. E. S., Lippitz, A., Koprinarov, I., Kuhn, G., Weidner, S., & Vogel, L. (1999). Chemical reactions at polymer surfaces interacting with a gas plasma or with metal atoms – their relevance to adhesion. *Surface and Coatings Technology, 119,* 772–782.

15. Krentsel, E., Fusselman, S., Yasuda, H., Yasuda, T., & Miyama, M. (1994). Penetration of plasma surface modification. II. CF_4 and C_2F_4 low temperature cascade arc torch. *Journal of Polymer Science Part A: Plasma Chemistry, 32*(10), 1839–1845.

16. Spatenka, P., Hladík, J., Kolouch, A., Pfitzmann, A., & Knoth, P. (2005). Plasma treatment of polyethylene powder-process and application. *48th Annual Technical Conference Proceedings of the Society of Vacuum Coaters, 48,* 95–98.

17. Horáková, M. Špatenka, P., Hladík, J., Horník, J., Steidl, J., & Polachova, A. (2011). Investigation of adhesion between metal and plasma-modified polyethylene. *Plasma Processing and Polymers, 8,* 983–988.

CHAPTER 2

PLASMA DYNAMICAL DEVICES: REVIEW OF FUNDAMENTAL RESULTS AND APPLICATIONS

ALEXEY GONCHAROV

Institute of Physics NAS of Ukraine, Kiev, 03650, Ukraine,
E-mail: gonchar@iop.kiev.ua

CONTENTS

ABSTRACT

The current status on ongoing research and development of the novel generation plasma dynamical devices based on the axial-symmetric cylindrical electrostatic plasma lens configuration and the fundamental plasma optical principles of magnetic electron isolation and equipotentialization magnetic field lines are briefly reviewed. The experimental, theoretical, and simulation researches have been carried out over recent years collaboratively between IP NASU (Kiev), LBNL (Berkeley, USA), and HCEI RAS (Tomsk). These researches enable detailed and accurate description and modeling of plasma-optical systems, as well as prediction of their novel high-technology applications. This chapter, in detail, describes the latest devices for focusing high-current negative charge particle beams and for filtering dense plasma flow from micro droplets.

2.1 INTRODUCTION

High-current neutralized charged particle beams (positive, negative ions and electrons) and energetic ion-plasma flows are used widely for fundamental investigations and high-technology applications; for example, in heavy fusion research, high current linear accelerators, spacecraft control systems, high dose ion implantation for material surface modification, and ion-plasma treatment.

The fundamental concept of the new generation plasma devices are based on the application of cylindrical plasma lens configuration and plasma-optical principles of magnetic insulation electrons, and maintain the magnetic field lines as equipotentials ("equipotentialization") for the control of electrostatic fields introduced into the plasma medium for manipulating non-magnetized ions. This plasma-optical concept was first described by Morozov [1, 2]. These early contributions also clarified the essential advantages of electrostatic plasma lens as compared with the more traditional electrostatic and magnetic lenses. The experiments have demonstrated mainly that the optical strength of the plasma lens can be up to two orders of the magnitude greater than for an Einzel lens, and up to four orders of the magnitude greater than for a magnetic lens under the same experimental conditions. This lens is a well-explored tool for focusing high-current, large area, and energetic heavy ion beams [3, 4], where the concern of beam space charge neutralization is critical. The crossed electric and magnetic fields inherent the plasma lens configuration provides the suitable method for establishing a stable plasma discharge at low pressure. In this way, using plasma lens configuration, a number of low maintenance, cost effective, and high reliability plasma generation devices using permanent magnets and possessing considerable flexibility towards spatial configuration were developed [5]. These kinds of devices are part of a larger class of plasma devices (plasma accelerators, magnetrons, jet propulsions, and magnetically insulated diodes) that use a discharge in crossed electric and magnetic fields with a closed electron drift for the production, formation, and manipulation of high current beams and ion-plasma flows. In these conditions, the variation of the magnetic field line configuration within device volume and the distribution of electric potential enable the formation and control of high current ion beams while maintaining their quasi-neutrality. This makes the practical applications of such devices more attractive. They can be operated as a stand-alone instrumentation, for example, for liquid crystal alignment on large-area substrates; or as part of an integrated processing system together with magnetron sputtering, for example, for deposition of spectrally-selective coatings on industrial glass. These devices can be applied both for fine ion cleaning, polishing and activation of substrates before deposition, including synthesis new nanomaterials and exotic coatings.

One particularly interesting result of these basic researches was observed as the essential positive potential at the floating substrate. This suggested us to the possibility of an electrostatic plasma lens for focusing high-current beams of negatively charged particles, electrons, and negative ions, based on the use of the dynamical cloud of positive space charge under magnetic isolation electrons [6, 7]. The experimental results demonstrate an attractive possibilities of an application in the positive space charged plasma lens with magnetic electron insulation for focusing and manipulating wide aperture high-current with no relativistic electron beams. For relatively low-current mode for which electron beam space charged less than positive space charged plasma lens, it realizes the electrostatic focusing on passing electron beam. In case of high-current mode when electron beam space charge much more than space charge plasma lens, the lens operates in plasma mode to create transparent plasma accelerating electrode and compensate space charge propagating the electron beam.

The original approach for effective additional elimination of micro droplets in a density flow of cathodic arc plasma is also described [8]. This approach is based on application of the cylindrical plasma lens configuration for introducing in a volume of propagating along axis dense low temperature plasma flow convergent toward axis energetic electron beam produced by ion-electron secondary emission from electrodes of plasma optical tool. The theoretical appraisals and experimental demonstrations that have been carried out at the Institute of Physics NASU provide confidence and optimism that proposed idea for removal and clearing the micro droplet component from dense, low-temperature metal plasma has the high practical potential for elaboration novel state-of-the-art plasma processing for the filtering of microdroplets (or their reduction to the nanoscale) from the dense plasma formed by erosion plasma sources like vacuum arc and laser produced plasma, without losses of plasma production efficiency.

2.2 BASIC PLASMA OPTICAL PRINCIPLES

The fundamental plasma optical principles are based on using of magnetically insulated cold electrons (i.e., transverse mobility $\ll$ parallel mobility) to provide space charge neutralization of the focused ion beam and maintain the magnetic field lines as equipotentials (or equipotentialization). This means that value of the magnetic field is those that following the inequalities are correct $\rho_{Be} \ll R \ll \rho_{Bi}$, where ρ_{Be} and ρ_{Bi} are the electron and ion larmor-radiuses, and R-typical system size, respectively.

The electrostatic plasma lens is an axially-symmetric plasma-optical device with a set of cylindrical ring electrodes located within the magnetic field region, with magnetic field lines connecting ring electrode pairs symmetrically about the lens mid-plane (Figure 2.1). The condition of equipotential magnetic field lines of length $l \gg R$, the lens radius, passing through the axial region of the system and crossing the outermost electrodes (which are grounded) follows from a model in which the lens volume is uniformly filled with cold background of the electrons of density n_e and energetic beam ions of density $n_b = I_b/(eQv_b\pi R^2)$, where eQ is the ion charge, with e as the electronic charge and Q as the ion charge state (1, 2, 3, ...), v_b is the beam ion velocity, and R is the beam radius. This condition can be given as [4]:

$$n_e - \frac{I_b}{eQv_b\pi R^2} = \pm \frac{\varphi_L}{e\pi R^2}$$

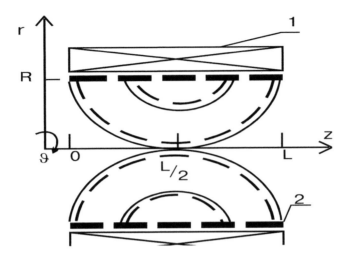

FIGURE 2.1 Schematic of the plasma lens. 1 – magnetic coils; 2 – cylindrical electrodes; dashed lines – equipotentials; solid lines – magnetic field lines.

where φ_L is the maximum electric potential on the ring electrodes. The plus sign corresponds to beam focusing and the minus sign corresponds to beam defocusing – the dispersive operational regime of the lens. Here, we assume Q to be the mean ion charge state, for simplicity. In order that equipotential magnetic field lines are to be generated, an electron density is needed, which is sufficient to compensate both the space charge due to the beam and the vacuum electric potential within the lens volume. It can be seen from the equation that the beam focusing can be obtained for low ion beam density, which is similar to the Gabor lens. In this case, the ion beam density does not play a role and the electrons are used only for space charge compensation and transformation of the external vacuum electric field in order to make the electric field lines transverse to the magnetic field lines for focusing low current positive ion beams. For a high current beam when, if un-neutralized, the beam could blow up under its own space-charge forces, the neutralization can be provided by electrons of sufficient density held within the beam by its space charge. This regime occurs when the beam potential parameter I_b/v_b exceeds significantly the maximum externally applied lens voltage, $I_b/v_b \gg \varphi_L$. In the high beam current regime, quasi-neutral plasma is formed within the lens volume consisting of cold magnetized electrons drifting across the magnetic field lines and fast beam ions that are affected to first approximation only by the radial electrostatic field of the lens. It is noted that the equipotential-ization condition follows from the steady-state hydrodynamic equation of motion of cold electrons, which in this case is

$$E = -1/c(v_d \times B),$$

where v_d is the drift velocity of cold electrons, B is the magnetic field within the lens volume, and c is the velocity of light. The macroscopic electrostatic field E can exist only in the presence of closed electron drift and an "insulating" magnetic field. Then, the electric field is perpendicular to the magnetic field, leading to magnetic field lines that are equipotentials.

The focal length F of this kind of electrostatic plasma lens is given by

$$F = \theta \varphi_b R / 2 \varphi_L$$

where, φ_b is the ion beam accelerating potential (i.e., φ_b is the ion source extractor voltage and the energy of the beam ions is $E_b = eQ\varphi_b$), φ_L is the maximum electric potential on the ring electrodes, and θ is a geometric parameter about unit. Importantly, it is noted that the focal length of the plasma lens does not depend on the ion charge-to-mass ratio; this is a conse-quence of the purely electrostatic optical system, as previously mentioned.

2.3   PLASMA LENS WITH POSITIVE SPACE CHARGE

Plasma-optical principles of introducing spatial over thermal E-fields in the plasma medium of an intense ion beam resulted in essential progress in focusing and manipulating high-current positive ion beams including heavy ion beams with current values in a range of amperes. The plasma lens configuration of crossed electric and magnetic fields provides an attractive method for maintaining a stable plasma discharge at low pressure.

This background work suggested us to the possibility of an electrostatic plasma lens for focusing high-current beams of negatively charged par-ticles, electrons, and negative ions. The lens is based on the main plasma optical principle meaning that electrons are magnetically insulated and non-magnetized, free positive ions create a controlled uncompensated space charge cloud. Such plasma lens with a positive space charge cloud for focusing high-current negatively charged particle beams can be a very attractive and efficient device. The original plasma lens with positive space charge cloud is produced by a toroidal plasma source like an anode layer accelerator (Figure 2.2). In this kind of systems, the electrons are separated from ions by relatively strong magnetic field in the discharge channel. The accelerated ions are weakly affected by the magnetic field owing to their mass. This system can be described by a set of equations:

$$(1/r)\partial\left(r\,\partial U/\partial r\right)/\partial r + \partial^2 U/\partial z^2 = 4\pi q_i n_i$$

$$M_i\frac{dv_i}{dt} = q_i E + \frac{1}{c}\left[v_i \times B\right]$$

$$V \cdot div(j_{out}) = S \cdot j_{in}$$

$$\oint_S E_s dS = \int_V \rho dV$$

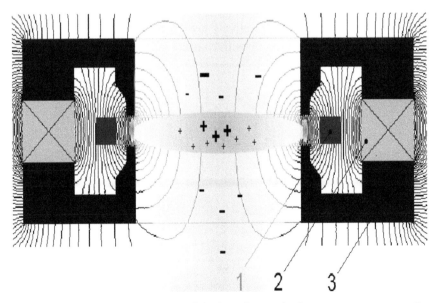

FIGURE 2.2 Simplified schematic of the lens. 1 – anode, 2 – permanent magnets, 3 – magnetic system with cathode.

This theoretical model is used for computer simulations of the steady-state potential distributions in formatting positive space charge cloud and for focusing passing electron beam.

Here the results of wide-aperture (6 cm) non-relativistic (up to 20 keV) intense (from 100 mA up to 100 A), repetitively a pulsed (pulse duration 120 μs) electron beam by focusing the positive space charge plasma lens [6, 7] are briefly described. The experiments have been carried out in the Tomsk (High Current Electronic Institute, SB RAS) with using plasma lens produced by Institute of Physics NAS of Ukraine. Maximum lens of anode potential is 2 kV; the lens discharge current is up to 100 mA; and the working gas Ar, pressure 10^{-5}–10^{-4} Torr.

One can see a typical electron focusing effect by plasma lens both in case of low-current mode and for high-current mode. A typical focusing effect for low-current mode is shown in Figure 2.3 and for high-current mode is shown in Figure 2.4.

These experimental results demonstrate an agreeable possibilities application positive space charged plasma lens with magnetic electron insulation for focusing and manipulating wide aperture, high-current, and no relativistic electron beams.

As mentioned earlier, for relatively low-current mode for which electron beam space charged less than positive space charged plasma lens it realize

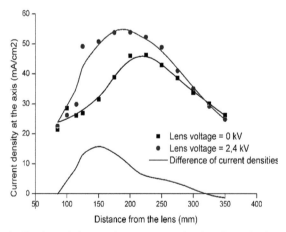

FIGURE 2.3 Distribution of electron beam current density along the beam path. Beam energy – 10 keV, beam current – 200 mA, Magnetic field at the lens center B(0,0) – 50 G, pressure p = 1 × 10–4 Torr. The lens discharge current – 5 mA. Bottom curve – current density difference j(UL=2.4 kV) – j(UL=0).

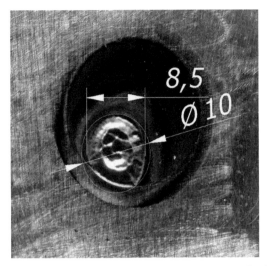

FIGURE 2.4 Imprint of the focused electron beam on the target. The dimensions are in mm. Beam current – 100 A; beam energy – 16 keV; lens voltage – 1 kV; lens current – 100 mA. Substrate is made of stainless steel.

electrostatic focusing passing along axis electron beam. This means that the effect of e-field exceeds the effect of magnetic field. In case of high-current mode when electron beam space charge much more than space charge plasma lens, the lens operates in plasma mode to create transparent plasma accelerating electrode and compensate space charge propagating electron beam. The magnetic field lens in this case is used for effective beam focusing. Under described experimental conditions, the maximal compression factor was up to 30× and beam current density at the focus was about 100 A/cm^2.

The obtained simulation results in closed to experiment conditions are in a good agreement with the experimental data.

2.4 PLASMA FILTER FOR CLEARING MICRO-DROPLETS

As said above, we suggest for the first time, a new more practical approach for the effective removal of the micro-droplets, as well as its conservation and incorporation into the plasma stream is produced by the erosion plasma sources (vacuum arc or laser produced plasma). This approach is based on application of the cylindrical plasma lens configuration for introducing in a volume of propagating along axis dense low temperature plasma flow convergent toward axis energetic electron beam produced by ion-electron secondary emission from electrodes of plasma optical tool. The first experiments and theoretical estimations [8] demonstrate the workability and an idea of application that the new plasma-optical tool based on plasma lens configuration with convergent and oscillating fast electrons for effective additional evaporation, destroying, and clearing of liquid metal droplets in a passing intense flow of dense metal plasma. The simplified schematic of a new plasma-optical system utilizing an electron beam for effective elimination of microdroplets in passing arc metal plasma flow is shown in Figure 2.5. Here **C** is the cathode, **A** is the anode, **D** is the aperture, **MC** is the magnetic coil, **M** is the magnetic field lines, Δ is the spatial layer, in which the strong radial electrical field is supported, and $\mathbf{C}_1$ is the hollow cylinder, from which internal surface generated electrons of secondary ion-electron emission.

The application of a voltage results in the formation of a spatial layer $\Delta \ll \rho_e \equiv eE_r/m_e\omega_{He}^2$ with a large electrical field E_r. This field is mainly directed along with the radius. The presented inequality can be written in more simple form:

$$\omega_{He}/\omega_{pe}(e\varphi_0/T_e)^{1/4}(10/9\pi)^{1/2} \ll 1,$$

where T_e is the plasma electron temperature, φ_0 is the near wall jump of electrical potential. ω_{He}/ω_{pe}-electron cyclotron and electron Langmuir

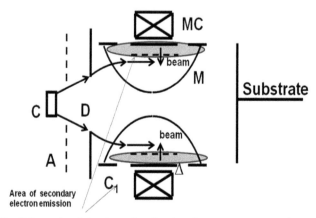

FIGURE 2.5   Schematic of clearing microdroplets from a vacuum arc plasma flow.

frequencies, respectively. From this inequality, it can be seen that in less magnetic field, the result is better. At the same time one needs to provide electron magnetic insulation $\rho_{Be} \ll R$. The calculations show the energy of the formatting electron beam enough for micro droplet elimination during propagating plasma flow through plasma tool.

The experimental prototype of the new plasma-optical device was elaborated, produced and tested in conditions of propagating cathode arc plasma flow. The experiments have been carried out under the following experimental conditions:

- DC discharge arc current: 60–90 A;
- Discharge voltage: 20 V at B = 0 G;
- Discharge voltage: 40 V at B = 360 G;
- Substrate bias = –200 V;
- Substrate area: 1 cm²;
- Additional gas argon with pressure up to 3×10^{-4} Torr;
- Total deposition time during these experiments for different conditions was the same 3 min;
- Residual pressure at vacuum chamber is 3×10^{-5} Torr;
- Distance between substrate and target (arc cathode) is 250 mm;
- Cathode diameter (Cu) is 20 mm.

In Figures 2.6 and 2.7, it is seen that the SEM photos (square of images 250 μm × 330 μm) demonstrate the experimental results on copper deposition onto stainless steel samples.

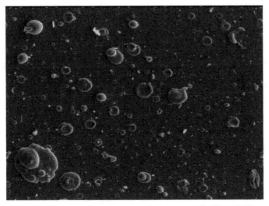

FIGURE 2.6   SEM photo of copper microdroplets B = 0 G, U = 0 V.

FIGURE 2.7   SEM photo of copper microdroplets B = 360 G, U = –900 V.

As menioned earlier, the presented experimental results and theoretical estimations mainly demonstrate an agreeable possibility suggested
approach for devising an effective micro droplets filters new generations.

2.5   CONCLUSION

The purpose of this review is to discuss an achieved contemporaneous
understanding level physical mechanisms determining the progress in
the development of new generation plasma dynamical devices based on
plasma lens configuration and fundamental plasma optical principles,
electron magnetic insulation, and equipotentialization magnetic field lines.
Along with early described plasma devices, two novel plasma optical tools
perspective for state-of-the-art applications for focusing and manipulating
negative charged intense particle beams and for filtering micro droplets
from propagating toward substrate dense plasmas flow formed by erosion
plasma sources were described.

ACKNOWLEDGMENT

This work is supported by the grant of 34-08-14 (Ukraine) and 14-08-
90400 (Russia).

KEYWORDS

- electron magnetic insulation
- erosion plasma sources
- filtering micro droplets
- high current charged beams
- plasma lens
- plasma optics

REFERENCES

1. Morozov, A. I. (1965). Focusing cold quasi-neutral beams in electromagnetic fields. *Dokl. Acad. Nauk. USSR 163*(6), 163–167.
2. Morozov, A. I. (2008). *Introduction to Plasmadynamics*. Fismatlit, Moscow.
3. Goncharov, A. A., & Brown, I. G. (2004). High-current heavy ion beams in the electrostatic plasma lens. *IEEE Trans. Plasma Sci. 32*, 80–83.
4. Goncharov, A. (2013). Invited review article: the electrostatic plasma lens, *Rev. Sci. Instrum. 84*, 021101.
5. Goncharov, A. A., & Brown, I. G. (2007). Plasma devices based on the plasma lens: a review of results and applications. *IEEE TPS 35*, 986–991.
6. Goncharov, A., Dobrovolskiy, A., Dunets, S., Evsyukov, A., Litovko, I., Gushenets, V., & Oks, E. (2011). Positive-space-charge lens for focusing and manipulating high-current beams of negatively charged particles. *IEEE TPS, 39*, 1408–1411.
7. Goncharov, A. A., Dobrovolsky, A. N., Dunets, S. N., Litovko, I. V., Gushenets, V. I., & Oks, E. M. (2012). Electrostatic plasma lens for focusing negatively charged particle beams. *RSI, 83*, 02B723.
8. Goncharov, A. A., Maslov, V. I., & Fisk, A. (2012). Novel plasma-optical device for the elimination of droplets in cathodic arc. *55th SVC Annual Techn. Conference Proceedings*, Santa Clara, CA, USA, 441–444.

CHAPTER 3

CONCEPTUAL DESIGN OF A PERMANENT RING MAGNET-BASED HELICON PLASMA SOURCE

ARUN PANDEY, DASS SUDHIR, M. BANDYOPADHYAY, and A. CHAKRABORTY

ITER-INDIA, Institute for Plasma Research, Bhat, Gandhinagar 382428, Gujrat, India, E-mail: arun.pandey@ipr.res.in, dass.sudhir@iter-india.org, arun.chakraborty@iter-india.org

CONTENTS

ABSTRACT

The Institute for Plasma Research (IPR) (India) has initiated a multi-driver based large size helicon negative ion source R&D program. The program is initiated through a single driver helicon plasma source having permanent ring magnet for the necessary axial magnetic field. A conceptual model for the design of a helicon plasma source is described along with the modeling of the design. Following the design by analysis approach, source parameters are optimized and expected source performances are simulated. These simulations and calculations are done using the computer code HELIC, with a view of development of a helicon plasma source using hydrogen gas. The magnetic field topology for the ring magnet is simulated with another code (BFieldM) and the results obtained are used in the simulation.

3.1   INTRODUCTION

Helicon plasma sources are very promising plasma sources due to their high ionization efficiency [1]. The physics behind such high efficiency is still a subject of investigation. However, due to having high plasma density ($\sim 10^{13}\,\mathrm{cm}^{-3}$) using low RF power ($\sim$ few kW), helicon based plasma sources are used in the fields of plasma processing [2] and space exploration [3].

Worldwide, many groups are continuously contributing great research works in this field over the past few decades [1–8]. The underlying physics of the power absorption and high-density plasma production by the helicon waves is still not well understood and the problem is still under investigation. Most of the experimental investigations are being carried out in the setup equipped with electromagnets. Only few laboratories are working on permanent magnet based helicon experiment [6, 9]. These permanent magnet based helicon source modules have been developed for plasma processing [6] and space application [9]. IPR also has an operational helicon plasma experimental setup with electromagnets. The present work shows the first attempt to build a negative hydrogen ion source for fusion related applications with a permanent magnet based helicon plasma source module. This chapter describes the conceptual design of that permanent magnet based helicon source, to be the setup in IPR.

The Institute for Plasma Research (IPR) is engaged in long term research and development (R&D) programs on fusion related technologies and neutral beam injector (NBI) system is one of them [10]. In large area, high plasma density sources are needed in NBI ion sources

[11]. Current technologies relies on either filament based or inductively coupled plasma (ICP) based ion sources [12]. However, the efficiency of such technologies is low. Due to high ionization efficiency against the power input, helicon source with a multi-driver configuration would be a promising candidate for a large size ion source. The present work is linked to the above-mentioned objective. IPR has initiated a multi-driver based large size helicon negative ion source R&D program. The program is initiated through a single driver helicon plasma source having permanent ring magnet for the axial magnetic field. The conceptual design activity is being carried out using two computer codes HELIC [13] and BfieldM [14]. HELIC computes the power deposition spectra to plasma for a given antenna configuration, magnetic field, and geometrical inputs. In helicon plasma, magnetic field topology inside the source is very important for wave excitation and source operation. Magnetic field distribution due to permanent ring magnets is calculated by BfieldM code. The chapter describes the conceptual design of a single driver permanent magnet based helicon plasma source and also its simulated plasma performance for hydrogen gas using these codes.

3.2 DESCRIPTION OF HELIC AND BfieldM CODE

HELIC is developed by Arnush [13], which is essentially a C++ program for the design of RF plasma sources with and without magnetic field. It predicts the power deposition spectra for given input parameters of plasma density, magnetic field, antenna configuration, and geometry. The program solves four coupled radial differential equations for each k_z to obtain two independent waves, and these are combined by applying the boundary conditions algebraically. It assumes the plasma to be cylindrically symmetric, and the d.c. magnetic field to be axially uniform and does not take into account the mode of coupling of antenna power to the plasma nor does it consider the plasma transport. Maxwell's equations for radially non-uniform plasma with the standard cold-plasma dielectric elements are manipulated in the code.

The basic schematic configuration showing antenna and magnetic field direction is shown in Figure 3.1. The program computes the following quantities: (a) wave variables: magnitudes, components, and phases of **B**,

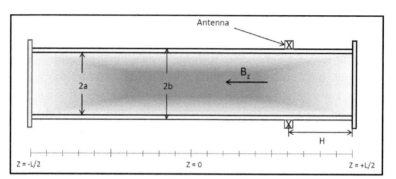

FIGURE 3.1   Geometry of HELIC calculations [14].

E, and **J** for a given azimuthal wavenumber m; (b) power spectra, S(k) P(k), P(r), P(z), where $k = k_z$, P(k) is the power deposition at various k for the given antenna and plasma parameters, S(k) is the response of the plasma at various k, P(r) is the radial profile of power deposition over a given range in z, P(z) is the axial profile of power deposition integrated over cross-section; (c) resistive loading R(n, B) for a range of n and B_0.

The radial and axial profiles of deposited power, P_r and P_z, respectively, are also computed as the output from HELIC. The user can also specify the type of endplate, insulating or conducting type. For the present case, the design is simulated for a configuration with a conducting endplate.

The calculation of the plasma loading is directly related to the power absorbed by the plasma from the antenna. Plasma can be simply approximated as an electrical load in the electrical circuit as shown in Figure 3.2.

The power balance equation gives,

$$P_{in} = P_{rf} \frac{R_p}{R_c + R_p} \tag{1}$$

If $R_c << R_p$, then $P_{in} \sim P_{rf}$; i.e., most of the power from the source goes into the plasma. Hence, in the simulation, the primary focus is on obtaining results corresponding to a high value of R_p, which correspondingly indicates highly efficient plasma in that density regime.

Magnetic field plays an important role in helicon plasma source configuration and its operation. One of the main features of the proposed conceptual design is the use of a permanent ring magnet for the excitation of helicon modes in plasma. NdFeB permanent ring magnet has 4.5 kG

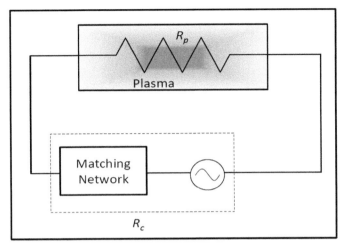

FIGURE 3.2   Schematic of the electrical system showing circuit resistance R_c and the plasma loading R_p.

surface magnetic field sufficient to replace normally used strong electro-magnet and help to get rid of electrical and corresponding cooling water interfaces for the helicon source and make the system simple. Thus, it becomes inevitable to know beforehand the topology of magnetic field lines, for the optimization of the design geometry. The typical field lines from a ring magnet are shown in the Figure 3.2 with respect to the relative position of the discharge tube. Note that the tube is kept at a distance from the magnet because of two reasons: (i) the field lines should extend throughout the discharge tube, which will not be possible in the strong field region close to the magnet; and (ii) the magnetic field should not vary much radially inside the tube.

This input file of BfieldM [15] code contains information about the geometry of the magnet, magnetization of the magnet, and the number of grid points for which the calculation is to be done. This code is exclusively for permanent ring magnet system only.

3.3   RESULTS

HELIC code estimates the power deposition spectra for given input parameters of plasma density, magnetic field, antenna configuration,

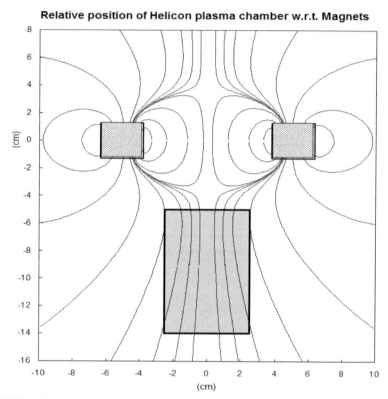

FIGURE 3.3 Magnet field lines from a ring magnet. A rectangular zone depicts the discharge tube location to be placed at some distance from the magnet on the axis.

and geometry. For the present case, uniform density plasma is assumed. For the geometry shown in Figure 3.1, with a = 5 cm, b = 5.5 cm, and a bounded plasma cavity, the following simulations were performed using HELIC. The separation between antenna and the endplate, shown as H, can also be varied. For simplicity, the antenna is placed at the center of the cavity ($z = 0$).

A loop antenna gives better results than other types of antenna, as it shall be shown later (Figure 3.10), hence the plasma performance is simulated with a single loop antenna and a conducting endplate. These input parameters of the antenna type, endplate material, and dimensions of plasma chamber were fixed only after optimization through simulations. Figure 3.4 shows the variation of power absorbed for different radial

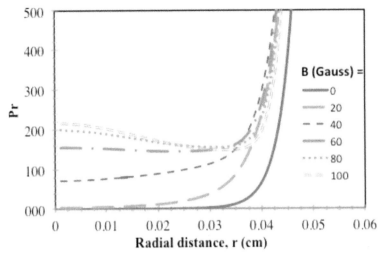

FIGURE 3.4   Excitation of the helicon modes and density peaking on the axis when the magnetic field Bz is increased. B = 0G is the pure inductively coupled plasma configuration.

distances from the center of the discharge tube for different axial magnetic field strength. It can be seen that the power absorbed at the center (near r = 0 cm), which is almost non-existent at lower fields and increases as the field is increased beyond 40G. This marks the beginning of the helicon mode excitation in plasma in presence of axial magnetic field. Therefore, axial magnetic field of value greater than 40G is necessary to start the conceptual design.

Therefore, understanding of the magnetic profile is important to start for the design. Typical field profiles for an available magnet ring of inner diameter 5 cm and outer diameter 10 cm of thickness 2 cm with magnetization is in thickness direction are shown in Figure 3.5.

The plot of the magnetic field data is calculated using BFieldM program. It shows that after a certain z value, the axial variation of B_z is relatively uniform and higher than 40G. This region can be chosen as the position of the discharge tube (as shown in Figure 3.1). As discussed in Section 3.2, the optimized condition for helicon discharge can be realized through R_p estimation. Figure 3.6 shows that the peak of plasma resistance, R_p for different plasma density n_e in presence of different axial magnetic field, shifts towards higher density value as magnetic field increases.

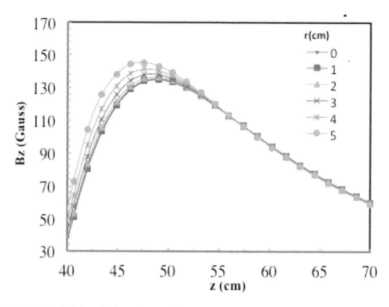

FIGURE 3.5 Axial variation of B_z at different radial positions.

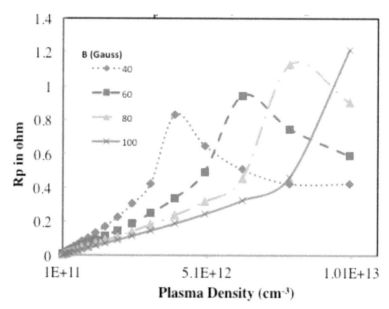

FIGURE 3.6 Variation of R_p for different plasma density conditions in presence of different axial magnetic field values.

However, at very large magnetic fields (>500G) R_p decreases with the increasing of magnetic field value (as shown in Figure 3.7).

To launch a helicon wave, the RF driving frequency should be greater than the ion cyclotron frequency but lesser than electron cyclotron frequency. Since the hydrogen gas is considered for the present design, the frequency should be more than ~1 MHz. A systematic study to understand the frequency dependency is shown in Figure 3.8. The commercially available RF generators are mainly of 13.56 MHz and the Figure 3.8 shows that frequency is sufficient to excite helicon wave inside the discharge. The figure clearly shows that R_p obtained at 2 MHz is low and thus, it would not be very useful to use a source with such low frequencies, for the present case. However, R_p variation with different frequencies can be altered by applying different magnetic field (as shown in Figure 3.9). From Figure 3.9, it can be concluded that we can get the same range of R_p with 13.56 MHz source instead of using a 27.12 MHz source, if the magnetic field is increased. The curves for these two frequencies are quite close to each other, while R_p value for 2 MHz source is very less, even at higher magnetic fields.

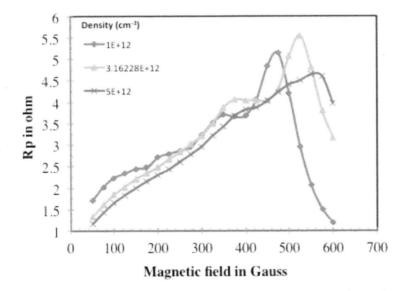

FIGURE 3.7 Variation of Rp with different magnetic field for different plasma density values.

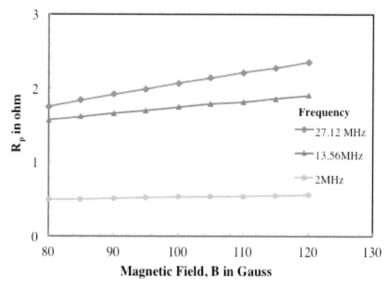

FIGURE 3.8   Variation of Rp with plasma density for different driving frequencies of RF power source.

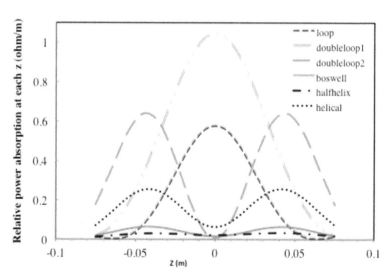

FIGURE 3.9   Variation of Plasma Resistance with magnetic field for different frequency values.

Antenna configuration, through which the power is going to be transferred to the plasma, plays a vital role in an ICP or helicon plasma source. Different antenna configurations are simulated. It is apparent from Figure 3.10 that a single or double loop antenna gives the maximum power absorption near the antenna location. In the Figure 3.10, the center of the antenna is placed at $z = 0$. For smaller plasma cavities, loop antenna works better than a half-helix or Boswell type antenna.

One end of plasma discharge tube is open to allow plasma to fill the expansion region, but the other end is closed by an endplate, either an insulator or a metal plate. Endplate material is important from the point of view of the wave reflections, taking place at the boundaries. Chen [16] has highlighted the importance of helicon wave reflection from the endplates and the interference of the waves. Figure 3.11 shows a comparative analysis of the plasma loading R_p obtained with half-helix and a loop type (single loop) antenna. The plot also compares the conducting and insulating type endplate configurations. The conducting endplate with the single loop antenna gives the maximum plasma loading as evident from the simulation results (Figure 3.11).

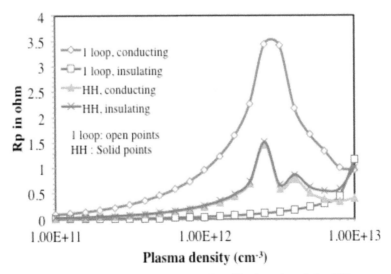

FIGURE 3.10   Variation of the power absorbed profile along the axis for different antenna types.

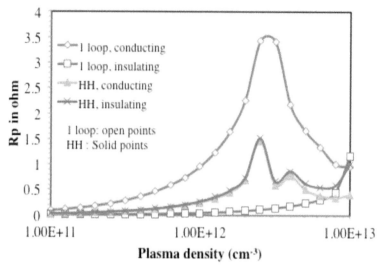

FIGURE 3.11   Helicon discharge comparison obtained with Half-Helix and Loop antenna with different endplate configurations.

3.4   PLASMA LOSSES

The stable operational regime for plasma density depends on a balance between coupled RF power P_{in} and plasma loss P_{out} mechanism [6, 14]. As shown by Shamrai [17], P_{out} varies linearly with plasma density. Due to different nonlinear physical processes, P_{in} vs. plasma density curve is not monotonic. A stable operating point is possible if the P_{in} and P_{out} curves cross each other.

Electrons are confined by the magnetic field of the helicon discharge as they are magnetized but the field is too weak to confine the ions. However, the electrons are lost through the sheaths at the ends of short tubes to maintain plasma neutrality. Plasma losses are determined by the ion losses, the electrons following at the same rate in order to preserve quasineutrality. The ions free-fall to the sheaths at the walls at low pressure conditions, and their loss rate is given by,

$$\tau = \frac{1}{2} n_e c_s \tag{2}$$

Hence, the loss rate is given by,

$$-\frac{dN}{dt} = \frac{1}{2} n_e c_s A;$$ (3)

where, A is the total surface area of interaction and C_s is the ion acoustic speed. In a steady state, each ion-electron pair that is lost carries away some energy, W, where,

$$W = E_c + W_i + W_e$$ (4)

where, $E_c (T_e)$ is the energy required to ionize each gas atom, including the energy lost in inelastic collisions. Each ion lost to the walls loses its energy gained in the pre-sheath (0.5 kT$_e$) and that gained in the wall sheath (5 kT$_e$), making a total of $W = 5.5$ kT$_e$. The losses accompanying the electron loss is approximately equal to 2 kT$_e$. Combining both the equations we can get power lost by the following relation, which shows linear dependency on plasma density.

$$P_{out} = \frac{dN}{dt} W_e = \frac{1}{2} n_e C_s A W_e$$ (5)

3.5 PROPOSED CONFIGURATION

Figure 3.12 shows the proposed permanent ring magnet based helicon plasma configuration for hydrogen plasma. This configuration is to initiate the program and will be optimized in terms of magnet geometry and its position, discharge tube dimension, and its endplate and antenna configuration. Later, similar configuration of multiple helicon sources will be arranged in desired matrix to create a large size plasma source to build a large area ion source suitable for NBI application.

3.6 SUMMARY

The design performances of the proposed helicon plasma source are studied using BFieldM and HELIC computer codes developed by Arnush [13].

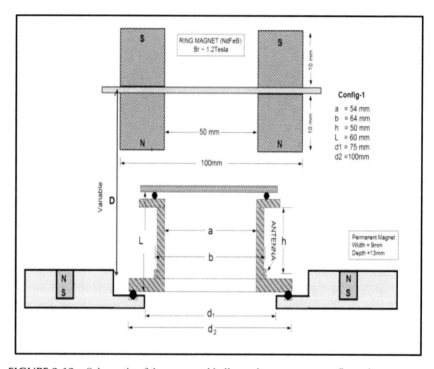

FIGURE 3.12   Schematic of the proposed helicon plasma source configuration.

In the simulation studies following parameters are examined: discharge gas, antenna type, RF generator frequency and its power level, magnetic field profile, and discharge tube geometry and its endplate material. The simulated results show that expected plasma density can be achieved as high as $\sim 10^{12}$ cm^{-3}.

KEYWORDS

- **HELIC**
- **helicon plasma**
- **helicon wave**
- **ion source**

REFERENCES

1. Degeling, A. W., & Boswell, R. W. (1997). *Phys. Plasmas 4.*
2. Boswell, R. W. (1984). *Plasma Phys. Control. Fusion 26*, 1147.
3. Chen, F. F., Evans, J. D., & Tynan, G. R. (2001). *Plasma Sources Sci. Technol. 10*, 236–249.
4. Chen, F. F. (1991). Plasma ionization by helicon waves, *Plasma Phys. Control. Fusion 33*, 339.
5. Chen, F. F., & Boswell, R. W. (1997). Helicons: the past decade, *IEEE Transactions on Plasma Science 6*, 25.
6. Francis F. Chen, & Humberto Torreblanca (2007). Large-area helicon plasma source with permanent magnets, *Plasma Phys. Control. Fusion, 49*, A81–A93.
7. Ellingboe, A. R., & Boswell, R. W. (1996). Capacitive, inductive and helicon wave modes of operation of a helicon plasma source, *Phys. Plasmas 3*, 2797.
8. Boswell, R. W., & Vender, D. (1995). An experimental study of breakdown in a pulsed Helicon plasma, *Plasma Sources Sci. Technol. 4*, 534.
9. Shinohara, S., Nishida, H., Tanikawa, T., Hada, T., Funaki, I., & Shamrai, K. P., Characterization of developed high-density helicon plasma sources and Helicon Electrodeless Advanced Thruster (HEAT) project, *Pulsed Power Conference (PPC), 2013 19th IEEE*, vol. 1, pp. 8, 16–21 June 2013.
10. Speth, E. (1989). Neutral beam heating of fusion plasmas, *Rep. Prog. Phys. 52*, 57–121.
11. Marcuzzia, D., et al. (2009). *Fusion Engineering and Design, 84*, 1253–1258.
12. Ian G. Brown (2004). *The Physics and Technology of Ion Sources*, Wiley-VCH.
13. Arnush, D. (2000). *Phys. Plasmas, 7*, 3042.
14. Torreblanca, H. (2008). Thesis, University of California, Los Angeles.
15. BfieldM, http://www.seas.ucla.edu/ltptl/PPTs/Instructions%20for%20BFieldM%20 program.pdf.
16. Chen, F. F. et al. (1997). *Plasma Phys. Control. Fusion 39*, A411.
17. Shamrai, K. P. (1998). *Plasma Sources Sci. Technol. 7*, 499.

CHAPTER 4

SOLITARY WAVE SOLUTIONS OF MODIFIED KADOMSTEV-PETVIASHIVILI EQUATION FOR HOT ADIABATIC DUSTY PLASMA HAVING NON-THERMAL IONS WITH TRAPPED ELECTRONS

APUL NARAYAN DEV,[1] RAJESH SUBEDI,[2] and JNANJYOTI SARMA[3]

[1]*Center for Applied Mathematics, Siksha 'O' Anusandhan University, Khandagiri, Bhubaneswar – 751030, Odisha, India E-mail: apulnarayan@gmail.com*

[2]*Department of Science and Humanities, College of Science and Technology, Rinchending, Phuentsholing – 450, Bhutan, E-mail: rajesh@cst.edu.bt*

[3]*Department of Mathematics, R. G. Baruah College, Guwahati – 781025, Assam, India, E-mail: jsarma_2001@yahoo.com*

CONTENTS

ABSTRACT

In this chapter, an investigation is presented on the properties of dust acoustic (DA) solitary wave propagation in adiabatic dusty plasma including the effect of the nonthermal positive and negative ions and trapped electrons. The reductive perturbation method has been employed to derive the lower order modified Kadomstev-Petviashivili (mK-P) and higher order mK-P for dust acoustic solitary waves in a homogeneous, unmagnetized, and collisionless plasma, whose constituents are trapped electrons, singly charged positive, and negative nonthermal ions and massive charged dust particles. The stationary analytical solution of the lower order mK-P and higher order mK-P equations are solved using well-known *tanh*-method. These solutions are numerically analyzed and the effect of various dusty plasma constituents DA solitary wave propagation is taken into account. It is observed that both the ions in dusty plasma play a key role in the formation of DA solitary waves and also the ion concentration and trapped electrons concentration controls the transformation of compressive potentials of the waves.

4.1   INTRODUCTION

The nonlinear wave phenomena in dusty plasmas have been widely studied theoretically and experimentally in the last few decades, since the presence of extremely massive charged dust particles plays an imperative role in understanding the electrostatic disturbances in space plasma environments as well as in laboratory plasma devices [31]. The presence of dust grains in two components electron-ion plasma is responsible for the appearance of new types of electrostatic waves including solitary or shock waves and has been reported by many researchers in both theoretical [19, 21, 32] and experimental point of views [1, 3]. One of those electrostatic waves is the low frequency dust-acoustic (DA) mode in unmagnetized dusty plasma whose constituents are charged dust fluid and Boltzmann distributed electrons and ions. It was Rao et al. [25] who first theoretically reported the existence of these low frequency DA solitary waves. They showed the formation of rarefactive type of DA solitary waves solution in dusty plasma and the predictions of DA solitary waves were conclusively verified by the laboratory experiment [4]. Lin and Duan [11] considering the nonthermal ion in dusty plasma derived a Korteweg-de Vries (K-dV) equation for DA wave and was reported that the nonthermal ions have very important effect on the propagation of DA solitary waves. In continuation with this, Mamun et al. [17] investigated the nonlinear DA waves in a two components unmagnetized dusty plasma consisting of a negatively charged cold dust fluid and isothermal electrons. In another work of Mamun [15] was reported that the adiabatic effect of inertia-less electron and ion fluids has significantly modified the basic properties of the DA solitary waves. In continuation of that a number of theoretical investigations [6, 13, 14, 18] have been made on DA solitary waves by assuming a three components unmagnetized dusty plasma consisting of a negatively charged cold dust fluid and inertialess isothermal electron and ion fluids. These works are only valid for a cold dust fluid and isothermal electrons and ions. It has also been confirmed from both theoretical and experimental observations that presence of negative ions in dusty plasma plays an important role in many aspects including charging of the dust particles. In the experimental work of Adhikary et al. [1], the formation of rarefactive dust ion acoustic (DIA) waves under influence of negative ions in dusty plasma was presented with their characteristic properties. In another study of Roy et al. [26] on the role of negative ions with dust charge fluctuation, estimating the dispersion relation for DA wave it was shown that the low temperature negative ions can reduce both the frequency and damping of the dust acoustic waves.

Some numerical investigations on linear and nonlinear DAW show a significant amount of ions trapping in the wave potential, which implies that there is a departure from Boltzmann ion distribution and one encounter vortex-like ion distribution in phase space [12, 15, 16, 33]. On the other hand, the nonlinear behaviors of electrostatic waves in plasma with this trapped state [28, 29] has received considerable attention and have been studied by a number of authors in the last few years in unmagnetized and magnetized plasmas [19]. Most of the studies have focused on deriving K-dV and K-P equations using reductive perturbation technique as well as Sagdeev potential approach [23]. Sayed and Mamun [27] studied the effects of dust temperature for hot adiabatic dusty plasma with Boltzmann distribution on DA solitary waves. They found that the effect of dust fluid temperature have significantly modified the basic properties (amplitude and width) of the solitary potential structures in warm dusty plasma. Mamun et al. [15, 16] investigated the effects of vortex-like and nonthermal ion distributions within the small amplitude regime by using modified K-dV equation and they concluded the possibility of coexistence of large amplitude rarefactive as well as compressive dust-acoustic solitary waves, whereas these structures appear independently when the wave amplitudes become infinitely small). Pakzad [22] studied the effect of DA solitary waves in warm dusty plasma with variable dust charged, two temperature nonthermal ions. Pakzad [24] and Dorranian and Sabetkar [9] reported that with increasing nonthermal ion population, the amplitude of solitary wave decreases, while the width of solitary waves increases. Recently, Dev et al. [7] also studied the effect of nonthermal electrons and vortex-like electrons distributions in compressive, rarefactive solitary wave and spiky solitary wave in clod dusty plasma by deriving the K-P equation and mK-P equations.

In 2014, Adikary et al. [2] derived the mK-dV and modified Burgers equation [8] in warm dusty plasma containing trapped electrons, nonthermal positive, and negative ions. So far it is a concern that the solitary wave in an unmagnetized warm dusty plasma containing trapped electrons, and nonthermal positive and negative ions have not yet been studied detail for different higher order nonlinearity. In this chapter, a detailed investigation of the propagation characteristics of dust acoustic wave in an unmagnetized warm dusty plasma containing trapped electrons, and nonthermal positive and negative ions is reported to derive the lower order modified K-P and higher order modified K-P equation. Here the effect of nonthermal ions as well as the trapped electrons is taken into account and their thermal effects are considered in the analytical treatment for the small amplitude solitary wave limit.

4.2 NORMALIZED EQUATIONS

In the present plasma model, we considered the propagation of dust acoustics waves in collisionless, unmagnetized warm dusty plasma consisting of non-isothermal electrons, and positive and negative nonthermal ions. The total charge neutrality at equilibrium requires that $n_{e0} = n_{p0} - Z_d n_{d0} - n_{n0}$, where n_{e0}, n_{p0}, n_{n0} and n_{d0} are the equilibrium values of electron, positive ions, negative ions, and dust number density, respectively. Z_d is the dust charged particles. In the present plasma system, we considered the electrostatic dust acoustic solitary wave having extremely low phase velocity, which is followed by the negatively charged massive dust particles. The pressure provides the restoring force and the inertia comes from the dust mass. The dynamics of the dust particles in one-dimensional dust acoustic wave in such a dusty plasma system can be described by the following basic equations as

$$\frac{\partial n_d}{\partial t} + \nabla.\left(n_d \bar{v}_d\right) = 0 \tag{1}$$

$$\left(\frac{\partial}{\partial t} + \bar{v}_d.\nabla\right)\bar{v}_d + \frac{1}{m_d n_d}\nabla p_d = -\frac{q_d}{m_d}\nabla\varphi \tag{2}$$

$$\left(\frac{\partial}{\partial t} + \bar{v}_d.\nabla\right)p_d + \gamma p_d \nabla.\bar{v}_d = 0 \tag{3}$$

$$\nabla^2\varphi = 4\pi e\left(n_e + n_n - n_p + Z_d n_d\right) \tag{4}$$

The nonthermal number density of positive ion n_p, and negative ion n_n can be described by the following relations,

$$n_p = n_{p0}\left(1 + \alpha\phi + \alpha\phi^2\right)\exp\left(-z_p\phi\right) \tag{5}$$

$$n_n = n_{n0}\left\{1 + \alpha\sigma_p\phi + \alpha\left(\sigma_p\phi\right)^2\right\}\exp\left(z_n\sigma_p\phi\right) \tag{6}$$

with $\alpha = 4\gamma_1/1 + 3\gamma_1$ and $\sigma_p = T_p/T_n$, where n_d is the number density of the negatively charged stationary dust particles in plasma, Z_d is the number of electrons residing on the dust surface at equilibrium, p_d is the pressure of the dust fluid, e is the electronic charge, m_p (m_n) is the positive (negative) ion mass, z_p (z_n) is the positive (negative) ion charge state, v_d is the dust fluid velocity, T_e (T_p) is the electron (positive ion) temperature, K_B is the Boltzmann constant, φ is the electrostatic potential, and γ_1 is the population of nonthermal ions in the plasma. The adiabatic index $\gamma = 5/3$ [$= (2 + D)/D$, where D is the number of orders of freedom] is due to the three-dimensional geometry of the system.

The trapped electrons n_e can be described by the following relations [5] as:

$$n_e = n_{e0}\left\{\exp(\beta\phi) - G(\beta\phi)\right\} \tag{7}$$

Here $G(\beta\phi) = \sum_{r=1}^{n}\frac{2^{r+1}b_r}{\prod(2r+1)}(\beta\phi)^{\frac{2r+1}{2}}$, $\beta = T_p/T_e$, $b_r = \left(1 - \gamma_2^r\right)/\sqrt{\pi}$, where $r = 1, 2, 3, \ldots$ and the parameter γ_2 is defined as $\gamma_2 = T_{ef}/T_{et}$, in which T_{ef} and T_{et} are the temperatures of free electrons and trapping electrons in the plasma, respectively. The parameter γ_2 determines the nature of the distribution function giving plateau if $\gamma_2 > 0$ and a dip if $\gamma_2 > 0$, and a hump shape formed if $\gamma_2 > 0$. However, $\gamma_2 = 1$ corresponds to the Maxwellian distribution of the electrons. In the present plasma system, the range of γ_2 will be considered as $0 < \gamma_2 < 1$ for the trapped electrons.

Now, considering N_d dust number density normalized by its equilibrium value n_{d0}, V_d is the dust-fluid velocity normalized by $c_{sd} = \left(Z_d K_B T_p/m_d\right)^{1/2}$, ϕ is the DA wave potential normalized by $k_B T_p\varphi/e$, the time variable T is normalized by $\omega_{pd}^{-1} = \left(m_d/4\pi n_{d0}Z_d^2 e^2\right)^{1/2}$, the space variable X is normalized by $\lambda_{Dd}^{-1} = \left(4\pi n_{d0}Z_d e^2/\kappa_B T_p\right)^{1/2}$, pressure p_d is normalized by $P_d = n_{d0}\kappa_B T_d$ and from Eqs. (1)–(4), we get

$$\left(\frac{\partial}{\partial T} + \nabla.\bar{V}_d\right)N_d = 0 \tag{8}$$

$$\left(\frac{\partial}{\partial T} + \bar{V}_d.\nabla\right)\bar{V}_d + \frac{\sigma_d}{N_d}\nabla P_d = \nabla\phi \tag{9}$$

$$\left(\frac{\partial}{\partial T}+\bar{V}_d.\nabla\right)P_d+\gamma P_d\nabla.\bar{V}_d=0 \tag{10}$$

$$\nabla^2\phi=p_1\phi-p_2\phi^{3/2}+p_3\phi^2-p_4\phi^{5/2}+\left(N_d-1\right) \tag{11}$$

where $p_1=\left\{\left(\mu_p-\mu_n-1\right)\beta+\mu_n\left(z_n+\alpha\sigma\right)-\mu_p\left(\alpha-z_p\right)\right\}$, $p_2=b_1\left(\mu_p-\mu_n-1\right)\beta^{3/2}$,

$p_3=\beta\left(\mu_p-\mu_n-1\right)/2+\mu_n\left(\left(z_n\right)^2/2+\alpha\sigma z_n+\alpha\left(\sigma\right)^2\right)-\mu_p\left(\left(z_p\right)^2/2-\alpha z_p+\alpha\right)$,

$p_4=\left(\mu_p-\mu_n-1\right)b_2\beta^{5/2}$ and $\mu_p=n_{p0}/Z_dn_{d0}$, $\sigma_d=T_d/Z_dT_p$, $\mu_n=n_{n0}/Z_dn_{d0}$,

$n_{e0}/Z_dn_{d0}=\mu_p-\mu_n-1=\mu_e$ are considered.

4.3 DERIVATION OF LOWER ORDER MODIFIED KADOMSTEV-PETVIASHIVILI (mK-P) EQUATION

Now, to derive the lower order mK-P equation for the propagation of small but finite amplitude DASW, we use the standard reductive perturbation technique in which the independent variables ξ and τ are stretched as

$$\xi=\varepsilon^{3/4}\left(X-V_0T\right), \eta=\varepsilon^{3/2}Y, \zeta=\varepsilon^{3/2}Z \text{ and } \tau=\varepsilon^{9/4}T \tag{12}$$

where, V_0 is the phase speed (normalized by c_{sd}) of the wave along x-direction and ε is a small nonzero constant measuring the weakness of the dispersion. The dependent variables N_d, V_d, P_d and ϕ can be expanded in power series of ε as

$$N_d=1+\varepsilon N_d^{(1)}+\varepsilon^{3/2}N_d^{(2)}+.... \tag{13}$$

$$V_{dx}=\varepsilon V_{dx}^{(1)}+\varepsilon^{3/2}V_{dx}^{(2)}+.... \tag{14}$$

$$P_d=1+\varepsilon P_d^{(1)}+\varepsilon^{3/2}P_d^{(2)}+.... \tag{15}$$

$$\phi=\varepsilon\phi^{(1)}+\varepsilon^{3/2}\phi^{(2)}+\varepsilon^2\phi^{(3)}.... \tag{16}$$

$$V_{dy,z} = \varepsilon^{7/4} V_{dy,z}^{(1)} + \varepsilon^{9/4} V_{dy,z}^{(2)} + \dots \tag{17}$$

Substituting the stretched co-ordinates and the expressions for N_d, V_d, P_d and ϕ into the normalized basic Eqs. (8)–(11) and equating the coefficients of lowest order of ε, we get,

$$V_d^{(1)} = \frac{3V_0 \varphi^{(1)}}{\left(5\sigma_d - 3V_0^2\right)} \tag{18}$$

$$N_d^{(1)} = \frac{3\varphi^{(1)}}{\left(5\sigma_d - 3V_0^2\right)} \tag{19}$$

$$P^{(1)} = \frac{5\varphi^{(1)}}{\left(5\sigma_d - 3V_0^2\right)} \tag{20}$$

Together with dispersion relation,

$$V_0 = \left[\frac{5}{3}\sigma_d + \frac{1}{\left\{\left(\mu_p - \mu_n - 1\right)\beta + \mu_n\left(z_n + \alpha\sigma\right) - \mu_p\left(\alpha - z_p\right)\right\}}\right]^{1/2} \tag{21}$$

From the next higher-order of ε, we derive a set of equation with the relation $N_d^{(1)}$, $V_d^{(1)}$, $P_d^{(1)}$, $\phi_d^{(1)}$ and $N_d^{(2)}$, $V_d^{(2)}$, $P_d^{(2)}$, $\phi_d^{(2)}$, substituting the values from Eqs. (18) to (21) and following by a straight forward elimination of $N_d^{(2)}$, $V_d^{(2)}$, $P_d^{(2)}$, $\phi_d^{(2)}$ finally we obtain the following lower order mK-P equation,

$$\frac{\partial}{\partial \xi}\left(\frac{\partial \phi^{(1)}}{\partial \tau} + A_2\left(\phi^{(1)}\right)^{3/2}\frac{\partial}{\partial \xi}\phi^{(1)} + B\frac{\partial^3 \phi^{(1)}}{\partial \xi^3}\right)$$
$$+ C\left(\frac{\partial^2 \phi^{(1)}}{\partial \eta^2} + \frac{\partial^2 \phi^{(1)}}{\partial \zeta^2}\right) = 0 \tag{22}$$

where the nonlinear coefficient A_1, the dispersion coefficient B, and the transverse coefficient C are given by

$$A_1 = \frac{p_2}{p_1}\frac{\left(3V_0^2 - 5\sigma_d\right)}{4V_0}, \quad B = \frac{\left(3V_0^2 - 5\sigma_d\right)}{6p_1V_0}, \quad C = \frac{V_0}{2}$$

Equation (22) represents the well-known lower order mK-P equation describing the nonlinear propagation of the DA waves in electronegative dusty plasma with nonthermal ions and trapped electrons. To find this solution at the end, we use the transformation $\chi = \left(l\xi + m\eta + n\zeta - U\tau\right)$, where l, m, n are the direction cosines along the x, y, z axes and let $\phi^{(1)}\left(\xi,\eta,\zeta,\tau\right) = \psi\left(\chi\right)$. Thus, from Eq. (22) one obtains

$$Bl^4\frac{d^2\psi}{d\chi^2} + A_1l^2\frac{3\psi^{2/3}}{2} - Ul\psi + C\left(m^2 + n^2\right) = 0 \qquad (23)$$

We now apply the *tanh*-method in which we define $x = \tanh\left(\chi\right)$, $\psi\left(\chi\right) = W\left(x\right)$. Then Eq. (23) becomes

$$\frac{2A_1l^2}{3}W^{3/2} + Bcl^4\left(1-z^2\right)^2\frac{d^2W}{dx^2} - Bcl^4Z\left(1-Z^2\right)\frac{dW}{dx} + C\left(m^2+n^2\right)W - UlW = 0$$
$$(24)$$

For the series solution of Eq. (24) we assume $W\left(x\right) = \sum\limits_{r=0}^{\infty}a_rx^{\delta+r}$ and then the leading order analysis of finite terms gives $r = 4$ and $\delta = 0$ so that $W(x)$ becomes $W\left(x\right) = a_0\left(1-x^2\right)^2$. Now substituting the value of $W(x)$ in Eq. (24), we can obtain the values of $a_0 = \phi_{m1}$. The required stationary solution of lower order mK-P equation is

$$\phi^{(1)} = \phi_{m1}\sec h^4\left(\frac{\chi}{\omega_1}\right) \qquad (25)$$

where $\phi_{m1} = \frac{225}{64}\left[\left\{Ul - C\left(m^2 + n^2\right)\right\}/A_1l^2\right]^2$ and $\omega_1 = 4\left[Bl^3/\left\{Ul - C\left(m^2 + n^2\right)\right\}\right]^{\frac{1}{2}}$ are the amplitude and width of the solitary waves, respectively, and l, m, n representing the direction cosines of the angle made by the propagation with x-axis, y-axis, and z-axis, respectively. The trapped effect enhanced the amplitude of lower order mK-P equation but not in width. Since ϕ_{m1} is always positive, therefore the solution of the lower order mK-P equation admits only compressive solitary waves.

4.4   DERIVATION OF HIGHER ORDER NEW MODIFIED KADOMSTEV-PETVIASHIVILI (MK-P) EQUATION

Now, if the nonlinear coefficient of the K-P equation vanishes, at $A_1 = 0$ for some critical values of the plasma parameters, then amplitude of the K-P equation $\phi_{m1} \to \infty$, i.e., the lower order mK-P equation fails to form the solitary waves. The failure forces us to look at another equation, which is suitable for describing the evolution of the system. Instead of the stretching coordinate used before we have to use higher stretching coordinates of the perturbation theory,

$$\xi = \varepsilon^{3/4}\left(X - V_0 T\right),\ \eta = \varepsilon^{3/2}Y,\ \zeta = \varepsilon^{3/2}Z \text{ and } \tau = \varepsilon^{9/4}T \qquad (26)$$

where, V_0 is the phase speed (normalized by c_{sd}) of the wave along x-direction and ε is a small nonzero constant measuring the weakness of the dispersion. The dependent variables N_d, V_d, P_d and ϕ can be expanded in power series of ε as before Eqs. (13)–(16) and

$$V_{dy,z} = \varepsilon^{7/4}V_{dy,z}^{(1)} + \varepsilon^{9/4}V_{dy,z}^{(2)} + ... \qquad (27)$$

Using the above new form of the stretched coordinate Eq. (26) and variable expansion parameter of Eqs. (13)–(16) and (27) in the normalized basic Eqs. (8)–(11) and collecting the coefficient of the terms of the lowest order ε, we obtain the same result as Eqs. (18)–(21). However, for the higher order of ε, yields the following relation

$$V_d^{(2)} = V_0\left(P_2\left(\varphi^{(1)}\right)^{3/2} - P_1\varphi^{(2)}\right) \qquad (28)$$

$$P^{(2)} = \frac{5}{3}\left(P_2\left(\varphi^{(1)}\right)^{3/2} - P_1\varphi^{(2)}\right) \qquad (29)$$

$$N_d^{(2)} = P_2\left(\varphi^{(1)}\right)^{3/2} - P_1\varphi^{(2)} \qquad (30)$$

For the higher order of ε, we derives a set of equation with the relation $N_d^{(1)}, V_d^{(1)}, P_d^{(1)}, \phi^{(1)}, N_d^{(2)}, V_d^{(2)}, P_d^{(2)}, \phi^{(2)}$ and $N_d^{(4)}, V_d^{(2)}, P_d^{(4)}, \phi^{(4)}$, substituting the value from Eqs. (18)–(21) and Eqs. (28)–(30), followed by a straightforward elimination of $N_d^{(2)}$, $V_d^{(2)}$, $P_d^{(2)}$, $\phi^{(2)}$ and $N_d^{(4)}$, $V_d^{(4)}$, $P_d^{(4)}$, $\phi^{(4)}$. Finally we obtain the new higher order mK-P equation with trapped electrons in following form,

$$\frac{\partial}{\partial \xi}\left(\frac{\partial \phi^{(1)}}{\partial \tau} + A_2 \left(\phi^{(1)} \right)^{3/2} \frac{\partial}{\partial \xi} \phi^{(1)} + B \frac{\partial^3 \phi^{(1)}}{\partial \xi^3} \right)$$
$$+ C\left(\frac{\partial^2 \phi^{(1)}}{\partial \eta^2} + \frac{\partial^2 \phi^{(1)}}{\partial \zeta^2} \right) = 0 \tag{31}$$

where the dispersion coefficient B and transverse coefficient C same as before and the nonlinear coefficient A_2 as

$$A_2 = \left(\frac{26 p_2 \sigma_d + 5V_0^2 p_2}{9V_0} - \frac{5\left(5\sigma_d - 3V_0^2\right)p_4}{12V_0 p_1} \right)$$

Repeating the same we obtained the solution of higher order new mK-P equation (31) as

$$\phi^{(1)} = \phi_{m2} \sec h^{\frac{4}{3}}\left(\frac{\chi}{\omega_2} \right) \tag{32}$$

where $\phi_{m2} = \left[35\left\{ U l - C\left(m^2 + n^2 \right) \right\} / 8A_2 \, l^2 \right]^{2/3}$ and $\omega_2 = 4/3\left[Bl^3 / \left\{ U l - C\left(m^2 + n^2 \right) \right\} \right]^{\frac{1}{2}}$ are the amplitude and width of the solitary waves, respectively, and l, m, n representing the direction cosines of the angle made by the propagation with x-axis, y-axis, and z-axis, respectively. Again, the trapped effects enhance the amplitude of higher order mK-P equation but not in width. The nonlinear coefficient A_2 can be positive, negative, or zero. For positive values of nonlinear coefficient A_2, the solution of higher order new mK-P equation gives compressive solitary wave and for negative values of nonlinear coefficient A_2, the solution of K-P equation gives imaginary form which does not give us the solitary waves.

4.5   RESULTS AND DISCUSSION

We analytically examine the dependence of the lower order mK-P, K-P, and higher order mK-P explaining the propagation of electrostatics solitary wave on various dusty plasma parameters. Here the dust particles in the plasma are considered as uniform in size and negatively charged, while the background plasma are singly charged positive and negative ions. Figures 4.1 and 4.2 shows the effect of the variation of positive ion density ratio μ_p and positive ion temperature ratio β on the amplitude of the lower order mK-P equation ϕ_{m1} and higher order mK-P equation ϕ_{m2}. Here, the other plasma parameters are considered as $m_p = 40 \times 1.6^{-27}$kg, $n_{e0} = 4 \times 10^{14}$m^{-3}, n_{p0} $= 5.4 \times 10^{14}$m^{-3}, $n_{n0} = 3.0 \times 10^{14}$ $n_{d0} = 1.2 \times 10^{10}$m^{-3}, $Z_{d0} = 1.5 \times 10^4$e, $T_e = 1.5$ eV, $T_i = 0.1$ eV [2]. Figure 4.1 clearly depicts that the amplitude of lower order mK-P equation increases with the enhancement of positive ions density μ_p and decreases as the ions temperature ratio β increases. Figure 4.2 clearly depicts that the amplitude of higher order mK-P equation increases with the enhancement of positive ions density μ_p and decreases as increases the ions temperature ratio β.

Figures 4.3 and 4.4 shows the effect of the variation of negative ion density ratio μ_n and positive ion temperature ratio β on the amplitude of the lower order mK-P equation and higher order mK-P equation ϕ_{m2}. Figure 4.3 clearly depicts that the amplitude of lower order mK-P equation increases with the enhancement of negative ions density μ_n. Figure 4.4 clearly depicts that the amplitude of higher order mK-P equation increases with the enhancement of negative ions density μ_n.

Figures 4.5 and 4.6 shows the variation of solitary wave profile $\phi^{(1)}$ for lower order mK-P equation and higher order mK-P equation with the variation of the nonthermal ion concentration γ_1 and spatial variable χ. Figure 4.5 clearly depicts that, for increasing values of the nonthermal ion concentration γ_1, the amplitude of compressive solitary wave is decreasing. Similarly, Figure 4.6 clearly depicts that, for increasing values of the nonthermal ion concentration γ_1, the amplitude of compressive solitary wave is decreasing. We conclude from these figure that the amplitude of the solitary waves are maximum for lower order mK-P equation and decreasing for the higher order nonlinear term of K-P equation.

Figures 4.7 and 4.8 shows the variation of solitary wave profile $\phi^{(1)}$ of lower order mK-P and higher order mK-P equation with the variation of the trapped electron concentration γ_2. These two figures show that the amplitude of compressive solitary waves increases with the enhancement of the trapped electrons concentration γ_2. It is seen that the amplitude of compressive solitary wave increases with the enhancement of the non-isothermal

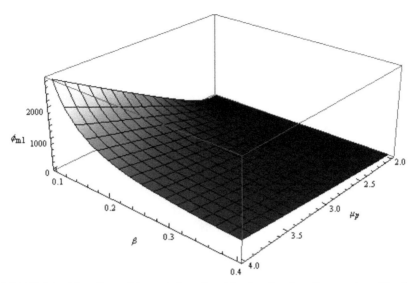

FIGURE 4.1   The effect of the variation of positive ion density ratio μ_p and positive ion temperature ratio β on the amplitude of the lower order modified K-P.

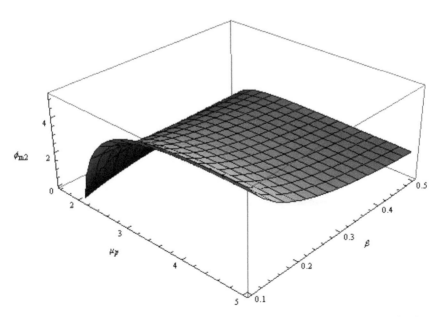

FIGURE 4.2   The effect of the variation of positive ion density ratio μ_p and positive ion temperature ratio β on the amplitude of the higher order modified K-P.

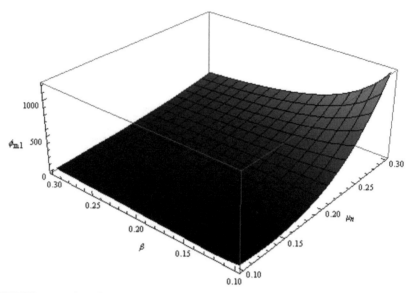

FIGURE 4.3 Plot of the amplitude of the lower order modified K-P with respect to negative ion density ratio μ_n and positive ion temperature ratio β.

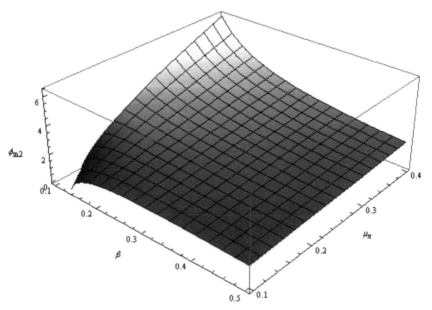

FIGURE 4.4 Plot of the amplitude of the higher order modified K-P with respect to negative ion density ratio μ_n and positive ion temperature ratio β.

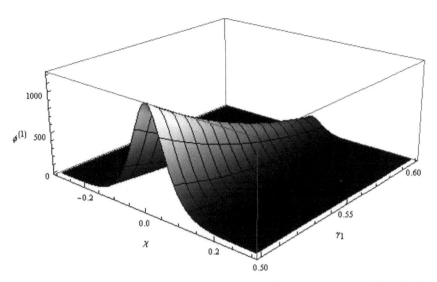

FIGURE 4.5 Depicts the variation of compressive solitary wave profile $\varphi(1)$ with the variation of the non-thermal ion concentration γ_1 for lower order modified K-P equation.

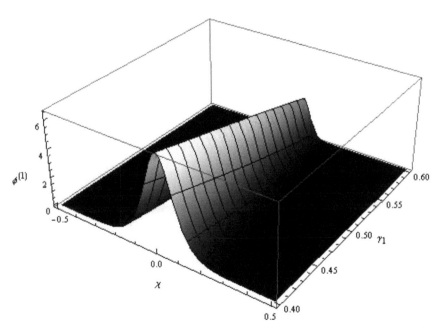

FIGURE 4.6 Depicts the variation of compressive solitary wave profile $\varphi(1)$ with the variation of the non-thermal ion concentration γ_1 for higher order modified K-P equation.

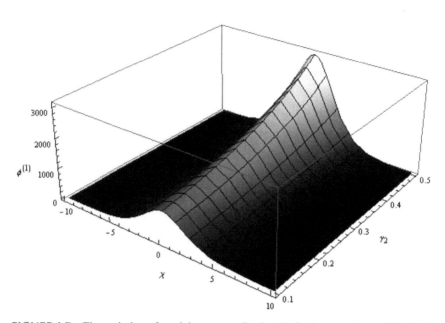

FIGURE 4.7   The variation of spatial wave amplitude $\varphi(1)$ for lower order modified K-P equation with the variation of the trapped electron concentration γ_2.

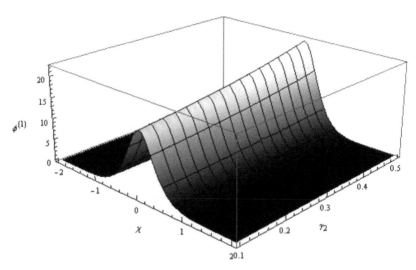

FIGURE 4.8   the variation of spatial wave amplitude $\varphi(1)$ for higher order modified K-P equation with the variation of the trapped electron concentration γ_2.

electron concentration γ_2. This may be because as trapped to free electrons temperature ratio is increases, i.e., as the system approaches towards thermal equilibrium or more precisely as nonlinearity of the system decreases, the solitary wave potential increases accordingly.

4.6 CONCLUSION

In this present work we have investigated the properties of DA solitary wave in warm adiabatic dusty plasma in presence of positive and negative nonthermal ions, and trapped electrons. The reductive perturbation method is employed to derive the lower order mK-P equation and higher order mK-P equation for dust acoustic solitary waves. The stationary analytical solution of the lower order mK-P and higher order mK-P solved by well-known *tanh*-method and these solutions are numerically analyzed and the effect of various dusty plasma constituents DA solitary wave propagation is taken into account. It is concluded that both positive and negative ions density ratio and ions temperature ratio play an important role in DA solitary waves. We also observed that the nonthermal ion concentration γ_1 and trapped electron concentration γ_2 play a key role in amplitude of solitary waves.

KEYWORDS

- dusty plasma
- reductive perturbation method
- solitary wave
- tan-hyperbolic method

REFERENCES

1. Adhikary, N. C., Deka, M. K., & Bailung, H. (2009). Observation of rarefactive ion acoustic solitary waves in dusty plasma containing negative ions, *Phys Plasmas*, *16*(6), 63701–63704.

2. Adhikary, N. C., Deka, M. K., Dev, A. N., & Sarma, J. (2014). Modified Korteweg–de Vries equation in a negative ion rich hot adiabatic dusty plasma with nonthermal ion and trapped electron, *Phys. Plasmas, 21*(8), 83703–83706.

3. Barkan, A., D'Angelo, N., & Merlino, R. L. (1994). Charging of Dust Grains in a Plasma, *Phys. Rev. Lett., 73*(12), 3093.

4. Barkan, A., Merlino, R. L., & D'Angelo, D. (1995). Laboratory observation of the dust acoustic wave mode, *Phys. Plasmas, 2*(6), 3563–3565.

5. Das, G. C., & Sarma, J. (1998). A new mathematical approach for finding the solitary waves in dusty plasma, *Phys. Plasmas, 5*(6), 3918–3925.

6. Dev, A. N. D., & Sarma, J. (2014). Solitary Wave Solution of Higher order Kadomstev-Petviashvili Equation for Complex Plasma, *International Journal of Technology, 4*(1), 13–19.

7. Dev, A. N., Sarma, J., Deka, M. K., Misra, A. P., & Adhikary, N. C. (2014). Kadomtsev-Petviashvili (KP) Burgers equation in dusty negative ion plasmas: Evolution of dust-ion acoustic shocks, *Commun. Theor. Phys. 62*(6), 875–880.

8. Dev, A. N., Deka, M. K., Sarma, J., & Adhikary, N. C. (2015). Shock Wave Solution in a Hot Adiabatic Dusty Plasma having Negative and Positive Nonthermal Ions with Trapped Electrons, *J. Korean Physical Society, 67*(2), 339–345.

9. Dorranian, D., & Sabetkar, A. (2012). Dust acoustic solitary waves in a dusty plasma with two kinds of nonthermal ions at different temperatures, *Phys. Plasmas, 19*(1), 013702.

10. Futana, Y., Machida, S., Saito, Y., Matsuoka, A., & Hayakawa, H. (2003). Moon-related nonthermal ions observed by Nozomi: Species, sources, and generation mechanisms, *J. Geophys. Res., 108*(A1), 1025–1010.

11. Lin, M. M., & Duan, W. S. (2007). Kadomstev-Petviashvili, mK-P and coupled K-P equation for two ion-temperature dusty plasma. *Chaos Solitons and Fractals, 23*(3), 929–937.

12. Lin, C., & Lin, M. M. (2007). Analytical study of the dust acoustic waves in a magnetized dusty plasma with many different dust grains, *Adv. Stud. Theor. Phys., 1*(12), 563–570.

13. Ma, J. X., & Liu, J. (1997). Dust-acoustic soliton in a dusty plasma, *Phys. Plasmas 4*(10), 253–257.

14. Mamun, A. A. (1999). Arbitrary amplitude dust-acoustic solitary structures in a three-component dusty plasma, *Astrophys. Space Sci. 268*, 1–17.

15. Mamun, A. A. (2008). Coexistence of positive and negative solitary potential structures in dusty plasma, *Phys. Lett. A, 372*(5), 686–689.

16. Mamun, A. A. (2008). Dust-acoustic solitary waves in an adiabatic hot dusty plasma. *Physics Letters A, 372* (6), 884–887.

17. Mamun, A. A., & Hassann, M. A. (2000). Effects of dust grain charge fluctuation on an obliquely propagating dust acoustic solitary potential in a magnetized dusty plasma, *J. Plasma Phys., 63*(2), 191–200.

18. Mamun, A. A., & Shukla, P. K. (2002). Electrostatic Solitary and Shock Structures in Dusty Plasmas, *Phys. Scripta T98*, 107–114.

19. Mamun, A. A., Shukla, P. K., & Cairns, R. A. (1996). Solitary Potentials in dusty plasmas, *Phys. Plasmas 3*(2), 702–704.

20. Mamun, A. A., Shukla, P. K., & Cairns, R. A. (1996). Effects of vortex-like and nonthermal ion distributions on non-linear dust-acoustic waves, *Phys. Plasmas, 3*(2), 2610–2614.

21. Misra, A. P., Chowdhury, K. R., & Chowdhury, A. R. (2007). Saddle-node bifurcation and modulational instability associated with the pulse propagation of dust ion-acoustic waves in a viscous dusty plasma: A complex nonlinear Schrödinger equation, *Phys. Plasmas 14*(11), 012110–6.

22. Pakzad, H. R. (2009). Solitary waves of the K-P equation in warm dusty plasma with variable dust charge, two temperature nonthermal ions. *J. Plasma Fusion Res., 8*(1), 261–264.

23. Pakzad, H. R., & Javidan, K. (2009). Solitons of the KP equation in dusty plasma with variable dust charge and two temperature ions: energy and stability, Indian, *J. Phys., 83*(9), 349–363.

24. Pakzad, H. R. (2010). Modified KP-Burger and KP-Burger equations in coupled dusty plasmas with variable dust charge and non-isothermal ions, *Ind. J. Phys., 84*(7), 867–879.

25. Rao, N. N., Shukla, P. K., & Yu, M. Y. (1990). Dust-acoustic waves in dusty plasmas, *Planet. Space Sci., 38*(4), 543–546.

26. Roy, B., Sarkar, S., Khan, M., & Gupta, M. R. (2005). Propagation of Dust Acoustic Waves in a Complex Plasma with Negative Ions, *Phys. Scripta, 71*(6), 644.

27. Sayed, F., & Mamun, A. A. (2007). Dust acoustics Korteweg-de Vries solutions in an adiabatic hot dusty plasma. *Phys. Plasmas, 14*(1), 14502–14503.

28. Schamel, H. (1972). Stationary solitary, snoidal and sinusoidal ion acoustic waves, *Plasma Phys. 14*(10), 905.

29. Schamel, H. (1975). Analytic BGK modes and their modulational instability, *J. Plasma Phys., 13*(1), 139–145.

30. Schamel, H., & Bujarbarua, S. (1980). Solitary plasma hole via ion–vortex distribution, *Phys. Fluids, 23*(8), 2498.

31. Shukla, P. K., & Mamun, A. A. (2002). Introduction to Dusty Plasma Physics. Bristol: Institute of Physics. IOP Publishing Ltd 2002, Institute of Physics Publishing Bristol and Philadelphia. London Institute of Physics Publishing, Dirac House, Temple Back, Bristol BS1 6BE, UK.

32. Shukla, P. K., & Silin, V. P. (1992). Dust ion-acoustic wave, *Phys. Scripta, 45*(5), 508–512.

33. Winske, D., Gary, S. P., Jones, E., Rosenberg, M., Chow, V. W., & Mendis, D. A. (1995). Ion heating in a dusty plasma due to the dust/ion acoustic instability, *Geophys. Res. Lett., 22*(15), 2069–2073.

CHAPTER 5

GLOBAL TRANSITION FROM DRIFT WAVE DOMINATED REGIMES TO MULTI-INSTABILITY PLASMA DYNAMICS AND SIMULTANEOUS FORMATION OF A RADIAL TRANSPORT BARRIER

SAIKAT CHAKRABORTY THAKUR,[1,2] CHRISTIAN BRANDT,[2] and GEORGE R. TYNAN[2]

[1]458, EBU-II, CER – UCSD, 9500 Gilman Drive, Mail Code 0417, La Jolla, CA 92093, USA, E-mail: saikat@ucsd.edu

[2]Center for Energy Research, University of California at San Diego, San Diego, CA 92093, USA

CONTENTS

ABSTRACT

Recent studies in the Controlled Shear Decorrelation eXperiment (CSDX) reported a sharp non-monotonic global transition (at a critical magnetic field of $B = 140$ mT, with all other source parameters kept constant) in the plasma dynamics during the route to fully developed broadband turbulence. For B < 140 mT, the plasma is dominated by density gradient driven resistive drift wave (RDW) instabilities, propagating in the electron diamagnetic drift direction. The resulting particle flux is radially outwards. For B > 140 mT, a new global equilibrium is achieved where we observe the simultaneous existence of three radially separated plasma instabilities. The density gradient region, still dominated by RDWs, separates the plasma radially into the edge region and the core region. The edge region is dominated by strong, turbulent, shear driven Kelvin-Helmholtz (KH) instabilities, while the core region shows coherent Rayleigh Taylor (RT) modes driven by azimuthal rotation. The RT modes at the core have very high azimuthal mode number, propagate in the ion diamagnetic drift direction and are associated with intense argon ion (Ar-II) emission. In this regime, the radial particle flux is directed outward for small radii and inward for large radii, thus forming a radial particle transport barrier leading to stiff profiles, decreased turbulence levels and increased core plasma density. Simultaneously, the Ar-II light emission from the core region increases by an order of magnitude leading to the formation of a very bright blue core. Blue cores have been previously observed in helicon plasma though its origin is hotly debated in the helicon source community. The radial extent of the inner RT mode and radial location of the particle transport barrier coincides with the radial extent of the inner blue core. Simultaneously, we find enhanced axial and azimuthal plasma flows in the core plasma, further helping in keeping the core and the edge distinctly separated. This new global equilibrium with simultaneous RT-DW-KH instabilities shows very interesting and rich plasma dynamics including intermittency, formation and propagation of blobs, formation of a radial particle transport barrier, inward particle flux going up against density gradients, etc. This transition also suggests that changes in the cross-field radial particle transport due to low frequency instabilities are crucial to helicon core formation.

5.1 INTRODUCTION

Helicon plasma sources are radio frequency (RF)-driven plasma sources noted for their exceptionally high ionization efficiencies. They can produce very high density ($> 10^{19}$ m^{-3}) but relatively low temperature plasmas (electron temperature $\sim$ few electron volts (eV); ion temperatures less than 1 eV) at low but finite magnetic fields and relatively low power. Compared to the other classes of rf-driven plasma sources like capacitive coupled plasma (CCP) and inductively coupled plasma (ICP), helicon sources can produce much higher plasma densities for similar power inputs. For example, a 1 kW helicon source typically produce plasma densities of $\sim 10^{19}$ m^{-3}, while for the same power CCPs typically produce plasma with densities $\sim 10^{16}$ m^{-3} and the corresponding plasma densities in ICPs typically reach $\sim 10^{17}$ m^{-3} [17]. An excellent review of the early studies of helicon sources and the subsequent research till the mid 1990s can be found in the papers [4, 9]. More recent progress in helicon sources is reviewed by Scime et al. [29], which is a summary of a mini-conference hosted by the American Physical Society – Division of Plasma Physics in 2007. Since the first experiments on the development of a helicon source in the late 1960s and early 1970s [2], the high density plasmas produced by helicon sources have found several applications like materials processing [25], etching of semiconductors [7], ion sources for inertial confinement fusion [15], plasma thrusters [45], studies of drift wave turbulence and zonal flows, etc. [5, 21, 22, 28].

Conventional rf-produced plasmas can typically operate in one or more of the following three different modes: capacitive, inductive, and helicon, depending on the source parameters, antenna design, and device geometry. Distinct transitions between these three modes have been observed by several groups working on helicon plasma [3, 8, 11–13, 16, 26, 30, 31]. The mode transitions are seen as distinct non-monotonic density jumps as the input RF power or the magnetic field is varied (with all other source parameters kept constant). Chapter 2 of Wiebold [41], with help of the references within, gives a very nice review of the details of the mode transitions between the capacitive, inductive, and helicon modes in RF plasmas. In most cases, these transitions have hysteresis between increasing and decreasing power or magnetic field. In some cases [12, 13], there is a direct capacitive-helicon transition, with increasing power. But for decreasing power both the traditional helicon-inductive and the inductive-capacitive transitions are observed.

In the capacitive mode, the plasma densities are very low ($\sim 10^{16}$ m^{-3} to 10^{17} m^{-3}), typically with hollow radial density profiles. Time-averaged plasma potentials can be as high as the rf-driving voltages and the ionization is localized under the leads of the antenna. The plasma emission is dominated by excited neutrals, so for argon plasma the color of this mode of operation is typically diffuse pink, the emission being in the wavelength range 600 nm to 850 nm. As the input RF power or the magnetic field is increased, the system goes through a distinct discrete upward jump in the plasma density ($\sim 10^{17}$ m^{-3} to 10^{18} m^{-3}), into the inductive mode. The plasma potentials are much lower than that in the capacitive mode. The intensity of the neutral light emission is stronger along with signatures of ion emission. In argon, the plasma looks brighter pink with some amount of blue argon ion emission mostly along the center. On further increasing the RF power to the plasma source or the magnetic field, the helicon mode can be achieved, experimentally seen as another discontinuous jump in the plasma density. In the helicon mode, one can achieve very high plasma densities ($\sim 10^{19}$ m^{-3}), centrally peaked density profiles and very low plasma potentials. The helicon mode is dominated by very bright blue argon ion emission, due to the strong argon ion emission lines in the range of 400 nm to 500 nm. In addition, since this is the only mode in which the helicon wave propagates, the helicon wave magnetic fields can be measured in this mode, unlike in the capacitive and inductive modes. Figure 5.2 of Franck [12] shows a very nice representation of the density profiles and the mode jumps between the capacitive, inductive, and helicon modes in a typical helicon device. Because of the nature of the light emission from the plasma at the three different modes, it is quite easy to distinguish between the capacitive, inductive, and helicon modes of operation of an RF plasma device. Figure 2.4 (in Ref. [41]) gives a nice picture of the visual representation of the three different modes in a typical helicon device. The light emission has been shown from spectroscopy to be related to the relative abundance of excited neutrals (Ar I) and the ions (Ar II) that reflects changes in the ionization fraction during the mode transitions in the RF source [6].

Since the first detailed studies of helicon waves in gaseous plasmas [2, 3], the jump to the helicon mode is associated with the formation of a *blue core*, a radially localized central region of very strong ion light emission. There are numerous studies [2, 3, 6, 13, 26, 29], where the appearance of the core is used as an identification of the helicon mode. The formation of the blue core is typically taken to be an indication of achieving the helicon mode. In other words, the appearance of the blue core is a proxy for measuring the detailed plasma characteristics to confirm propagating helicon waves. It is generally believed that this is due to the peaked central densities and the strong blue Ar II emission that the core region looks brighter. The formation of the visual bright core is not just restricted to argon helicon plasma, but has been observed for other working gases like xenon [40] and krypton [20] in a helicon source as well.

Here we describe new experimental results that help in understanding the plasma dynamics behind the formation of the helicon core. We point out in this study that it is not necessary for core formation to accompany the inductive – helicon transition. A few previous studies have shown that the helicon mode exists even without the formation of the core [27, 42], mostly in the $m = 0$ (antenna having only circular coils, without the helical connections, [33]) helicon antenna. Moreover, in many cases the word "core" is very loosely used and even any central brightening of the plasma emission is thought to be a "core." In this chapter, we present experimental results that clearly distinguish between the inductive – helicon mode transition in an RF heated, argon plasma, and the formation of the classic "blue core." For certain source parameters, helicon plasma (shown by the discrete jump to high densities with increasing power and magnetic field, strong Ar-II emission, peaked central densities, low plasma potential, etc.) can occur without the formation of a distinct core. For such conditions, the plasma is dominated by low frequency resistive drift wave (RDW) instabilities driven by the radial density gradient. Note that in the helicon mode, the central densities are very peaked and the corresponding density gradient is sufficient to drive resistive drift waves [5, 21, 28]. The RDWs propagate in the electron diamagnetic drift direction and the resulting particle flux is radially outwards for all radii.

Initially, we are already in the helicon mode, but without the bright core. As we increase the magnetic field in the chamber, the plasma goes through a discrete and discontinuous change as an intense sharp core is formed at a particular value of the magnetic field. A new global equilibrium state is achieved where we find the simultaneous existence of three radially separated plasma instabilities [33]. The density gradient region, still dominated by RDWs, separates the plasma radially into the edge region and the core region. The edge region is dominated by strong, turbulent, shear driven Kelvin-Helmholtz (KH) instabilities, while the core region shows coherent Rayleigh-Taylor (RT) modes driven by the centrifugal forces due to azimuthal rotation. The RT modes are ion-dominated modes that propagate in the ion diamagnetic drift direction and are associated with enhanced light emission. The particle flux is directed outward for small radii and inward for large radii, thus forming a radial particle transport barrier which leads to another slight increase in the core plasma density. Simultaneously, the Ar II emission from the core region increases by an order of magnitude. The radial extent of the inner RT mode and radial location of the particle transport barrier is the same as the radius of the blue core. This new equilibrium with the RT – RDW – KH instabilities leads to the formation of the very stable, strong, and enhanced blue core. Just prior to this new global equilibrium state with the enhanced blue core, the system undergoes incomplete intermittent transitions between the two equilibrium states, leading to the visual perception of a centrally bright helicon core in a time-averaged sense. This is the first time that the development of the helicon core is shown to be associated with changes in radial transport.

5.2 EXPERIMENTAL APPARATUS

The experiments were carried out in an upgraded version of the CSDX, a cylindrical magnetized helicon plasma device [5, 33] devoted to the study of drift wave instabilities and zonal flows. The CSDX device is 2.8 m long with a vacuum chamber radius of 0.1 m. One end of the chamber mates with a pyrex glass bell jar which supports the helicon antenna while the other end is attached to a 1000 liters turbo-molecular pump to achieve a base pressure of 10^{-7} mTorr. The working gases are introduced through a calibrated MKS mass flow controller at the first port available downstream of the antenna (see Figure 5.1). The effective pressure in the chamber can also be controlled by a butterfly valve located in front of the turbo-molecular pump. The chamber pressure is measured by two Baratron gauges, one located at the port of the gas feed and the other located just in front of the butterfly valve at the end of the chamber. Plasmas are typically created at neutral gas pressures of 1 to 10 mTorr. The chamber is surrounded by 28 magnetic field coils that can provide an axially uniform, magnetic field of up to 240 mT. Radial variation of the magnetic field within the chamber is negligible and there is no detectable magnetic ripple between any two coils. In the last 50 cm of the chamber, the magnetic field decays by ~ 30%.

Argon plasma is produced by a 0.15 m diameter, $m = 1$ helical antenna (antenna having two circular coils connected by helical conducting straps, [33]), driven by a 13.56 MHz, 5 kW RF power supply. The present plasma source has been recently upgraded from a 1.5 kW, 10 cm diameter, $m = 0$ (antenna having only circular coils, without the helical connections, [33]) helicon antenna [5]. The reflected power is less than 20 W. Typical average electron, ion and neutral temperatures in CSDX are 4 eV, 0.6 eV [34, 35] and 0.5 eV, respectively, with plasma densities in the order of $10^{19} \, \text{m}^{-3}$.

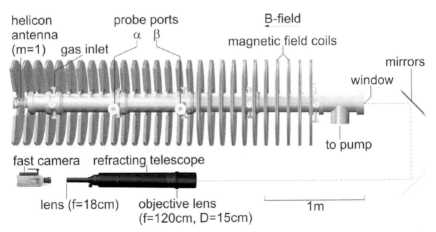

FIGURE 5.1 A schematic description of the CSDX device along with the magnets, the positioning of the steering mirrors, telescope and fast framing camera set up. The probes are inserted at the port position α, about 100 cm downstream from the source (taken as z = 0 cm) and about 80 cm downstream of the gas feed and at the port position β which is further 80 cm downstream.

Previously we also found that the plasma dynamics in CSDX is sensitive to the end boundary conditions [10, 36], as parallel currents can leak through finite end sheaths and short out the effective perpendicular ion polarization currents, which are associated with the turbulent Reynolds stress and the resulting radially sheared azimuthal E x B flow. Hence, here we operated CSDX with insulating boundary conditions to prevent currents from flowing to and through the walls. In this experiment, the end of the vacuum chamber ended in a glass window to allow imaging the full azimuthal plasma cross-section. The inner walls of the chamber which extended beyond the end of the array of magnets opposite to the plasma source (~0.5 m of the chamber near the pumps) were covered with Teflon insulation. On the source side, the inner wall of the flange that mates the glass bell jar ($r = 0.075$ m) to the stainless steel chamber ($r = 0.01$ m) was also covered by an insulating layer of boron nitride. Finally, for $r < 0.075$ m on the source end, the plasma encountered the end of the glass bell jar. Thus, it was ensured that all possible regions of open magnetic field lines could possibly come in contact with the chamber walls were insulated for this experimental study.

The experiments described in this chapter were performed at the following canonical conditions: the neutral gas pressure was 4.1 mTorr at the gas feed location and 3.2 mTorr at the end of the machine in front of the butterfly valve at a constant gas flow rate of 25 sccm (standard cubic centimeters per minute), and the power input to the antenna was kept constant at 1.6 kW, while the magnetic field (B) was varied from 40 mT to 240 mT. Even at the lowest B (40 mT), the plasma is already in the helicon mode of operation. As B is increased, the plasma characteristics change monotonically until 140 mT, where we find a sharp discrete change in the plasma characteristics and simultaneously observe the formation of the intense blue core. The magnetic field at which the blue core and all the global transition characteristics described later occur change slightly with power, pressure and gas flow rates but the features of the global transition leading to the blue core formation are qualitatively similar.

5.3   DIAGNOSTICS USED

Multi-tip Langmuir probes and fast imaging are used as the primary diagnostics. The plasma produced is relatively cold (a few eV) and the moderate heat fluxes to the probes allow taking long sets of data for good statistics. The probes were inserted into the chamber at two locations: α (~0.8 m) and β (~1.6 m) downstream of the gas feed (see Figure 5.1). The end window allows using a fast framing camera to record light intensity fluctuations. A time-averaged electron temperature and plasma potential are measured by an rf-compensated single tip swept Langmuir probe [32]. The probe was swept at 360 Hz and for each evaluation, we averaged over 100 sweeps. For density fluctuation measurements, we use Langmuir probes biased to a fixed voltage (typically −100 V) to collect ion saturation current. We measured the floating potential by using probes with a high resistance ($R = 100$ kΩ) placed in series, thus ensuring almost no current collected by the probe tip. In the absence of strong electron temperature fluctuations the floating potential fluctuations are interpreted as plasma potential fluctuations for comparison with RDW theories [5]. Similarly, the measured ion saturation current fluctuations are interpreted as density fluctuations. Reynolds stress and particle flux measurements are done with a 4-tip probe, which has been previously used for detailed studies of the DW turbulence and zonal flow interactions [42]. The Nyquist frequency of the probe data is 250 kHz, which is above the ion cyclotron frequency in our experiment and well above the observed low frequency ($f < 25$ kHz) phenomena that are observed during these experiments.

A 4-tip Mach probe is used to measure the bulk plasma velocity. The tips, placed symmetrically around the probe axis and separated by insulators, are biased negatively to collect ion saturation currents. The flows are interpreted based on the ratio (R_M) of the ion saturation currents collected by 180° opposite probe tips. The measurements are then fitted to the equation $R_M(\theta) = \exp[K \sin(\delta\alpha)/(\delta\alpha(M_{//} \cos \theta + M_{\perp} \sin \theta))]$, where θ is the angle between the magnetic field and the line connecting the 180° opposite probes $\delta\alpha$ is the acceptance angle of each probe tip (~90° in our probe design), $K = 1.34$, (using Hutchinson's model of ion collection [14]) and the parallel and perpendicular Mach numbers ($M_{//}$ and $M_{\perp}$, respectively) are the only fitting parameters. The methods have been compared against both time delay estimation (TDE) methods and Doppler shifted measurements of the ion fluid velocity using laser-induced fluorescence (LIF) across a wide range of B fields in CSDX [37, 38, 43].

For imaging the dynamics in the azimuthal cross-section of the plasma column, a fast framing camera is used to record the light intensity. A refractive telescope is used to focus the visible light from the plasma onto a Phantom V 710 high speed camera to reduce parallax effects. Two mirrors increase the optical path length, as shown schematically in Figure 5.1. The light intensity fluctuations have been previously found to be correlated with the ion saturation current fluctuations and is hence taken to be a proxy for density fluctuations [1, 19, 24]. A visible light from the plasma is steered using the two large mirrors (see Figure 5.1) on to a 1.2 m, $f/8$ Celestron C6-RGT telescope placed (~7 m) away from the focal plane having a 180 mm, $f/4$ lens as the objective. The image is then focused onto the camera sensors with a 25 mm, $f/1.4$ C-mount lens. The parallax is reduced due to the long focal distance which allows the lines of sight to be aligned to the background magnetic field to within an error of ~ ± 0.6°. Moreover, the contribution of the parallel dynamics on the imaging measurements is minimized since the depth of field around the image plane was made much less than the total length of the machine. For the present optical arrangement, each pixel images a volume comprising of $\delta x = \delta y = 1.5$ mm in the focal plane ($r_{plasma}/\delta x \sim 50$) and of depth $\delta z \sim 10$ cm ($L/\delta z \sim 30$). To understand the relative contribution of ions and neutrals, we use two interchangeable filters in front of the camera. From basic spectroscopy studies in argon helicon plasma in CSDX, it is known that the light emitted is dominated by singly ionized argon (Ar II) emission in the range 420–520 nm and neutral argon emission (Ar I) which is in the range 650–850 nm. Therefore, here we use commercially available filters to look at the ion emission (FWHM from 410 to 490 nm) and the neutral emission (long-pass filter for wavelength > 650 nm) separately. We also used narrow-band filters (at 488 nm for Ar II and 750 nm for Ar I) in previous experiments [19] to collect only one dominant argon ion or neutral transition, but the collected light intensity was smaller and had to be integrated for a longer time to get any reliable measurement, thus reducing the time resolution. For this experiment, the image sequences of 5000 frames were recorded at 210,500 frames per second with an exposure time of 1 to 2 μs at a spatial resolution of 128 × 128 pixels covering an area slightly larger than the whole plasma cross-section (~110 pixels). This allows us to study the plasma dynamics with sufficiently high spatial and temporal resolution.

5.4   EXPERIMENTAL RESULTS

In this section, we describe the experimental details before and after the formation of the core in the helicon mode and in the process study the inherent physical mechanism involved in the transition. As mentioned before, the helicon blue core is formed as the magnetic field in CSDX is increased. First, we show that we are in the helicon mode even at the lowest B field of operation. Figure 5.2 shows the ion saturation current measured for $B = 40$ mT as the RF power is increased. The mode transitions observed are very similar to those seen in Refs. [12, 13]. We find a direct capacitive to helicon mode transition (the density increases by about two orders of magnitude) at ~ 920 W of forward power in the increasing direction. But while decreasing the RF power, we find strong hysteresis; the mode first enters the inductively coupled phase and then ~ 620 W, we find the inductive to capacitive back transition.

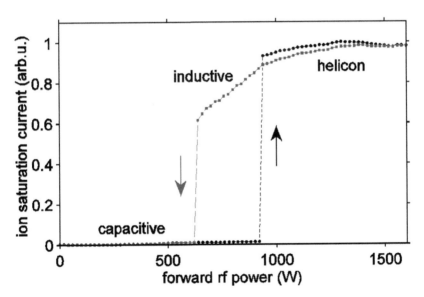

FIGURE 5.2   Capacitive, inductive, and helicon mode transitions for CSDX: the ion saturation current (measured at $r = 2$ cm) as a function of forward power to show the mode transitions, at B = 40 mT. The increasing RF power is denoted by black, while decreasing power is denoted by blue (and by arrows).

For 25 sccm flow of the neutral argon gas at 1.6 kW of RF power, even at 40 mT, we find all the standard helicon mode characteristics: sharp mode transition leading to increase in densities to $\sim 10^{19}$ m^{-3}, the bright blue Ar II dominated plasma with centrally peaked density profiles with low plasma potentials. A subsequent increase of the RF power does not bring substantial change in the system, as the RF heated plasma source is already in the helicon mode. Here, we find that even though we have centrally peaked plasma densities [5, 37], the center of the plasma is bright blue, but there is no substantially bright centrally localized region that looks like a core. So this is similar to the conditions when helicon plasma can form without the "blue core" [27, 44].

Once the device is in the helicon mode, further increase in the magnetic field initially changes the plasma characteristics monotonically, until 140 mT; where we see a sudden discrete change in all the plasma characteristics and the simultaneous appearance of the helicon blue core. Further increase in the magnetic field does not lead to any more discrete changes and the blue core mode remains stable.

As mentioned before, the visual changes during the operation of a helicon source, due to the changes in the light emission (in the visible range), aid in the operation and the identification of the mode transitions. Even at the lowest magnetic field of 40 mT, bright blue Ar II emission dominated the plasma emission, but visually we did not find a distinct core; even though the light emission from the plasma peaked at the center and trailed off towards the larger radii as the plasma density goes down by more than an order of magnitude at the edge. As the magnetic field was increased, we found a preferential brightening in the central region, which was around 100–130 mT and did have a visual effect of a very broad fuzzy core-like appearance. If we had not continued the experiments for even higher magnetic fields, we might have been misled into thinking that is the core similar to what is typically seen in most helicon devices. However, at 140 mT, the global plasma appearance changed completely and a very sharp distinct core was formed at the center.

To get a better quantitative understanding of this phenomenon, we use two-dimensional imaging of the plasma cross-section using the fast camera. Figure 5.3 shows the time averaged two-dimensional profiles of the Ar I (excited argon neutral) emission as we change the magnetic field.

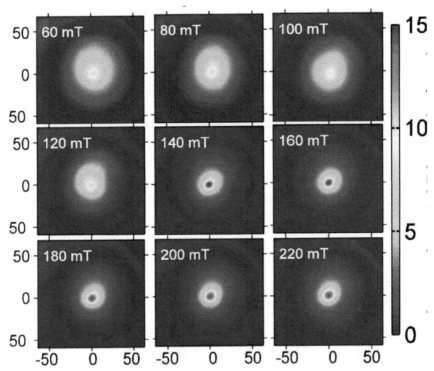

FIGURE 5.3 2-D profiles of time averaged argon neutral (Ar I) light emission. The units of the spatial scale are in mm. The antenna is 50 mm in radius (for comparison).

Similarly, Figure 5.4 shows the time averaged two-dimensional profiles of the Ar II (argon ion) emission for the same magnetic fields. Note that the antenna has a radius of 0.05 m. So for 40 mT, these profiles looks very much like the two-dimensional density cross-section of the plasma in a helicon mode as shown in Franck et al. [13]. But, we clearly see the formation of the very intense sharp helicon core at 140 mT. The emitted light intensities remain similar for all values of the magnetic field larger than 140 mT. Between 40 mT and 130 mT, the light intensity monotonically increases (for both Ar I and Ar II light) followed by a sudden sharp change in the light emission pattern. Between 130 mT and 140 mT, the intensities of both Ar I and Ar II go up discretely (a factor of two for Ar I and a factor of 10 for Ar II) and the radial profiles become very narrow. This is the appearance of the very sharp helicon blue core. We note that the central part was saturated after the formation of the core. The Ar II emission was so bright for $B > 140$ mT, to prevent complete saturation and damage to the detectors of the camera, we had to turn the aperture of the camera down. In spite of that, as we see from Figure 5.4, the center looks saturated for $B > 140$ mT. With respect to that, the Ar II emission seems weak for $B < 140$ mT since we plotted the intensities of all the B fields using the same color scheme. But we do get a clear picture of the formation of a sharp, spatially localized, very intense core at 140 mT.

To understand the physical features that lead to the formation of this core, we investigated the underlying instabilities associated with this global transition. We base our observations from both probe measurements of density and potential fluctuations and also the light intensity fluctuations from the fast framing camera. In Thakur [37, 38], we outlined the methodology used and given here the details of the plasma dynamics during the transition from nonlinear-coupled drift Eigen modes to broadband

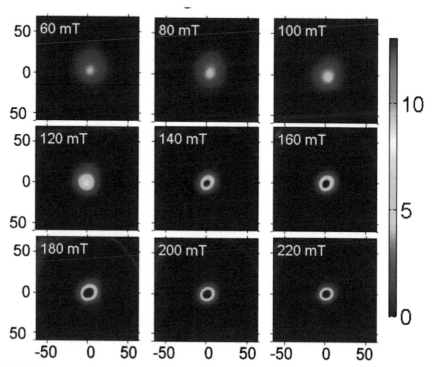

FIGURE 5.4 2-D profiles of time averaged argon ion (Ar II) light emission. The units of the spatial scale are in mm. The antenna is 50 mm in radius (for comparison).

multi-instability plasma turbulence. Here, to stress on the core formation, we show and compare the plasma characteristics from four sets of magnetic field values: 80 mT (plasma is in helicon mode but far away from the helicon core formation), 120 mT (approaching the transition to the core formation), 140 mT (the first magnetic field for which we see the stable core formation), and 160 mT (after the core is formed).

We show the radial profiles of the mean plasma density and the floating potentials as solid white lines over-plotted on Figure 5.5. For 80 mT and 120 mT, the peak plasma densities near the center are very similar and the profiles are Gaussian. The density profile at 120 mT is slightly narrower than that at 80 mT, most probably due to the confining effects of a larger magnetic field. But for 140 mT, we see a drastic change in the shape of the profile. The plasma can sustain a non-Gaussian profile with very steep density gradients. At 160 mT, after the helicon core has formed, the core plasma density increases by about 15–20%, and the profiles relax back to being Gaussian. We also see the corresponding floating potential profiles. For 80 mT, the floating potential is mostly flat at ~ −10 V from the center to about ~0.09 m and then increases to become zero at the chamber wall. For larger values of B, while the floating potential at the center remains almost the same, a potential well (~−30 V) is formed in the radial region of $0.05 < r < 0.08$ m. This dip in the mid-radii increases with B and thus leads to a very steep potential gradient near the edge of the plasma ($0.08 < r < 0.1$ m). This potential is well and therefore the corresponding edge electric field is the strongest for 140 mT and then relaxes back, as seen for 120 mT and 160 mT. These subtle differences become important when we investigated the fluctuation properties. These mean profiles lead to instabilities that play a crucial role in the formation of the helicon core.

The radial profiles of the frequency spectra (fast Fourier transform) of the density and the potential fluctuations are also shown in Figure 5.5. At 80 mT, the plasma is dominated by several nonlinearly interacting but discrete Eigen modes over a weak broader background. The density and the potential fluctuations of both the peaks at around the location of the maximum density gradient (shown as a white solid line) and have similar intensities. Moreover, the phase between the density and potential fluctuations are near zero at this region. All of these are signatures of standard resistive drift wave (RDW) instabilities, as it has been studied previously

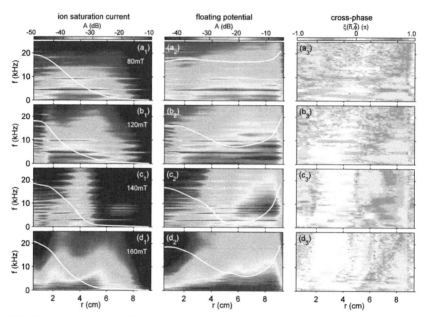

FIGURE 5.5 Radial profiles of the density and potential frequency spectra and cross-phases with magnetic field: radially resolved frequency spectra obtained from fluctuations of ion saturation current (column 1: a1 – d1) and floating potential (column 2: a2 – d2), and cross-phase between density and potential (column 3: a3 – d3). In columns 1 and 2 the time-averaged radial profiles of plasma density and floating potential are plotted superimposed as a solid white line (the scales of the X-axis for columns 1 and 2: density profiles 0 to 1.5×10^{19} m^{-3} and floating potential profiles -34 V to 0 V).

in an older version of CSDX (with a 0.05 m diameter helicon antenna, with maximum magnetic field of 100 mT) [5, 22]. For higher magnetic field, we start to see new features in the potential spectra at the edge, in addition to the in-phase density and potential fluctuations around the location of the density gradient. We find strong potential fluctuations at the edge where the mean potential gradients are strong, and far away from the density gradients and which has no counterpart in the density fluctuations. At 140 mT, the edge potential fluctuations become very strong, and the phase difference between density and potential deviates strongly from zero. These are signatures of a shear-driven Kelvin Helmholtz (KH) instability. Meanwhile at 140 mT, the in-phase density and potential fluctuations are localized to only near the strong density gradient area, signifying drift waves are still present at that location. We also find another signature of deviation from drift waves near the center of the plasma. Using fast imaging (see Figures 5.6 and 5.7, and the corresponding discussions), we identify that as a rotation induced Rayleigh Taylor (RT) instability (also known as gravitational instability, where the rotation gives the centrifugal force term that acts like gravity) . More details of the identification of the individual instabilities, including comparison to theoretical linear growth rates, are given in Refs. [37, 38]. For 160 mT, the signatures are similar to that for 140 mT, except

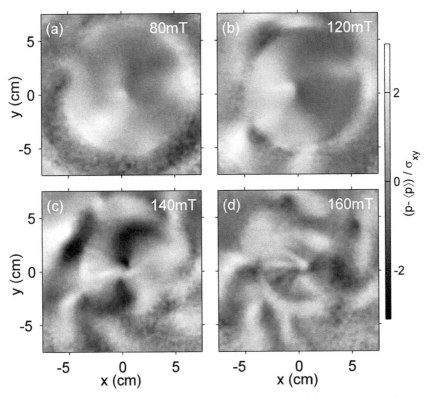

FIGURE 5.6 Snapshots of camera movies recording light of ArI emission lines for the different magnetic fields. Each pixel is normalized to its standard deviation σxy to enhance contrast. Note the formation of a quasi-coherent structure with a very high mode number (m > 15) near the plasma center for 140 mT (here it is still strongly modulated by the m = 2 drift wave) and 160 mT. This is the only mode that propagates in the ion diamagnetic drift direction, consistent with an ion dominated Raleigh-Taylor mode. All other modes propagate in the electron diamagnetic drift direction, consistent with drift waves.

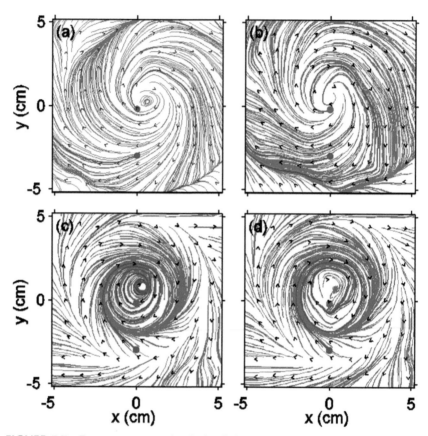

FIGURE 5.7 Frequency-averaged velocity fields obtained from time delay estimated velocimetry of the camera movies are shown in streamline plots. The frequency is averaged over 3–10 kHz. Here, we have (a) = 80 mT, (b) = 120 mT, (c) = 140 mT and (d) = 160 mT.

that there is larger scatter in the plots, most probably signifying stronger nonlinear interaction as we see from the broadening of the spectra.

In Figure 5.6, we show snapshots of camera movies recording the Ar I emission for the different magnetic fields. Since the densities and the light emission intensity at the edge is low, each pixel is normalized to its standard deviation σ_{xy} to enhance visibility. For 80 mT and 120 mT, we find modes with low azimuthal mode number, m (typically $m < 5$). These modes propagate in the electron diamagnetic drift direction and have been previously identified as resistive drift waves. For 140 mT, we find the formation of a quasi-coherent structure with very high mode number (m > 15) near the plasma center (though still strongly modulated by a super-imposed m = 2 drift wave). For 160 mT, the high mode number waves are more stable and robust. This is the only mode that propagates in the ion diamagnetic drift direction, consistent with an ion dominated RT mode. All other modes outside of the inner radii rotate in the electron diamag-netic drift direction, consistent with drift waves. Moreover, for B > 140 mT, at the edge, we also see very strong shear driven waves being born, growing in amplitude and spatial extent and finally breaking off. These waves produce intermittent bursts of plasma that go from the edge all the way towards the location of the inner mode.

To summarize, for B < 130 mT, we find standard drift wave instabilities with low m modes propagating in the electron diamagnetic drift direction. As a coincident with the appearance of the blue core, for B > 140 mT, we find the simultaneous presence of three different instabilities, the ion domi-nated high mode number RT mode at the center, the RDW at the density gradient region in the mid-radii, and very strongly turbulent KH at the edge.

In addition to the formation of these three instabilities for B > 140 mT, we find sharp changes in the global behavior of the radial particle flux of the plasma in response to these low frequency instabilities. In Figure 5.7, we show streamline plots of the averaged velocity, calculated from the Ar I light emission camera data, using standard time delay estimation (TDE) methods of inferring velocities [23, 43]. The TDE methods were applied to 25 msec long movies (averaged by windowing for 5 msec) to get the plasma velocity vectors at each point. For 80 mT and 120 mT, the veloc-ity fields indicate that the plasma spirals outward, as represented by the two colored dots (red dot taken near the center and the blue dot taken ~ 0.03 m). For 80 mT, the rate at which the plasma spiral out is faster than that at 120 mT. However, we see a very distinct change in the behavior for 140 mT and 160 mT. We find that the plasma starting from the red dot near the center remains trapped inside a circle of radius ~ 0.02 m. On the other hand, for plasma starting at the blue dot, it is pulled inward toward the circle of radius ~ 0.02 m, but takes a really long time to penetrate the inner radius of ~ 0.02 m. It seems as if for r > 0.02 m, the inward radial velocities dominate, but once it is near the edge of the circle, the azimuthal velocity is dominant and the inward radial motion almost stops.

In Figure 5.8, we show the radial profiles of the particle transport as the magnetic field is varied from the lowest values of helicon source operation (40 mT) to the maximum magnetic values that our current system permits (240 mT). This is calculated from the fluctuating density and electric fields measured by the Langmuir probes. Thus, the particle flux is measured by a completely different set of diagnostics than that showed in Figure 5.7. In this figure, red signifies positive particle flux in the outward direction, while blue represent negative particle flux which is directed inwards. For low magnetic fields we see that the particle flux is positive which means that the plasma propagates out toward the larger radii, consistent with a picture of diffusive transport down the density gradient (the helicon mode has a centrally peaked radial density profile). As we increase the magnetic field, near the B-field threshold at 140 mT, when the strong KH at the edge develops, we find negative particle flux, in which the particles move inward (consistent with the velocity fields calculated from the fast camera movies). This is also consistent with the existence of a strongly sheared $E \times B$ flow in this regime. With further increase in B we find a distinct separation of regions with inward and outward particle flux. The inner radial location where a possible RT instability dominates, we see

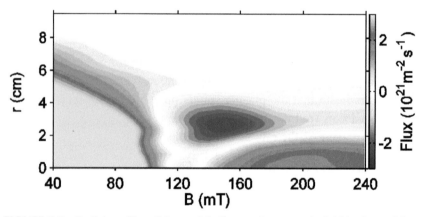

FIGURE 5.8 Radial profiles of the particle flux, as the magnetic field is changed from low values (40 mT) to high values (240 mT). Note the change in the characteristics at 140 mT. Blue signifies inward particle flux, while red signifies outward particle flux. At 140 mT, we observe the formation of a radial particle barrier (the thin white region in between the red and blue regions, showing no crossing of particles). This coincides with the radial location and the source parameters for the formation of the helicon blue core.

outward particle flux followed by a very narrow layer (the thin white area in between the blue and the red regions in Figure 5.8, for $B > 140$ mT) where there is almost zero radial particle flux. For larger radii, where KH dominates, we find inward particle transport.

Thus from the Figures 5.7 and 5.8, it seems that in the simultaneous presence of the three instabilities, the plasma develops a particle transport barrier, where the width of the transport barrier is consistent with the radial extent of the core RT-dominant region. This width is also the same as that of the helicon blue-core, as seen on the camera. The appearance of this radial particle transport barrier also coincides with the appearance of the blue-core mode with respect to B (~140 mT). For the same values of B, enhanced axial velocity at the core and an enhanced solid-body-like azimuthal rotation in the core is observed which may also have an effect on the radial particle transport. The formation of the transport barrier, coincident with the formation of the blue core is confirmed by both probe measurements and the velocity streamline plots extracted from the camera movies independently as shown in Figures 5.7 and 5.8.

5.5   DISCUSSIONS

Let us summarize the observations from the experiment. As the magnetic field is increased from low values, we observe the appearance of a very intense, sharp, and well-defined helicon core at 140 mT (see Figures 5.3 and 5.4). Along with the appearance of the core, we observe many phenomenological changes in the device. For $B < 140$ mT, the plasma is dominated by the resistive drift wave instability with low azimuthal mode number waves ($m < 5$) propagating in the electron diamagnetic drift direction. But for $B > 140$ mT, we find the simultaneous existence of three different plasma instabilities: a very high azimuthal mode number ($m > 15$), ion dominated RT mode rotating in the ion diamagnetic drift direction localized to the central core region, resistive drift waves in the mid-radial region where the density gradient is strong, and very strong turbulent shear driven KH instability at the plasma edge for larger radii (see Figures 5.5 and 5.6). Other than the RT mode, all azimuthal rotation is in the electron diamagnetic drift direction.

The radial localization of the instabilities and their appearance in parameter space is governed by the availability of the free energy sources that are responsible for driving these classic instabilities, for example, density gradients in case of RDW, strong rotation for the RT and strong shear for the KH [19]. In this experiment the mean profiles are consistent with the origin of these instabilities. The RDW is always found to be at the region with the strongest density gradient for all magnetic fields. For $B > 140$ mT, the KH is found to be near the edge region, consistent with the steep potential profiles that provide the strongest E x B shear. Previous studies in earlier versions of CSDX (with max $B = 100$ mT) have shown that the strong density gradients lead to nonlinearly interacting drift waves that lead to weak turbulence [5, 21]. The drift wave turbulence could nonlinearly drive mean azimuthally symmetric large-scale $m = 0$ shear flows, called zonal flows in CSDX [22, 39]. It seems that as the magnetic field is increased further, these zonal flows become so strong that they can drive the edge localized shear driven instabilities that we find for $B > 140$ mT. Simultaneously, the inner RT mode becomes prominent due to the enhanced solid-body-like rotation observed in the plasma core, which is the source of the centrifugal term that drives the RT waves.

Also from Figures 5.7 and 5.8, we find that the radial particle flux undergoes drastic changes at 140 mT, coincident with the formation of the helicon core. Before the core formation, for $B < 140$ mT, the particle flux is directed radially outward, down the density gradient for all radii. But for $B > 140$ mT, the plasma within the inner core is radially directed outward, while the plasma at larger radii is directed inward. This creates a very well defined radial particle barrier, whose radial location is the same as that of the radial extent of the helicon core, and is also the same as the radial extent of the inner RT ion dominated mode driven by the inner solid-body-like plasma rotation.

From the experimental evidence we see that the helicon blue core formation is related to the formation of the radial particle transport barrier that occurs due to the presence of three different inherent plasma instabilities in the system. For the parameters where the blue core is formed, the drift waves at the mid-radii essentially forms a barrier between the inner ion dominated RT modes and the outer strongly turbulent KH instability. We find this from both probe-based studies and from watching the light intensity movies from fast imaging. We find that the outer region and the inner regions of the plasma maintain completely separate phenomena due to the lack of inter-mixing of the energy sources that drive the separate instabilities in the two radially separated regions. These taken together, we believe, lead to the formation of the radial transport barrier, and thus the helicon core (Figures 5.3 and 5.4 together with Figures 5.7 and 5.8).

KEYWORDS

- blue core
- drift wave turbulence
- helicon source
- Kelvin Helmholtz
- plasma instabilities
- Rayleigh Taylor
- transport barrier

REFERENCES

1. Antar, G. Y., Yu, J. H., & Tynan, G. R. (2007). The origin of convective structures in the scrape-off layer of linear magnetic fusion devices investigated by fast imaging. *Physics of Plasmas, 14*, 022301.
2. Boswell, R. W. (1974). A study of waves in gaseous plasma. PhD Thesis, Flinders University, Adelaide, Australia (Also available online at www.heliconrefs.com).
3. Boswell, R. W. (1984). Very efficient plasma generation by whistler waves near the lower hybrid frequency. *Plasma Physics and Controlled Fusion, 26*, 1147.
4. Boswell, R. W., & Chen, F. F. (1997). Helicons – the Early Years. *IEEE Transactions on Plasma Science, 25*, 1229–1244.
5. Burin, M. J., Tynan, G. R., Antar, G. Y., Crocker, N. A., & Holland, C. (2005). On the transition to drift turbulence in a magnetized plasma column. *Physics of Plasmas, 12*, 052320.
6. Celik, M. (2011). Spectral measurements of inductively coupled and helicon discharge modes of a laboratory argon plasma source. *Spectrochimica Acta Part B, 66*, 149.
7. Chabert, P., Proust, N., Perrin, J., & Boswell, R. W., (2000). High rate etching of 4H–SiC using a SF6/O2helicon plasma. *Applied Physics Letters, 76*(16), 2310.

8. Chen, F. F., & Chevalier, G. (1992). Experiments on new RF plasma sources for etching and deposition. *Journal of Vacuum Science and Technology, A 10*, 1389.

9. Chen, F. F., & Boswell, R. W. (1997). Helicons – the Past Decade. *IEEE Transactions on Plasma Science, 25*, 1245–1257.

10. D'Ippolito, D. A., Russell, D. A., Myra, J. R., Thakur, S. C., Tynan, G. R., & Holland, C. (2012). Effect of parallel currents on drift-interchange turbulence: Comparison of simulation and experiment. *Physics of Plasmas, 19*, 102301.

11. Ellingboe, A. R., & Boswell, R. W. (1996). Capacitive, inductive and helicon-wave modes of operation of ahelicon plasma source. *Physics of Plasmas, 3*(7), 2797–2804.

12. Franck, C. M., Grulke, O., & Klinger, T. (2003). Mode transitions in helicon discharges. *Physics of Plasmas, 10*(1), 323–325.

13. Franck, C. M., Grulke, O., Stark, A., Klinger, T., Scime, E. E., & Bonhomme, G. (2005). Measurements of spatial structures of different discharge modes in a helicon source. *Plasma Sources Science and Technology, 14*, 226.

14. Hutchinson, I. H. (2005). Ion collection by a sphere in a flowing plasma: 3. Floating potential and drag force. *Plasma Physics and Controlled Fusion, 47*, 71–87.

15. Jung, H. D., Park, M. J., Kim, S. H., & Hwang, Y. S. (2004). Development of a compact helicon ion source for neutron generators. *Review of Scientific Instruments, 75*(5), 1878.

16. Kaeppelin, V., Carrère, M., & Faure, J. B. (2001). Different operational regimes in a helicon plasma source. *Review of Scientific Instruments, 72*, 4377.

17. Lieberman, M. A., & Lichtenberg, A. J. (2005). Principles of Plasma Discharges and Material Processing. John Wiley & Sons, New York, USA.

18. Light, A. D., Thakur, S. C., Brandt, C., Sechrest, Y., Tynan, G. R., & Munsat, T. (2013). Direct extraction of coherent mode properties from imaging measurements in a linear plasma column. *Physics of Plasmas, 20*, 082120

19. Light, M., Chen, F. F., & Colestock, P. L. (2001). Low frequency electrostatic instability in a helicon plasma. *Physics of Plasmas 8*, 4675.

20. Magee, R. M., Galante, M. E., Gulbrandsen, N., McCarren, D. W., & Scime, E. E. (2012). Direct measurements of the ionization profile in krypton helicon plasmas. *Physics of Plasmas 19*, 123506.

21. Manz, P., Xu, M., Thakur, S. C., & Tynan, G. R., (2011a). Nonlinear energy transfer during the transition to drift-interchange turbulence. *Plasma Physics and Controlled Fusion, 53*, 095001.

22. Manz, P., Xu, M., Fedorczak, N., Thakur, S. C., & Tynan, G. R. (2011b). Spatial redistribution of turbulent and mean kinetic energy. *Physics of Plasmas 19*, 012309.

23. Fedorczak, N., Manz, P., Thakur, S. C., Xu, M., Tynan, G. R., Xu, G. S., & Liu, S. C. (2012). On physical interpretation of two-dimensional time-correlations regarding time delay velocities and eddy shaping. *Physics of Plasmas 19*, 122302.

24. Oldenbürger, S., Brandt, C., Brochard, F., Lemoine, N., & Bonhomme, G. (2010). Spectroscopic interpretation and velocimetry analysis of fluctuations in a cylindrical plasma recorded by a fast camera. *Review of Scientific Instruments, 81*, 063505

25. Perry, A. J., Vender, D., & Boswell, R. W. (1991). The application of the helicon source to plasma processing. *American Vacuum Society, 9*(2), 310–317.

26. Rayner, J. P., & Cheetham, A. D. (1999). Helicon modes in a cylindrical plasma source. *Plasma Sources Science and Technology, 8*, 79–87.

27. Sakawa Y, Takino T and Shoji, T. (1998). Control of antenna coupling in high-density plasma production by m = 0 helicon waves. *Applied Physics Letters 73*, 1643.

28. Schröder, C., Grulke, O., Klinger, T., Naulin, V. (2005). Drift waves in a high-density cylindrical helicon discharge. *Physics of Plasmas, 12*, 4.

29. Scime, E. E., Keesee, A. M., & Boswell, R. W. (2008). Mini-conference on helicon plasma sources. *Physics of Plasmas, 15*, 058301.

30. Shinohara, S., & Yonekura, K. (2000). Discharge modes and wave structures using loop antennae in a helicon plasma source. *Plasma Physics and Controlled Fusion, 42*, 41–56.

31. Shoji, T., Sakawa, Y., Nakazawa, S., Kadota, K., & Sato, T. (1993). Plasma production by helicon waves. *Plasma Sources Science and Technology, 2*, 5.

32. Sudit, I. D., & Chen, F. F. (1994). RF compensated probes for high-density discharges. *Plasma Sources Science and Technology, 3*, 162.

33. Thakur, S. C. (2010). Understanding plasmas through ion velocity distribution function measurements, PhD dissertation, West Virginia University, Morgantown, USA, http://ulysses.phys.wvu.edu/~plasma/pdf/Chakraborty%20Thakur_Saikat_dissertation.pdf.

34. Thakur, S. C., McCarren, D., Lee, T., Fedorczak, N., Manz, P., Scime, E. E., Tynan, G. R., & Xu, M. (2012). Laser induced fluorescence measurements of ion velocity and temperature of drift turbulence driven sheared plasma flow in a linear helicon plasma device. *Physics of Plasmas, 19*, 082102.

35. Thakur, S. C., McCarren, D., Lee, T., Fedorczak, N., Manz, P., Scime, E. E., Tynan, G. R., Xu, M., & Yu, J. H. (2012). Comparison of azimuthal ion velocity profiles using Mach probes, time delay estimation, and laser induced fluorescence in a linear plasma device. *Review of Scientific Instruments, 83*, 10D708.

36. Thakur, S. C., Xu, M., Manz, P., Fedorczak, N., Holland, C., & Tynan, G. R. (2013). Suppression of drift wave turbulence and zonal flow formation by changing axial boundary conditions in a cylindrical magnetized plasma device. *Physics of Plasmas, 20*, 012304.

37. Thakur, S. C., Brandt, C., Cui, L., Gosselin, J. J., Light, A. D., & Tynan, G. R. (2014). Multi-instability plasma dynamics during the route to fully developed turbulence in a helicon plasma. *Plasma Sources Science and Technology, 23*, 044006.

38. Thakur, S. C., Brandt, C., Light, A., Cui, L., Gosselin, J. J., & Tynan, G. R. (2014). Simultaneous use of camera and probe diagnostics to unambiguously identify and study the dynamics of multiple underlying instabilities during the route to plasma turbulence. *Review of Scientific Instruments, 85*, 11E813.

39. Tynan, G. R., Holland, C., Yu, J. H., James, A., Nishijima, D., Shimada, M., & Taheri, N. (2006). Observation of turbulent-driven shear flow in a cylindrical laboratory plasma device. *Plasma Physics and Controlled Fusion, 48*, S51–S73.

40. West, M. D., Charles, C., & Boswell, R. W. (2009). High density mode in xenon produced by a Helicon Double Layer Thruster. *Journal of Physics D: Applied Physics, 42*, 245201.

41. Wiebold, M. D. (2011). The effect of radio-frequency self bias on ion acceleration in expanding argon plasmas in helicon sources. PhD Thesis, University of Wisconsin – Madison, USA (2011).

42. Yan, Z., Yu, J. H., Holland, C., Xu, M., Müller, S. H., & Tynan, G. R. (2008). Statistical analysis of the turbulent Reynolds stress and its link to the shear flow generation in a cylindrical laboratory plasma device. *Physics of Plasmas, 15*, 092309.

43. Yu, J. H., Holland, C., Tynan, G. R., Antar, G., & Yan, Z. (2007). Examination of the velocity time-delay-estimation technique. *Journal of Nuclear Materials, 363–365,* 728–732.

44. Yun, S., Cho, S., Tynan, G. R., & Chang, H. (2001). Density enhancement near lower hybrid resonance layer in m=0 helicon wave plasmas. *Physics of Plasmas 8,* 358.

45. Ziemba, T., Slough, J., & Winglee, R. (2005). High power helicon propulsion experiments. *Space Technology and Applications International Forum (STAIF), 746* (1), 965–975.

CHAPTER 6

MAGNETIC DRIFT AND ITS EFFECT ON CROSS-FIELD DIFFUSION PROCESS

P. HAZARIKA,[1] B. K. DAS,[1] M. CHAKRABORTY,[1]
and M. BANDYOPADHYAY[2]

[1]Centre of Plasma Physics-Institute for Plasma Research,
Nazirakhat, Sonapur – 782402, Kamrup, Assam, India,
E-mail: hazarikaparismita@rediffmail.com, bdyt.ds@rediffmail.com,
monojitc@yahoo.com

[2]ITER-India, Institute for Plasma Research, A-29, GIDC,
Electronic Estate, Sector-25, Gandhinagar – 382025, Gujarat, India,
E-mail: mbandyo@yahoo.com

CONTENTS

ABSTRACT

In order to study the effect of $E \times B$ and diamagnetic drift on cross-field plasma transport process, an experiment is carried out in double plasma device (DPD), and presented in this chapter. In the experimental configuration, the directions of both drifts are same and perpendicular to transverse magnetic field (TMF) as well as the DPD chamber axis. The TMF divides the plasma chamber into two distinct regions *viz.* source and target region on the basis of electron temperature. Plasma is produced in the source region by filament discharge method and then allowed to diffuse to the target region through the TMF. In order to study the electrically grounded and electrically biased side wall effect on different plasma parameters metallic plates are inserted in the TMF plane in a direction perpendicular to the TMF. Data are acquired by Langmuir probe and compared for different source configuration in terms of metallic plate bias.

6.1 INTRODUCTION

The transport of charged particles across magnetic field is a crucial issue in fusion grade negative ion sources and has been studied for the last three decades to have a better understanding over it. The chief purpose of the application of transverse magnetic field (TMF) in such negative ion source is to reduce the electron temperature upto a certain level so that it satisfies the condition for negative ion formation in the extraction region [1–3]. The production mechanism of negative ions in a conventional ion source is given elsewhere [2]. The electron cooling process by magnetic field is achieved since magnetic field selectively allows the passage of only the cold electrons and confines the hot electrons simultaneously which create two distinct regions (hot and cold) inside the ion source [4]. Although the cross-field diffusion of charged particle is a composite process, optimization of plasma density in the extraction part of ion source is crucial for enhancement of negative ion formation. The application of magnetic field often complicates the ion source operation due to the anisotropic behaviour of the charged particles as well as different drift associated with the field. Several theoretical models have been proposed so far to understand the cross-field diffusion process of the charged particle, but unfortunately very few of them was able to give the proper explanation of actual diffusion. Classical treatment of cross-field diffusion often exhibits very poor explanation due to the discrepancies over theoretical and experimental results and it has been observed in number of experiments that the actual cross-field diffusion often shows higher value than the theoretically predicted one. Such anomalous cross-field diffusion of charged particles has been observed in PIC simulations [5–7] and was well explained in a number of experiments through instabilities [8] as well as the short circuit effect [9]. Different models stress on different mechanism of filter operation and tries to explain this anomalous type diffusion. Another fact is that the orientation of the magnetic field produces magnetic drift which significantly influences cross-field diffusion. It has been observed that such magnetic drift in the filter region affects the extracted ion beam uniformity. In a typical negative ion source, the $E \times B$ drift as well as diamagnetic drift can produce significant drain of charged particle along the radial direction which affects on the overall cross-field diffusion process.

In the present experimental configuration, an axial electron flux exists from source to the target region and it gives rise to an axial electron current, J_{axial} (along the axis of the chamber), which is perpendicular to the magnetic field (B) generated by the transverse magnetic channels. This mutually perpendicular J_{axial} and B combines and is subjected to Lorentz force. This force leads to the formation of a current which is perpendicular to both J_{axial} and B, i.e., in $J_{axial} \times B$ direction. The axial current density J_{axial} consist of two terms, one is proportional to the plasma potential (plasma temperature) gradient term which gives rise to $E \times B$ drift current and other is proportional to electron pressure gradient (arises due to density gradient) term which gives rise to diamagnetic drift current term. It was suggested that between these two drift terms in a standard negative ion source the dominant contribution to the $J_{axial} \times B$ comes from the diamagnetic drift current term [10]. So throughout the manuscript we have given main importance to diamagnetic drift current term and neglected the $E \times B$ drift term. Previous 2D simulation results of electron transport through the magnetic filter also showed that the presence of chamber wall in the $J_{axial} \times B$ direction plays an important role in the transport across the filter [11].

In magnetized plasma the cross-field diffusion flux can be given by:

$$\Gamma = n\vec{v}_\perp = \pm n\mu_\perp \vec{E} - D_\perp \nabla n + \frac{n(\vec{v}_E + \vec{v}_D)}{\left[1 + \left(v^2/\Omega_c^2\right)\right]} \qquad (1)$$

where $E \times B$ drift velocity is,

$$\vec{v}_E \approx \frac{\vec{E} \times \vec{B}}{B^2} \qquad (2)$$

And the diamagnetic drift velocity term is

$$\vec{v}_D \approx -\frac{\vec{\nabla}p \times \vec{B}}{enB^2} \qquad (3)$$

where n is the plasma density, $\mu_\perp$ is the cross-field mobility, E is the electric field, $D_\perp$ is the cross-field diffusion coefficient, v is the collision frequency, Ω_c is the cyclotron frequency, p is the pressure, and e is the electronic charge. It is to be noted that both the drift are in same direction. But due to charge independent nature of $E \times B$ drift both electrons and ions will move in the same direction. On the other hand due to diamagnetic drift electrons and ions will move in the opposite direction as this drift depends on charge. To control the diamagnetic drift, two metallic plates are inserted in the $J_{axial} \times B$ direction (perpendicular to both TMF and the chamber axis) and biased the plates in both polarity configurations (applied electric field in a direction parallel to drift current direction and opposite to drift current direction).

6.2 EXPERIMENTAL SET-UP

The experiment is performed in a double plasma device (DPD), which is constructed with two identical multi-cusp magnetic cages. The discharge chamber shown in Figure 6.1(*a*) is 1.1 m long and 0.38 m in diameter. The multi-cusp cages (diameter 0.25 m; length 0.32 m) are placed inside the discharge apparatus and are separated from each other by two magnet channels which produce the TMF. The channels of the magnetic cages consist of fourteen vacuum-sealed rectangular tubes. Each tube consists of small permanent magnets of dimension 0.025 m × 0.025 m × 0.025 m (surface field strength 0.12 Tesla). The field strength at the centre of this transverse magnet channels separation is ~2.4 × 10⁻⁴ Tesla. The photographs of the experimental chamber along with the multi-cusp cage are shown in Figure 6.1(b) and 6.1(c).

The magnetic field profile between the TMF channels and along the axis of the chamber are shown in Figure 6.2(a) and 6.2(b).

The separation distance between the two TMF channels is ~0.15 m. The channels are kept insulated from each other and also from the magnetic

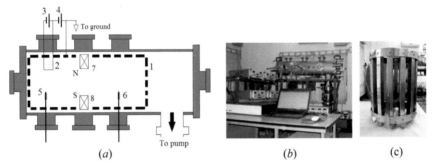

FIGURE 6.1 (a) Schematic diagram of the experimental chamber: (1) multi-dipole magnetic cage, (2) filament, (3) filament voltage power supply (VF), (4) discharge voltage power supply (VD), (5,6) Langmuir probes, and (7,8) transverse magnet channels. (b) The photograph of the vacuum chamber; and (c) the photograph of the multi-dipole magnetic cage.

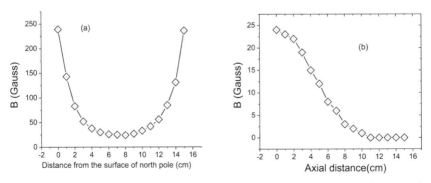

FIGURE 6.2 (a) Magnetic field profile between the TMF channels and (b) axial magnetic field profile.

cages to ensure deliberate electrical connection to the ground or to some power supplies according to experimental configuration requirements. At first the cylindrical vacuum chamber is evacuated with the help of a rotary and diffusion pump combination. Once the ultimate chamber pressure of ~10^{-6} mbar is achieved, hydrogen gas is injected until the chamber pressure is raised up to ~5×10^{-4} mbar. Plasma is produced in the chamber by igniting discharge with the help of five numbers of filaments. Each filament is of length 0.03 m and diameter 2×10^{-4} m, located inside the source cage. Discharge voltage (V_D) and discharge current (I_D) is fixed at 80 V and 1 A, respectively. The produced plasma in the source region diffuses into the target region through the TMF.

The effect of diamagnetic drift is studied by inserting two metallic plates (dimension: length ~ 7 cm, width ~ 2 cm) along the radial direction of the chamber in such a way such that each of them face each other in the TMF plane. The backsides of the plates are insulated so that only the front side draws current. An electric field E_l is generated along the DPD axis due to the potential gradient between source and target region. The TMF channels are insulated to reduce Simon's short-circuit current [9] on the TMF surface. An electric field (E_{app}) is applied between the two plates by biasing them externally with a power supply in either polarity such that applied electric field is in both parallel as well as opposite to the direction of the electric field generated due to charge polarization (E_2). The schematic cross-sectional view of the TMF region with the plates is shown in Figure 6.3. Experimentally drawn currents for two polarity cases are used to determine diamagnetic drift current and compared with the theoretical one using relation (3). To understand the effect of sidewall on the plasma transport data in the TMF region are acquired both in presence and absence of metallic plates. Data collection by Langmuir probe and corresponding analysis is done by hidden advanced Espion Langmuir probe system. The dimension of the probe is as follows: length 0.01 m and diameter 1×10^{-4} m. In order to measure the plasma flow velocity in the source, TMF and target regions, a planer Mach probe having length 0.013 m and diameter 0.008 m is used. Two tips of the Mach probe are separated from each other at a distance of 0.003 m by a ceramic insulator. In the TMF region this flow velocity is measured both along and across magnetic field direction.

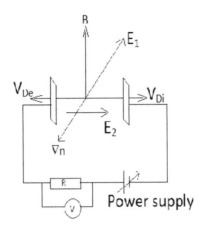

FIGURE 6.3   Schematic diagram of the TMF region.

6.3   EXPERIMENTAL RESULTS AND DISCUSSION

In our experimental chamber due to 2.4×10^{-3} TMF at the centre of the TMF region the electron gyro frequency (Ω_e) is found to be $\sim 10^8$–10^9 Hz, ion gyro frequency $(\Omega_i) \sim 10^5$ Hz and coulomb collision frequency (v_c) is $\sim 10^6$ Hz and electron-neutral $(e\text{-}n)$, (the most dominant binary collision) collision frequency (v_{en}) is $\sim 10^5$ Hz. Since in our experimental chamber $\Omega_i \sim v_c \sim v_{en} \sim v$ ions are unmagnetized and due to $\Omega_e \gg v$ electrons are highly magnetized. Percentage of ionization in our chamber is $\sim 0.5\%$ and due to these weakly ionization ions are assumed to be at room temperature throughout the whole experiment.

In our experimental setup near TMF region and along the axis of the chamber there are gradients of (i) plasma density and (ii) plasma potential due to plasma temperature variation created by the TMF. The density gradient and electric field generated by the gradient of potential in presence of transverse magnetic field generates (a) $(E_{axial} \times B)$ drift (b) diamagnetic drift in the same direction, perpendicular to both the axial and TMF direction. The $(E_{axial} \times B)$ drift is charge independent, but the diamagnetic drift is charged dependent and therefore, an electric field E_2 is expected to be developed across the TMF plane due to polarization of charges. This electric field E_2 combine with TMF generated magnetic field can produce another $(E_2 \times B)$ drift along the axis of the chamber. This axial $(E_2 \times B)$

drift combining with other drift originally present in our chamber may generate unstable electrostatic waves in the axial direction may be a cause of enhanced electron transport from source to the target region [12].

Two metallic plates are inserted in the filter region in order to measure the value of diamagnetic drift current experimentally. To study the effect of extra plasma loss area, a single piece of SS plate is installed near the chamber wall to observe the change in radial plasma distribution and the data are collected by the probe both in presence and absence of the SS plate. Figures 6.4 and 6.5 are the radial plasma density and temperature profile with and without SS plate.

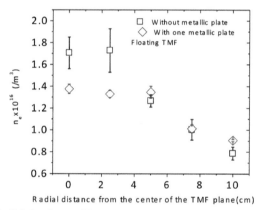

FIGURE 6.4   Radial density profile.

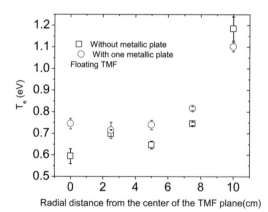

FIGURE 6.5   Radial temperature profile.

From Figure 6.4 it is observe that without the SS plate plasma density falls uniformly towards the chamber wall. When the SS plate is inserted radially in the TMF zone (placed very close to the chamber wall, the surface of this plate lies at the edge of the vertical TMF plane), plasma density also shows a fall from centre to the wall. But a depletion of plasma density has been observed at the centre after insertion of the plate whereas near the chamber wall it resembles with the previous case, i.e., without SS plate. The electron temperature decreases from chamber wall and almost saturates at the bulk plasma in both cases (Figure 6.5). The depletion of plasma density after the insertion of the metallic plates is due to the presence of electron absorbing surface. The current collected by two plates in either polarity configuration is shown in Figure 6.6. The difference of current in two polarity cases is linked to the unidirectional drift current. The drift current is calculated from this experimental data and is shown in Figure 6.7. From the figure it is observed that the drift current decreases with increasing plate biasing voltage.

Figure 6.8 shows the axial electric field profile from the source to target region for both the electric field configuration which increases with plate biasing voltage.

In order to investigate whether the nature of the drift is diamagnetic or $E_2 \times B$ drift dominating, experimentally calculated diamagnetic drift velocity value is compared with theoretically obtained $E_2 \times B$ and diamagnetic drift velocity value for both the polarity configurations (Figure 6.9). From the Figure 6.9 it is observed that experimentally observed velocity

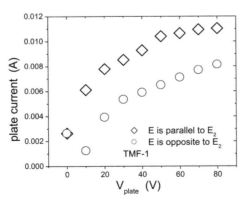

FIGURE 6.6   Plate biasing voltage (V_{plate}) vs. plate current.

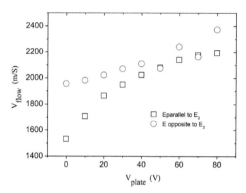

FIGURE 6.7   Experimental drift current vs. plate biasing voltage.

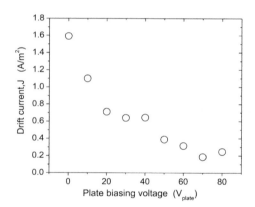

FIGURE 6.8   Axial electric field vs. plate voltage.

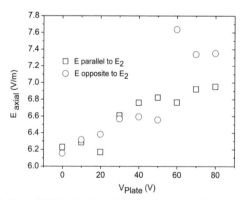

FIGURE 6.9   Variation of drift velocity with plate biasing voltage.

value is much lower than that of theoretically calculated diamagnetic drift velocity value. This indicates that along with diamagnetic drift some other phenomenon also taking place in our chamber which lowers our experimentally observed drift velocity value.

Axial plasma flow profile with increasing plate biasing voltage (V_{Plate}) in the source region is shown in Figure 6.10. From the figure an increasing trend of flow velocity in the source region along with increasing V_{plate} is observed. The monotonically increased axial electric (as shown in Figure 6.8) field may be responsible for accelerating plasma in the source region. A comparison between the measured axial flow velocity with the help of Mach probe and the flow velocity due to this axial electric field for both the polarity configurations is shown in Figure 6.11.

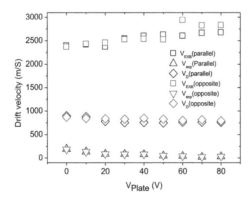

FIGURE 6.10   Variation of flow velocity with V_{plate}.

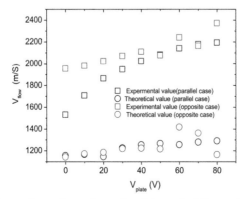

FIGURE 6.11   Comparison of experimental and theoretical flow velocity with V_{plate}.

From the figure it is seen that our experimentally observed velocity shows a higher value as compared to theoretically calculated value but the order for both the value is same and both the velocity value increases with V_{plate}. The variation of plasma density in the source region with respect to plate biasing voltage (V_{plate}) is shown in Figure 6.12. Density starts to decrease with increasing plate biasing voltage for both polarity cases. But in both the cases at each biasing voltage, density is found to be approximately same.

Figure 6.13 is the variation of plasma density in the target region. Target density is found to increase monotonically with V_{plate} for both the electric field configuration. Density gradient calculated from Figures 6.12

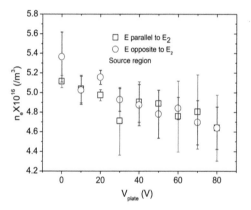

FIGURE 6.12 Density variation in the source region with V_{plate}.

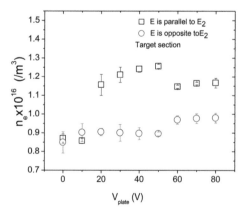

FIGURE 6.13 Density variation in target region with V_{plate}.

and 13 is shown in Figure 6.14. From the figure it is observed that density gradient is reducing with increasing bias voltage but is unaffected by bias polarity direction.

Figure 6.15 is the correlation between plasma gradient and both theoretical and experimental drift current. From the figure it is seen that density gradient is proportional to diamagnetic drift.

6.4 CONCLUSION

The diamagnetic drift in a negative ion source type chamber may be a prime candidate for modifying cross-field plasma transport. The purpose

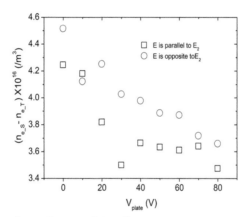

FIGURE 6.14 Density gradient vs. plate voltage.

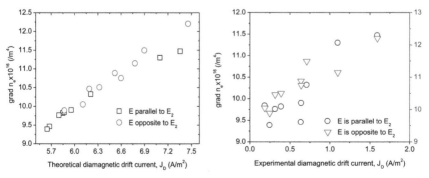

FIGURE 6.15 Variation of density gradient with theoretical and experimental diamagnetic drift current.

of this work is to get an idea about how plasma drifts affect the cross-field plasma transport and how the presence of metallic boundary in the J_{axial} $\times B$ direction influences the plasma parameters in the target region. It is observed that if we have a proper control on the plasma drifts, plasma density in the target region can be enhanced. A detailed analysis of this effect along with the effect of various instabilities on cross-field transport will be reported in future.

KEYWORDS

- cross-field plasma transport
- double plasma device
- plasma drift
- transverse magnetic field

REFERENCES

1. Djermanov, I., Kolev, S. T., Lishev, S. T., Shivarova, A., & Tsankov, T. S. (2007). Plasma behaviour effected by a magnetic filter. *J. Phys.: Conf. Ser. 63*, 1–5.
2. Holmes, A. J. T., McAdams, R., Proudfoot, G., Cox, S., Surrey, E., & King (1994). Intense negative ion sources at Culham Laboratory (invited). *Rev. Sci. Instrum. 65*, 1153–1158.
3. Hagelaar, G. J. M., & Oudini, N. (2011). Plasma transport across magnetic field lines in low temperature plasma sources. *Plasma Phys. Control. Fusion, 53*, 1–12.
4. Nakano, T., Mori, S., Tauchi, Y., Ooharo, W., & Fukumasa, O. (2009). Relationship between production and extraction of D⁻/H⁻ negative ions in a volumen negative ion source. *J. Plasma Fusion Res. Series, 8*, 789–793.
5. Ohi, K., Naitou, H., Tauchi, Y., & Fukumasa, O. (2001). Bifurcation in asymmetric plasma divided by a magnetic filter. *Plasma Phys. Control. Fusion, 43*, 1615–1624.
6. Naitou, H., Ohi, K., & Fukumasa, O. (2000). Beam instability excited by the magnetic filter. *Rev. Sci. Instrum, 71*, 875–876.
7. Riz, D., & Pamela, J. (1998). Modeling of negative ion transport in a plasma source (invited). *Rev. Sci. Instrum, 69*, 914–919.
8. Santhosh Kumar, T. A., Mattoo, S. K., & Jha, R. (2002). Plasma diffusion across inhomogeneous magnetic fields. *Phys. Plasmas, 9*(7), 2946–2953.
9. Simon, A. (1955). Ambipolar Diffusion in a Magnetic Field. *Phys. Rev., 98*(2), 317–318.

10. Chaudhury, B., Boeuf, J. P., Fubiani, G., & Claustre, J. (2012). Currents through a magnetic filter in a low temperatura plasma from a Particle-In-Cell monte Carlo Collisions model. *ESCAMPIG XXI*, 10–14.

11. Fubiani, G., & Boeuf, J. P. (2014). Plasma asymmetry due to the magnetic filter in fusion-type negative ion sources: Comparisons between two and three-dimensional particle-in-cell simulations. *Phys. Plasmas, 21*, 073512-8.

12. Pal, A. R., Chutia, J., & Bailung, H. (2004). Observation of instability in presence of $E \times B$ flow in a direct current cylindrical magnetron discharge plasma. *Phys. Plasmas, 11*(10), 4719–4726.

CHAPTER 7

ON THE EFFECT OF BASE PRESSURE UPON PLASMA CONTAINMENT

G. SAHOO,[1] R. PAIKARAY,[2] S. SAMANTARAY,[2,3] P. DAS,[2] J. GHOSH,[4] and A. SANYASI[4]

[1]Stewart Science College, Cuttack, Odisha, 753001, India, E-mail: gsahoo@iopb.res.in

[2]Ravenshaw University, Cuttack, Odisha, 753003, India

[3]Christ College, Cuttack, Odisha, 753001, India

[4]Institute for Plasma Research, Gandhinagar, Gujarat, 382428, India

CONTENTS

ABSTRACT

There is always competition between different transport mechanisms in plasma. To simulate tokamak scrape-off-layer (SOL) like situation experimentally, a tabletop experiment (CPS) is set up in the plasma research laboratory of Ravenshaw University, Cuttack. It is worth noting that matter and energy can be transported effectively across magnetic field lines in a tokamak SOL region and cause damage to the walls. This convective transport is a major issue needs to be addressed. In our system convective transport of plasma in form of blob from bulk plasma produced from a gas injected washer plasma gun, is simulated by adjusting the base pressure of the system (CPS). It is well known that mean free path of charged particles decrease with increase in base pressure. After coming out from plasma gun the plasma is shaped into a structure having finite diameter. It is observed from probe as well as fast imaging data that when the classical ion-neutral and neutral-neutral mean free path is smaller than that of diameter of plasma plume convective transport of plasma in form of blob, is observed. It is well known that classical ion-neutral and neutral-neutral mean free path are basically neutral density (base pressure) dependent, when temperature of plasma does not change appreciably. It is because at low base pressure where mean free path of charged species/neutrals are higher than plasma dimension, there is no source of energy formation/transfer, whereas at higher base pressure where mean free path of charged species/neutrals are smaller than that of plasma dimension collisionality in plasma increases and charge particles and fast neutrals in plasma produces fresh ions/excited species in the plasma plume that sustains plasma even after the energy source (plasma gun) is switched off. This increased lifetime of plasma (more than pulse width of pulse forming network (PFN)) provided space for convective transport in form of blobs, similar to that of SOL region detachment phenomenon in tokamaks.

7.1 INTRODUCTION

The experimental and theoretical investigations confirm the cross-field transport of matter and energy in scrape-off-layer (SOL) of tokamak-like devices [1–6]. This reduces the efficiency of the device and cause damage to the wall also. Addressing the issue of cross-field transport in a tabletop device is still a challenge. Plasma guns are very good source to produce moving plasma and have a wide range of applications in plasma science and nuclear technology [7–19]. In this experiment, a gas injected washer plasma gun [17–19] is used to make a compact plasma system [20]. This is used to study the cross-field dynamics of plasma blobs. External parameters like input power, pulse width of input voltage, and base pressure is supposed to play crucial role in the process of ionization and recombination. Earlier experiments in CPS device [8] have revealed that input power and pulse width has no role on plasma containment. The role of base pressure upon plasma containment and transport is very interesting. It is observed that when base pressure attains certain critical value the lifetime of plasma is increased appreciably and plasma structure is contained for more than 1 ms even if the pulse width of the voltage supplied to the source is ~ 140 μs. It is observed from probe as well as fast imaging technique. The possible explanation of this strange experimental finding is reported here. However, the theoretical work in this field is still very open and in near future a complete explanation is expected to be achieved by scientific community. CPS device is a table-top experiment and due to its compact nature it is easy to handle and it is possible to run it continuously for hours. It gives excellent results to understand the basic plasma physics. Probe, spectroscopy, and fast imaging are a wide range of diagnostics that are associated with the CPS device to explore the dynamics of expanding and moving plasma structures.

7.2   EXPERIMENTAL SETUP

The experimental set up for plasma experiments at Ravenshaw University, i.e., compact plasma system (CPS) [20] consists of plasma chamber, pulse forming network (PFN) [21], plasma gun, gas feed system, diagnostic tools, data acquisition system, and data analysis software. The plasma chamber, under discussion is having major radius 50 cm and minor radius 30 cm. The PFN is capable of producing square wave pulse ~140 μs. By changing the stages of PFN the pulse width can be changed. The gas fed system is designed to maintain desired base/background/ambient pressure in the plasma chamber. Langmuir probe, emission spectroscopy technique

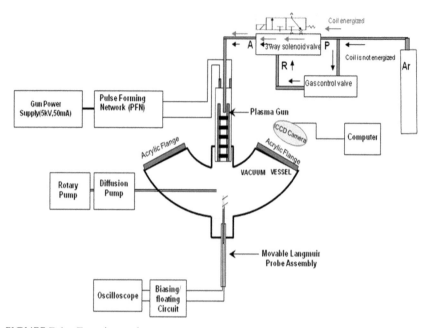

FIGURE 7.1   Experimental setup.

and fast imaging were carried out for plasma diagnostics. The schematic diagram of the setup is given in Figure 7.1.

7.3   RESULTS AND DISCUSSION

The base/background/ambient pressure is increased inside the chamber by injecting an argon gas into it through the gas-fed network as shown in Figure 7.1. The different diagnostic tools are used to measure plasma parameters are given in the following subsections.

7.3.1   PROBE MEASUREMENTS

Electric probes (Langmuir probes) [20] are used for plasma diagnostics. The ion saturation current of plasma is measured with Langmuir probe for different base pressure and is shown in Figure 7.2. It is observed that plasma stays for longer time at higher base pressure (> 1 mb). This is an indication of better containment of plasma at higher base pressure. Since, the pulse duration of PFN that energies plasma gun is 140 µs at higher base pressure the long-lived plasma structure is a matter of interest for investigators. The spatial variation of density shows that the plasma density remains almost constant up to 8 cm from plasma gun at higher base pressure (~ 1 mb or more). The trend is shown in Figure 7.3.

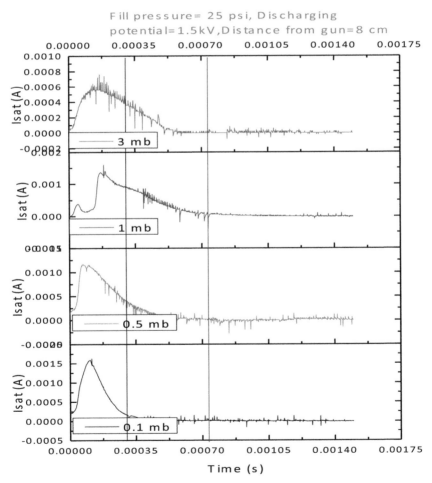

FIGURE 7.2 Ion saturation profile at a distance 8 cm from plasma gun for different operating conditions.

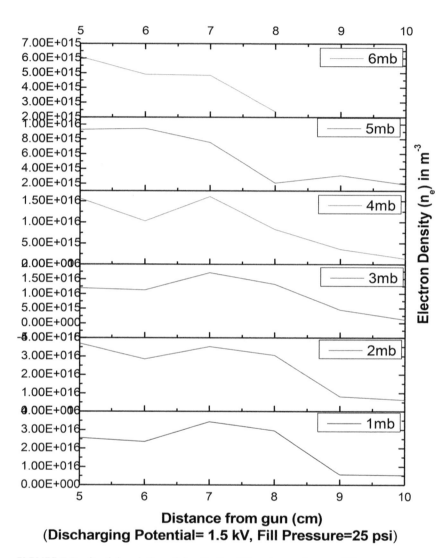

FIGURE 7.3   Spatial variation of density for different operating conditions.

7.3.2   SPECTROSCOPIC MEASUREMENTS

A compact spectrometer (USB 4000, Ocean Optics) is used to measure the electron temperature of plasma. Different excited/ionized states of plasma (produced from gun) were observed in spectroscopic signature. It has a high resolution and a fast integration time with sensitivity of 60 photons per count at 600 nm. This spectrometer was controlled by spectra suit software that is operated on windows operating systems.

At lower base pressure no prominent line of argon is noticed in the signature, whereas at higher base pressure a number of argon lines are observed. Assuming the plasma is in partially local thermodynamic equilibrium, the electron temperature of plasma can be estimated using ratio of intensity of spectral lines. Here from the line spectra H_α and H_β (due to impurity hydrogen present in the system) electron temperature are calculated.

The value of electron temperature estimated using ratio of intensity of spectral line method for different base pressure value is expressed in Figure 7.4. Even from Figure 7.5, it is noticed that at low base pressure (~0.1 mb) no line spectra of argon is noticed, whereas at high base pressure a number of argon lines are present in the spectroscopic signature. Argon excited states represent neutral argon atoms having more energy than that of normal neutral argon atom in ground state configuration. So, it may be concluded from spectroscopic signature that at higher base pressure a number of fast neutrals are present in bulk plasma. This is shown in Figure 7.6.

From the spectroscopic data it is found that below base pressure 3 mb argon lines at wavelengths 696.5438 nm and 706.7218 nm are not excited. When the base pressure is increased up to 1 mb or more argon lines at wavelengths 738.398 nm, 751.4652 nm, 772.42 nm, and 801.4786 nm are excited. Argon lines at wavelengths 763.5106 nm, 794.8176 nm, and 811.5311 nm are excited even if the base pressure is around 0.5–0.75 mb. At higher base pressure number of neutral atoms inside and outside plasma increases appreciably. As more lines are excited at higher base pressure clearly collision plays an important role in the process of excitation of argon.

The proposed mechanism of excitation due to collision may be written as

$$e+ Ar^0 \rightarrow Ar^m + e \qquad (1)$$

$$e + Ar^m \rightarrow Ar^* (4p) + e \qquad (2)$$

$$Ar^*(4p) \rightarrow Ar^*(4s) + h\upsilon \qquad (3)$$

As the base pressure increases, the collision frequency increases and the ion-neutral mean free path decreases. At pressure 1–6 mb the ion-neutral

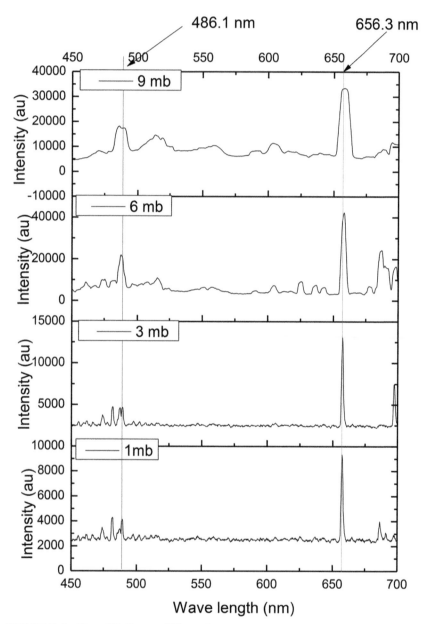

FIGURE 7.4 H$_\alpha$ and H$_\beta$ lines at different base pressures.

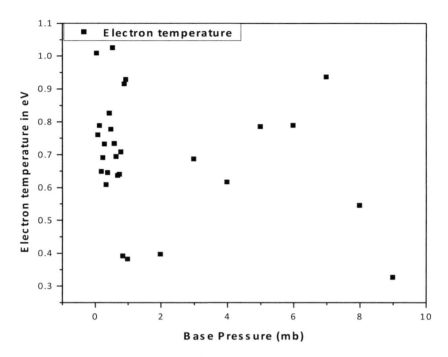

FIGURE 7.5 Electron temperature of plasma from H lines.

mean free path is ~5.7–0.85 cm, whereas the collision frequency is ~10^7–10^8 Hz [22]. The diameter of bulk plasma is more than 10 cm.

At base pressure 1 mb or above there is appreciable number of ion-neutral collisions. Considering all these facts the atomic process in the plasma is written as:

$$A^+ + e = A^* + h\upsilon \tag{4}$$

A fast ion of Ar picks up one electron from a slow atom (*Ar*) and becomes an excited atom (Ar*) which is having greater energy than slow neutrals in argon leaving behind slow Ar ion (*Ar$^+$*). In this process recombination along with ionization takes place. At the end of this process excited argon atom having upper level configuration $3s^2 3p^5 (^2P^0_{1/2}) 4p$ is created.

The process leaves behind a slow ion, which has enough probability to recombine with a free electron in plasma, to become an excited argon atom and emit electromagnetic radiation. The process can be written as

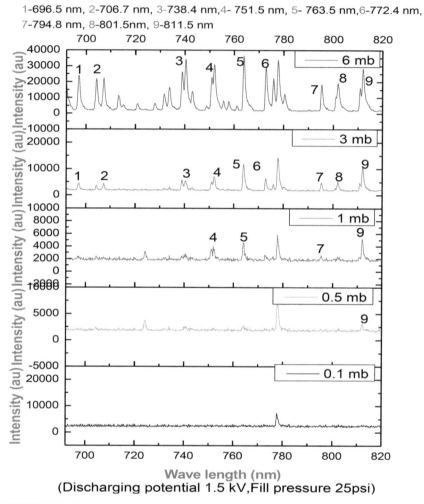

1-696.5 nm, 2-706.7 nm, 3-738.4 nm,4- 751.5 nm, 5- 763.5 nm,6-772.4 nm,
7-794.8 nm, 8-801.5nm, 9-811.5 nm

FIGURE 7.6   Signature of a number of argon lines at higher base pressure.

$$A + e = A^+ + e + e \tag{5}$$

Since, the energy difference between the energy levels of participating species usually do not match in this process, it is difficult to say which excited state is created in this process. Since the probability of radiative recombination is very high if there are slow neutrals in plasma, this process is a dominant process at higher pressure (~1 mb) or more.

If one takes into consideration the energy difference between the energy levels of some of the excited atoms it is found that it is ~ 2–3 eV in some cases [23]. This is close to the line averaged temperature at pressure ~0.8 mb. Therefore, three-body recombination is also a possible process at comparatively lower pressure 0.5–0.75 mb, where some small intensity Ar lines are observed (Ar line corresponding to 763.5106 nm and 811.5311 nm, etc.).

7.3.3 FAST IMAGING RESULTS

At 1 mb or above the confinement is appreciable, which is confirmed in imaging experiments (Mega Speed Camera, 20,000 fps). The signal received by the camera was in form of movie. Then, using the camera software the frames are being extracted. An origin software was used to extract data from the frames. The exposure time of the camera is 50 μs and frame rate is 9100 frames per second (fps). The discharging potential of gun is 1.5 kV. The plasma structure evolution at base pressure 2 mb is shown in Figure 7.7. Convective transport of plasma in form of blob is observed in the frame corresponding to the time 490 μs and afterwards [25]. However, plasma lifetime is appreciable comparing the input signal width from PFN.

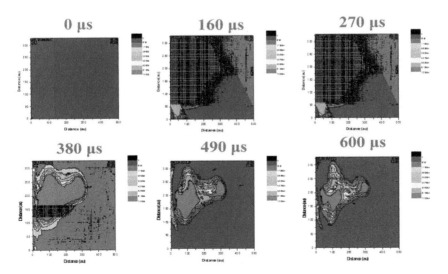

FIGURE 7.7   Images of plasma structure at base pressure 2 mb.

7.4 CONCLUSION

It is observed that base pressure plays a vital role for plasma containment and transport channel. Increasing background slow neutrals, by increasing base pressure up to 1 mb or above plasma lifetime is increased. Since at high pressure the ionization is a multi-step recombination, and it contributes towards a better containment. The spatial variation of electron density measured with Langmuir probe was shown as a flat top upto distance of ~ 8 cm from the plasma gun at higher base/background/ambient pressure. At plasma edge, i.e., beyond 8 cm, the density profile falls rapidly and at this region plasma transport in form of blobs is observed [24].

ACKNOWLEDGMENTS

The work is funded by Board of Research for Fusion Science and Technology (BRFST), Govt. of India.

KEYWORDS

- blob
- convective transport
- plasma gun
- plasma plume
- scrape-off-layer
- tokamaks

REFERENCES

1. Endler, M. (1999). Turbulent SOL transport in stellarators and tokamaks. Journal of nuclear materials, 266, 84–90.
2. Naulin, V. (2007). Turbulent transport and the plasma edge. Journal of nuclear materials, 363, 24–31.

3. Lipschultz, B., Bonnin, X., Counsell, G., Kallenbach, A., Kukushkin, A., Krieger, K., & Rognlien, T. (2007). Plasma–surface interaction, scrape-off layer and divertor physics: implications for ITER. Nuclear Fusion, 47(9), 1189.

4. Garcia, O. E. (2009). Blob transport in the plasma edge: a review. Plasma and Fusion Research, 4, 019–019.

5. D'Ippolito, D. A., Myra, J. R., & Zweben, S. J. (2011). Convective transport by intermittent blob-filaments: Comparison of theory and experiment. Physics of Plasmas, 18(6), 060501.

6. Sahoo, G., Paikaray, R., & Samantaray, S. (2014). Imaging of coherent plasma structures, 'blobs', carrying out matter and energy from bulk plasma-a review. International Journal *of Advanced Research*, 2(11), 438–442.

7. Bostick, W. H. (1956). Experimental study of ionized matter projected across a magnetic field. Physical Review, 104(2), 292–299.

8. Sahoo, G. (2014). PhD thesis submitted to Ravenshaw University.

9. Alidieres, M., Aymar, R., Jourdan, P., Koechlin, F., & Samain, A. (1963). Behavior of a Plasma Produced by a Button-Type Source in Presence of a Magnetic Field. The Physics of Fluids, 6(3), 407–417.

10. Steinhaus, J. F., Oleson, N. L., & Barr, W. L. (1965). Investigation of a plasma from an occluded gas cold plasma source. The Physics of Fluids, 8(9), 1720–1730.

11. Himura, H., Saito, Y., Sanpei, A., Masamune, S., Takeuchi, N., & Shiono, T. (2006). Optimized method of producing washers of titanium hydride for plasma gun using occluded hydrogen gas. Review of scientific instruments, 77(7), 073506,1–5.

12. Little, P. F., & Avis, B. E. (1966). Performance of a coaxial plasma gun. Journal of Nuclear Energy. Part C, Plasma Physics, Accelerators, Thermonuclear Research, 8(1), 11–20.

13. Cheng, D. Y. (1970). Plasma deflagration and the properties of a coaxial plasma deflagration gun. Nuclear Fusion, 10(3), 305–318.

14. Asai, T., Itagaki, H., Numasawa, H., Terashima, Y., Hirano, Y., & Hirose, A. (2010). A compact and continuously driven supersonic plasma and neutral source a. Review of Scientific Instruments, 81(10), 10E119, 1–3.

15. Baranga, A. B. A., Fisher, A., & Tzach, D. (1985). Small simple hydrogen plasma gun. Review of scientific instruments, 56(7), 1472–1474.

16. Voronin, A. V., & Hellblom, K. G. (1999). A titanium hydride gun for plasma injection into the T2-reversed field pinch device. Plasma physics and controlled fusion, 41(2), 293–302.

17. Jain, K. K., John, P. I., Punithavelu, A. M., & Rao, P. P. (1980). Gas injected washer plasma gun. Journal of Physics E: Scientific Instruments, 13(9), 928–930.

18. Osher, J. E. (1982). Plasma target output from a magnetically augmented, gas-injected, washer-stack plasma gun. Review of Scientific Instruments, 53(11), 1685–1692.

19. Sahoo, G., et al. (2011). IEEE XPLORE (International Conference on Multimedia Technology (ICMT-2011)), 6465–6467.

20. Sahoo, G., Paikaray, R., Samantaray, S., Patra, D. C., Sasini, N. C., Ghosh, J., & Sanyasi, A. (2013). A Compact Plasma System for Experimental Study. In Applied Mechanics and Materials (Vol. 278, pp. 90–100).

21. Sahoo, G., Paikaray, R., Samantaray, S., Patra, D. C., Sasini, N., Tripathy, S., & Sanyasi, A. K. (2013, June). A pulse forming network (PFN) for compact plasma system (CPS) at Ravenshaw University, India. In AIP Conference Proceedings(Vol. 1536, No. 1, pp. 1290–1291).

22. Huba, J. D. (2011). NRL Plasma Formulary (NRL/PU/6790-11-551).

23. G. Sahoo et al.,(2012) "Spectroscopic measurements of plasma blob produced by washer plasma Gun", Asian Journal of Spectroscopy, Special Issue., pp. 231–238.

24. Sahoo, G., et al. (2014). Kathmandu University Journal of Science, Engineering and Technology (KUSET), 10(II), 50–57.

25. Sahoo, G., Paikaray, R., Samantaray, S., Das, P., Ghosh, J., Chowdhuri, M. B., & Sanyasi, A. K. (2014). Base pressure plays an important role for production of plasma blob in argon plasma. Journal of Physical Science and Application, 4(6),348–357.

CHAPTER 8

ION-ACOUSTIC DRESSED SOLITONS IN ELECTRON-POSITRON-ION PLASMA WITH NONISOTHERMAL ELECTRONS

PARVEEN BALA and TARSEM SINGH GILL

¹Department of Mathematics, Statistics & Physics, Punjab Agricultural University, Ludhiana – 141004, India, E-mail: pravi2506@gmail.com

²Department of Physics, Guru Nanak Dev University, Amritsar – 143005, India, E-mail: tarsemgill50@gmail.com

CONTENTS

ABSTRACT

Using the standard reductive perturbation technique (*RPT*), ion-acoustic dressed solitons have been studied in an electron-positron-ion plasma with the nonisothermal distribution of electrons. To the lowest order, a modified Schamel Korteweg-de Vries (*KdV*) equation associated with (*1+1/2*) nonlinearity, also known as *Schamel-mKdV* model, has been derived. *RPT* is further extended to include the contribution of higher order nonlinear and dispersion terms. Using renormalization method, a stationary solution resulting from higher order perturbation theory has been found. Results of numerical computation for such contribution are shown in the form of graphs in different parameter regimes and a comparison with earlier investigations has been made. Such study is relevant to understand physics of astrophysical and cosmic plasmas, where *e-p* jets enter in cold interstellar environment.

8.1   INTRODUCTION

An electron-positron (*e-p*) plasma is considered not only as a building block of our early universe [43], but also an omnipresent constituent of a number of astrophysical environment, such as active galactic nuclei [35], pulsar magnetosphere [20, 34], solar atmosphere [20, 53], fire balls producing gamma-ray bursts, and at the center of our galaxy [3]. The *e-p* plasmas are also observed in laboratory experiments where positrons can be used as a probe to study the particle transport in tokamak plasmas [21, 52, 54]. Processes of *e-p* pair production can occur during ultra intense short laser pulse propagation in plasma [12]. Since most of the astrophysical and laboratory plasmas contain ions besides electrons and positrons, it is relevant there to discuss wave motions in electron-positron-ion (*e-p-i*) plasma.

The presence of positron components in electron-ion (*e-i*) plasma reduces the number density of ions and subsequent restoring force on electron fluid thereby leading to the modification of linear as well as nonlinear wave structures. The ion-acoustic wave (*IAW*) is an ion timescale phenomenon and this mode does not exist in *e-p* plasma. Further, the presence of ions leads to the existence of several low frequency waves, which otherwise do not propagate in *e-p* plasmas. However, nonlinear waves in *e-p-i* plasma behave quite differently. Over the last many years, there have been considerable interests among the researchers to study the characteristics of propagation of linear and nonlinear waves in *e-p* plasma [50] and *e-p-i* plasma [1, 2, 10, 13, 14, 17, 19, 25, 28, 31, 39–42, 45, 49, 55, 57, 60].

The observations made by space and laboratory plasmas have shown that particle distributions play a crucial role in characterizing the physics of nonlinear waves. Earlier investigations on ion-acoustic solitons (*IASs*) were based on particle distributions obeying Maxwellian distribution. Such distributions exist for the macroscopic ergodic equilibrium distributions. Both linear and nonlinear properties are influenced by velocity distribution of the particle constituents of the plasma. Moreover, they add considerable increase in richness and variety of wave motion that can exist in plasma and further influence the conditions required for the formation of these waves. In practice, the particles may not follow a Maxwellian distribution and based on the data, but the particle distributions are better modeled by velocity distributions and having flat top with high energy tails. The two most commonly used non-Maxwellian type distributions are nonthermal and nonisothermal particle distributions. The former one, associated with the particle flows resulting from the force fields present in the space and astrophysical plasmas, has abundance of superthermal particles. The second one, nonisothermal particle distribution is due to the formation of phase space holes caused by the trapping of electrons in a wave potential and is not only observed in space plasmas [47, 48] but also in laboratory [22, 44].

Moreover, the plasmas excited by an electron beam evolve towards a coherent trapped particle state rather than developing into turbulent one as has been confirmed by experiments [30]. As a matter of fact, trapping can occur and contribute even for infinite small amplitude [18]. Earlier, as well as recent investigations [16, 18, 32, 33, 46–48] have been reported on the study of *IASs* with vortex-like distributions. The presence of trapped and free particles can significantly modify the characteristics in collisionless plasmas [46, 47]. It may be mentioned here that the presence of hot ions do not change the phase velocity of ion-acoustic wave very much as $T_i << T_e$, T_p. Further, the ion-acoustic wave does exist even when the cold ions are considered. In spite of the fact that the hot e-p pair form most of the astrophysical and cosmic plasmas, a minority of cold electrons and heavy ions may not be ruled out. For example, outflows of e-p plasma from pulsars entering cold interstellar environment, low density e-i plasma leads to the formation of two electron temperature e-p-i plasma. In such plasma, electrons having different temperatures and higher density than positron have greater chance of being trapped by nonlinear *IAWs* [1]. On the other hand, positrons have much higher temperature cannot be trapped by the wave, hence remain free.

The Korteweg-de Vries (*KdV*) and modified Korteweg-de Vries (*mKdV*) model equations describe the small amplitude IASs and include lowest order nonlinearity and dispersion. These evolution equations are associated with quadratic and cubic type of nonlinearity. However, for particle described by vortex-like distribution, the nonlinearity is the (*1+1/2*) type and resulting equation is *Schamel-mKdV* equation. As the wave amplitude increases, the width and velocity of a soliton deviates from the one predicted by *KdV* or *mKdV* model. In such cases, more accurate results can be obtained by inclusion of higher order nonlinear and dispersive effects. For this purpose, the higher order approximation of reductive perturbation theory (*RPT*) has been considered as a powerful tool and applied in many situations [11] as well as in different plasma systems [5, 7, 9, 15, 16, 24, 26, 27, 36–38, 51, 59]. Higher order effects are shown to modify the solitary wave amplitude and may also result in shape deformation [11, 16]. This motivated us to investigate a more generalized case which incorporates several features studied by Tran [56] to account for discrepancies observed in experimental observations and prediction of theoretical models. In the present research work, ion-acoustic dressed solitons have been studied by introducing higher order nonlinear and dispersive effects in an electron-positron-ion plasma with non-isothermal distribution of electrons. The organization of the manuscript is as follows: In Section 8.2, evolution equations governing the dynamics of the *e-p-i* plasma have been given followed by the derivation of *Schamel-mKdV* equation. In Section 8.3, the stationary solution under appropriate conditions are obtained using renormalization technique developed by Kodama and Taniuti [26] and Kodama [27]. Section 8.4 is devoted to the discussion of the present research work and lastly conclusion is made for future direction for experimental *IASs* for such model.

8.2 FORMULATION OF THE PROBLEM

The *e-p-i* plasmas occur frequently in the astrophysical space environment and *IAWs* play an important role in such system. We consider a collisionless and unmagnetized plasma model consisting of cold ions and positrons with free and trapped electrons. In this model, the positrons are assumed as Boltzmann distributed and electrons are considered to obey nonisothermal distribution. The nonlinear behavior of the ion-acoustic waves may be described by the following set of normalized fluid equations:

$$\frac{\partial n}{\partial t} + \frac{\partial (nu)}{\partial x} = 0$$

$$\frac{\partial u}{\partial t} + u\frac{\partial u}{\partial x} + \frac{\partial \phi}{\partial x} = 0 \qquad (1)$$

$$\frac{\partial^2 \phi}{\partial x^2} = \mu n_e - (1-\mu)n_p - n$$

In the weakly nonlinear limit, these hot free positrons are isothermal following the Maxwellian distribution as $n_p = e^{-\delta\phi}$, where, $\mu = 1/(1-p)$, $p = n_{po}/n_{eo}$ and $\delta = T_{ef}/T_p$. In the above, n, n_p and n_e are the normalized ion, positron and electron number densities. u and ϕ are the ion fluid velocity and electrostatic potential respectively. Also, in Eq. (1), number densities of ions n, positron n_p, and electron n_e are normalized by unperturbed electron number density n_{eo}. Ion fluid velocity (u), electrostatic potential (ϕ), time (t), and space coordinate (x) have been normalized with respect to the ion-acoustic speed, $C_s = (K_B T_{ef}/m)^{1/2}$, $K_B T_{ef}/e$, inverse of ion plasma frequency in the mixture $\omega_{pi}^{-1} = \lambda_D/C_s$ and Debye length $\lambda_D = (\varepsilon_0 K_B T_{ef}/n_{eo} e^2)^{1/2}$, respectively. Here K_B is the Boltzmann constant and T_{ef} is the constant free electron temperature.

To study the effect of nonisothermal electrons on the characterization of nonlinear ion-acoustic waves in non-relativistic plasma, we use vortex like electron distribution of Schamel [47], which solves the electron Vlasov equation. Thus,

$$f_{ef} = \frac{1}{\sqrt{2\pi}}\exp\left(-\frac{1}{2(v^2 - 2\phi)}\right), |v| > \sqrt{2\phi}$$

$$f_{et} = \frac{1}{\sqrt{2\pi}}\exp\left(-\frac{1}{2\beta(v^2 - 2\phi)}\right), |v| \leq \sqrt{2\phi} \qquad (2)$$

Here f_{ef} (f_{et}) represents the free (trapped) electron distribution. The velocity v is normalized to the electron thermal velocity C_s and $\beta = T_{ef}/T_{et}$ [the ratio of the free electron temperature (T_{ef}) to trapped electron temperature (T_{et})] as non isothermal parameter. This parameter determines the number of trapped electrons. It has been assumed that the velocity of non-linear ion-acoustic waves is small in comparison with the electron thermal velocity. Also, it becomes obvious from this distribution that $\beta = 1$ ($\beta \to 0$) represents a Maxwellian (flat-topped) distribution, whereas $\beta < 0$ represents a vortex-like excavated trapped electron distribution. The electron distribution functions (2) can be readily integrated over the velocity space to get electron number density. In small amplitude limit, we can expand n_e and it is found that n_e is same for both $\beta \geq 0$ and $\beta < 0$. We find

$$n_e = 1 + \phi + \frac{1}{2}\phi^2 - \frac{4}{3}b\,\phi^{3/2} \tag{3}$$

where $\beta = (1 - \beta)/\sqrt{\pi}$ measures the deviation from isothermality. In order to derive nonlinear dynamical equations for the ion-acoustic waves from Eqs. (1) and (3), we must find an appropriate co-ordinate from where the wave can be described smoothly. For this purpose, we need to know the thickness and nonlinear velocity v_0 of the wave which can be taken from the equilibrium theory using vortex like electron distribution [23, 46, 47]. Thus, we find $\Delta \propto \varepsilon^{1/4}$ and $(v_0 - 1) \propto \varepsilon^{1/2}$, where ε is small parameter measuring the weakness of nonlinearity. This immediately leads to the following stretched co-ordinates $\xi = \varepsilon^{1/4}(\xi - \lambda t)$ and $t = \varepsilon^{3/4} t$, where λ is the phase velocity of ion-acoustic waves and ε measures the size of perturbation amplitude.

Furthermore, the dependent variables are expanded as power series in ε about their equilibrium values as:

$$\begin{pmatrix} n \\ p \\ u \\ \phi \end{pmatrix} = \begin{pmatrix} 1 \\ 1 \\ 0 \\ 0 \end{pmatrix} + \sum_{r=1}^{\infty} \varepsilon^{(r+1)/2} \begin{pmatrix} n^{(r)} \\ p^{(r)} \\ v^{(r)} \\ \phi^{(r)} \end{pmatrix} \tag{4}$$

Substituting the stretched coordinates and the perturbation relation (4) to the basic set of equations (1) and following the usual reductive perturbation theory [58], the first order equations yield:

$$n^{(1)} = \mu - (1-\mu)\delta\phi^{(1)}, \; u^{(1)} = \lambda\mu - (1-\mu)\delta\phi^{(1)}, \; p^{(1)} = 3\mu - (1-\mu)\delta\phi^{(1)}$$

and the phase velocity $\lambda = \dfrac{\sqrt{(1-p)}.}{\sqrt{(1-\delta p)}}$.

Here the phase velocity is a function of positron density p and the temperature ratio of free electron and positron, i.e., $\delta = T_{ef}/T_p$. It may be mentioned here that the phase velocity is independent of nonisothermal parameter β. To the next order in ε, we obtain a system of equations in the second order-perturbed quantities. On solving this system, a *Schamel-mKdV* equation is obtained as given by:

$$\frac{\partial\phi^{(1)}}{\partial\tau} + A\sqrt{\phi^{(1)}}\,\frac{\partial\phi^{(1)}}{\partial\xi} + B\frac{\partial^3\phi^{(1)}}{\partial\xi^3} = 0 \tag{5}$$

Here

$$A = \frac{b(1-p)}{\lambda(1-\delta p)^2} \text{ and } B = \frac{(1-p)^2}{2\lambda(1-\delta p)^2} \tag{6}$$

8.3 HIGHER ORDER SOLUTIONS

The second order quantities $n^{(2)}$, $u^{(2)}$ and $p^{(2)}$ can be expressed in terms of $\phi^{(1)}$ and $\phi^{(2)}$ as:

$$n^{(2)} = \frac{1}{\lambda^2}\phi^{(2)} - \frac{4\mu b}{3}(\phi^{(1)})^{3/2} + \frac{\partial^2\phi^{(1)}}{\partial\xi^2}$$

$$u^{(2)} = \frac{1}{\lambda^2}\phi^{(2)} - \left(\frac{2A}{3\lambda^2}\frac{4\mu b}{3}\right)(\phi^{(1)})^{3/2} + \left(\frac{B}{\lambda^2} - \lambda\right)\frac{\partial^2\phi^{(1)}}{\partial\xi^2} \tag{7}$$

$$p^{(2)} = \frac{3}{\lambda^2}\phi^{(2)} - \frac{4A}{3\lambda^3}(\phi^{(1)})^{3/2} - \frac{6B}{\lambda^3}\frac{\partial^2\phi^{(1)}}{\partial\xi^2}$$

To the next order in ε, we get the equation

$$\frac{\partial\phi^{(2)}}{\partial\tau} + A\frac{\partial(\sqrt{\phi^{(1)}}\phi^{(2)})}{\partial\xi} + B\frac{\partial^3\phi^{(2)}}{\partial\xi^3} = S(\phi^{(1)}) \tag{8}$$

where, the source term $S(\phi^{(1)})$ is given by:

$$S(\phi^{(1)}) = (2BC - 4\mu^2 b^2 D)\phi^{(1)} \frac{\partial \phi^{(1)}}{\partial \xi} - D \frac{\partial^5 \phi^{(1)}}{\partial \xi^5}$$

$$-\frac{3\mu b D}{\sqrt{\phi^{(1)}}} \frac{\partial^2 \phi^{(1)}}{\partial \xi^2} \frac{\partial \phi^{(1)}}{\partial \xi} + \frac{\mu b D}{2(\phi^{(1)})^{3/2}} \left(\frac{\partial \phi^{(1)}}{\partial \xi} \right)^3 - 4\mu b D \sqrt{\phi^{(1)}} \frac{\partial^3 \phi^{(2)}}{\partial \xi^3} \qquad (9)$$

where $D = B^3 X$. The coefficients C and X occurring here are given by,

$$C = \frac{\mu + (1-\mu)\delta^2}{2} + \frac{3}{\lambda^4} \text{ and } X = \frac{1}{\lambda^4} - \frac{2\lambda^3}{B} \qquad (10)$$

The basic set of equations (1) is reduced to a nonlinear $mKdV$ equation (5) in terms of $\phi^{(1)}$ and a homogeneous differential equation (8) in terms of $\phi^{(2)}$, for which the source term of equation (8) is described by a known function $\phi^{(1)}$. To solve two equations (5) and (8) analytically, the renormalization method developed by Kodama and Taniuti [26] is employed. Using this method, Eqs. (5) and (8) are modified as:

$$\frac{\partial \widetilde{\phi}^{(1)}}{\partial \tau} + A\sqrt{\widetilde{\phi}^{(1)}} \frac{\partial \widetilde{\phi}^{(1)}}{\partial \xi} + A \frac{\partial^3 \widetilde{\phi}^{(1)}}{\partial \xi^3} + \delta v \frac{\partial \widetilde{\phi}^{(1)}}{\partial \xi} = 0 \qquad (11)$$

$$\frac{\partial \widetilde{\phi}^{(2)}}{\partial \tau} + A \frac{\partial (\widetilde{\phi}^{(2)} \sqrt{\widetilde{\phi}^{(1)}})}{\partial \xi} + B \frac{\partial^3 \widetilde{\phi}^{(2)}}{\partial \xi^3}$$

$$+ B \frac{\partial^3 \widetilde{\phi}^{(2)}}{\partial \xi^3} + \delta v \frac{\partial \widetilde{\phi}^{(2)}}{\partial \xi} = S(\phi^{(1)}) + \delta v \frac{\partial \widetilde{\phi}^{(1)}}{\partial \xi} \qquad (12)$$

where $\widetilde{\phi}^{(1)}$ and $\widetilde{\phi}^{(2)}$ refers to renormalization variables.

Let us have the stationary solution by defining a new variable ψ as

$$\psi = \xi - (v + \delta v)\tau \qquad (13)$$

where, the parameter v is related to the Mach number $M = V/C_s$ by, $v + \delta v = M - 1 = \Delta M$, where V is the soliton velocity [6, 8, 29]. Under this transformation, Eqs. (11) and (12) becomes

$$B\frac{\partial^3\widetilde{\phi}^{(1)}}{\partial\psi^3} + A\sqrt{\widetilde{\phi}^{(1)}}\frac{\partial\widetilde{\phi}^{(1)}}{\partial\psi} - v\frac{\partial\widetilde{\phi}^{(1)}}{\partial\psi} = 0 \tag{14}$$

$$A\frac{\partial^3\widetilde{\phi}^{(2)}}{\partial\psi^3} + 2bA\frac{\partial(\widetilde{\phi}^{(2)}\sqrt{\widetilde{\phi}^{(1)}})}{\partial\psi} - v\frac{\partial\widetilde{\phi}^{(2)}}{\partial\psi} = S(\phi^{(1)}) + \delta v\frac{\partial\widetilde{\phi}^{(1)}}{\partial\psi} \tag{15}$$

It may be mentioned that Eqs. (14) and (15) are *ODEs*, which are integrated under the boundary conditions:

$$\widetilde{\phi}_1 = \widetilde{\phi}_2 = \frac{\partial\widetilde{\phi}_1}{\partial\psi} = \frac{\partial\widetilde{\phi}_2}{\partial\psi} = \frac{\partial^2\widetilde{\phi}_1}{\partial\psi^2} = \frac{\partial^2\widetilde{\phi}_1}{\partial\psi^2} = 0 \text{ as } |\psi|\rightarrow\infty. \tag{16}$$

Using above boundary conditions for $\widetilde{\phi}^{(1)}$ and $\widetilde{\phi}^{(2)}$, Eqs. (15) and (16) can be integrated with respect to the variable ψ and their derivatives upto second order yield

$$\frac{\partial^2\widetilde{\phi}^{(1)}}{\partial\psi^2}\left(\frac{2A}{3B}\widetilde{\phi}^{(1)1/2} - \frac{v}{B}\right)\widetilde{\phi}^{(1)} = 0 \tag{17}$$

$$\frac{\partial^2\widetilde{\phi}^{(2)}}{\partial\psi^2}\left(\frac{A}{B}\widetilde{\phi}^{(1)1/2} - \frac{v}{B}\right)\widetilde{\phi}^{(2)} = \frac{1}{B}\int_{-\infty}^{\psi}\left(S(\widetilde{\phi}^{(1)}) + \delta v\frac{\partial\widetilde{\phi}^{(1)}}{\partial\psi}\right)d\psi \tag{18}$$

The one soliton solution of Eq. (17) is given by

$$\widetilde{\phi}^{(1)} = \phi_0 \sec h^4\left(\psi W^{-1}\right) \tag{19}$$

where, the amplitude (ϕ_0) and width (W) of the solitary wave are given by

$$\phi_0 = \left(\frac{15v}{8A}\right)^2 \text{ and } W = \sqrt{\frac{16B}{v}} \tag{20}$$

It may be mentioned that B is assumed to be positive to ensure the reality of solutions. The Eq. (19) shows the existence of compressive solitons only. The source term of Eq. (12) becomes

$$\int_{-\infty}^{\psi}\left(S(\widetilde{\phi}^{(1)}) + \delta v\frac{\partial\widetilde{\phi}^{(1)}}{\partial\psi}\right)d\psi = \frac{\phi_0}{B}\left[(\delta v - BXv^2)\sec h^4(\psi W^{-1})\right.$$
$$\left. + BC\phi_0 \sec h^8(\psi W^{-1})\right] \tag{21}$$

In order to cancel the secular terms in $S(\tilde{\phi}^{(1)})$, we have to put

$$\delta v = BXv^2 \tag{22}$$

To solve Eq. (18), we define a new independent variable

$$\mu = \tanh h(\psi W^{-1}) \; (23) \tag{23}$$

Using this transformation Eq. (18) thereby becomes

$$\frac{d}{d\mu}\left[(1-\mu^2)\frac{d\tilde{\phi}^{(2)}}{d\mu}\right] + \left[30 - \frac{16}{1-\mu^2}\right]\tilde{\phi}^{(2)} = F(1-\mu^2)^3 = T(\mu) \tag{24}$$

where

$$F = \frac{(15)^4}{(8A)^4} 16BCv^3 \tag{25}$$

The two independent solutions of Eq. (24) are given by associated Legendre functions of the first- and second-kind and these are given by

$$P_5^4 = 945\mu(1-\mu^2)^2 \tag{26}$$

$$Q_5^4 = \frac{945}{2}\mu(1-\mu^2)\ln\frac{1+\mu}{1-\mu} - 334(1-\mu^2)^2$$

$$+975\mu^2(1-\mu^2)+630\mu^4 +264\frac{\mu^6}{1-\mu^2}+48\frac{\mu^3}{(1-\mu^2)^2}$$

By using the method of variation of parameters, particular solution of Eq. (18) can be written as

$$\tilde{\phi}_p^{(2)}(\mu) = q_1(\mu)P_5^4 + q_2(\mu)Q_5^4 \tag{27}$$

where,
$$q_1(\mu) = -\frac{F}{954 \times 384}\int Q_5^4(\mu)(1-\mu^2)^3 d\mu \tag{28}$$

$$q_2(\mu) = \frac{F}{954 \times 384}\int P_5^4(\mu)(1-\mu^2)^3 d\mu$$

The complementary solution of Eq. (18) is given by

$$\tilde{\phi}_c^{(2)}(\mu) = c_1(\mu)P_5^4 + c_2(\mu)Q_5^4 \tag{29}$$

Here, the first term is the secular one which can be eliminated by renormalizing the amplitude. Also, $c_2 = 0$, as a result of vanishing boundary conditions for $\tilde{\phi}^{(2)}(\psi)$ as $|\psi| \to \infty$. The solution of Eq. (24) is

$$\tilde{\phi}^{(2)}(\mu) = \tilde{\phi}_p^{(2)}(\mu) = \frac{F}{6}(1-\mu)^2 - \frac{F}{12}(1-\mu)^3 \tag{30}$$

In terms of the variable ψ, the stationary solution for the potential of ion-acoustic wave is given by

$$\tilde{\phi}(\psi) = \tilde{\phi}^{(1)}(\psi) + \tilde{\phi}^{(2)}(\psi) = \phi_0 \sec h^4\left(\psi W^{-1}\right) \tag{31}$$

$$+ \frac{F}{6}\sec h^4\left(\psi W^{-1}\right) - \frac{F}{12}\sec h^6\left(\psi W^{-1}\right)$$

where v and modified width are given by:

$$v = \Delta M(1 - X\Delta M) \text{ and } W' = \sqrt{\frac{16B}{\Delta M}}\left(1 + \frac{1}{2}X\Delta M\right) \tag{32}$$

8.4   DISCUSSION OF NUMERICAL RESULTS

It is worth mentioning here that underlying the principle rule of reductive perturbation technique [58], for inclusion of higher effects, the condition to be satisfied as:

$$\left|\tilde{\phi}^{(2)}\right|/\left|\tilde{\phi}^{(1)}\right| \leq 1 \tag{33}$$

This requirement has its origin in the proper ordering in ε. This means we have to choose appropriate solitary excitation velocity. It may be mentioned that increase in β allows the inequality (33) to be satisfied for lower values of v.

In order to highlight the importance of higher order effects, in a comparison of $\tilde{\phi}^{(1)}$, the soliton solution of Eq. (17) with $\tilde{\phi} = \left(\tilde{\phi}^{(1)} + \tilde{\phi}^{(2)}\right)$ the second order solution given by Eq. (18) becomes important. Both the cases of trapped electron parameter viz., $\beta<1$ and $\beta>1$ have been treated with other set of parameters as: $p = 0.1$, $\delta = 1$ and $v = 0.01$. The results are shown in the graphs plotted in Figure 8.1 for $\beta = 0.1(<1)$, and Figure 8.2 for $\beta = 1.5(>1)$. It is observed that the effect of second order nonlinearity leads to an increase in amplitude of dressed solitons (shown by dotted lines). This is in consistent with the finding of earlier investigations [11, 15, 16].

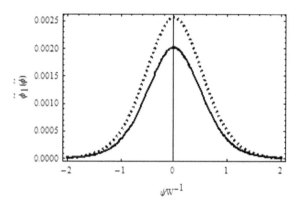

FIGURE 8.1   For $\beta<1$, the plot showing the comparison of KdV soliton solution (solid curve) and dressed soliton solution (dotted curve) as a function of ψW^{-1} with $p = 0.1$, $\delta = 1$, $v = 0.01$, and $\beta = 0.1$.

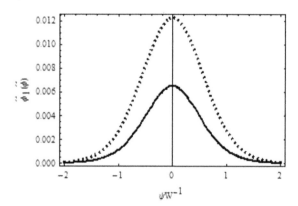

FIGURE 8.2   For $\beta>1$, the plot showing the comparison of KdV soliton solution (solid curve) and dressed soliton solution (dotted curve) as a function of ψW^{-1} with $p = 0.1$, $\delta = 1$, $v = 0.01$, and $\beta = 1.5$.

To illucidate the role of higher order effects on velocity of dressed solitons, a numerical calculation of the soliton velocity (ΔM) as a function of $\phi_0^{(t)} = \phi_0^{(1)} + \phi_0^{(2)}$ has been done for three different range of parameters (p, β) and results are shown in Figure 8.3. Here, the solid curve corresponds to $p = 0.0$, $\beta = 0.1$ (i.e., zero positron density) and dotted curve for $p = 0.1$, $\beta = 0.1$, i.e., for finite positron density. It is observed that the introduction of finite positrons leads to decrease in soliton velocity ΔM. However, further decrease in soliton velocity, ΔM is observed with trapped electrons (β) as obvious from dotted and dashed curves (for $p = 0.1$, $\beta = 0.2$). However, for $\beta > 1$, ΔM has been observed to increase with β while other behavior remains the same as that for $\beta < 1$. This trend is shown in Figure 8.4, where plot of ΔM as a function of $\phi_0^{(t)}$ is shown for three different values of $\beta(>1)$.

It is further mentioned that the width ratio of high and lower order solitons, i.e., W'/W is independent of nonisothermal parameter β and the width ratio decreases with positron density. However, the amplitude ratio shows increasing trend with these parameters. Nonisothermality plays an important role here, as for $\beta < 1$, the amplitude ratio $\phi_0^{(t)}/\phi_0$ increases while for $\beta > 1$, the ratio has been found to decrease (not shown).

It is interesting to note the plot of width-amplitude relation, i.e., $W'^2 \times \phi_0^{(t)}$ for dressed solitons as shown in Figure 8.5, where a plot of $W'^2 \times \phi_0^{(t)}$ as a

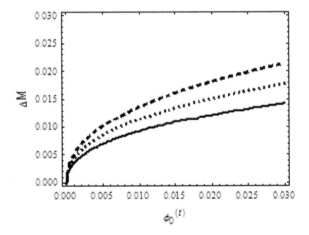

FIGURE 8.3 Variation of ΔM as a function of peak amplitude of dressed soliton $\phi_0(t)$ for three different sets of parameters (p, β). Here the solid curve is for p = 0.0, β = 0.1, the dotted curve for p = 0.1, β = 0.1, and the dashed curve for p = 0.1, β = 0.2 with δ = 1.

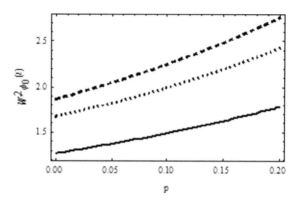

FIGURE 8.4 For $\beta > 1$, the variation of ΔM as a function of higher order peak amplitude $\phi_0(t)$ for three different values of β with $p = 0.1$ and $\delta = 1$. Here the solid curve is $\beta = 1.5$, the dotted curve for $\beta = 1.6$, and the dashed curve for $\beta = 1.7$ with $\delta = 1$.

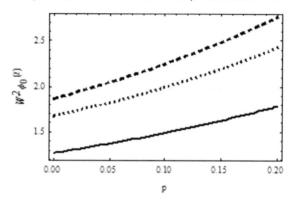

FIGURE 8.5 Variation of width-amplitude relation of dressed soliton, i.e., $W'2x\phi_0(t)$ as a function of relative positron concentration p for three sets of parameters (β, v). Here the solid curve is for $(\beta = 0.1, v = 0.01)$, the dotted curve is for $(\beta = 0.2, v = 0.01)$, and the dashed curve for $(\beta = 0.1, v = 0.011)$ and $\delta = 1$.

function of relative concentration of positrons (p) for three sets of parameters have been given. A slight increase in the width-amplitude product is observed with p and nonisothermal parameter $\beta(<1)$ as is clear from solid curves (for $\beta = 0.1$, $v = 0.01$) and dotted curve (for $\beta = 0.2$, $v = 0.01$). It is mentioned that soliton velocity also plays a crucial role in shaping the behavior of dressed solitons as obvious from dashed curves for $\beta = 0.2$, $v = 0.011$. An opposite trend has been observed for $\beta > 1$ (see Figure 8.6), where the width-amplitude relation is found to decrease with nonisothermal parameter.

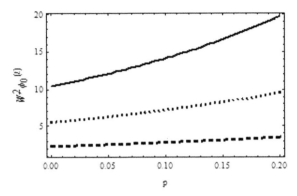

FIGURE 8.6   For $\beta>1$, variation of width-amplitude relation of dressed soliton, i.e., $W'2x\phi_0(t)$ as a function of relative positron concentration p for three different values of β with $\delta = 1$, and $v = 0.01$. Here the solid curve is for $\beta = 1.5$, and the dotted curveis for $\beta = 1.6$ and dashed curve for 1.7.

These observations stimulate one's interest in experimental plasma physics to study *IASs* in *e-p-i* plasmas. In standard *e-i* plasma with trapped electrons [56], it is possible to devise a mechanism for introducing positrons in *DP* machine. All predictions observed in the present *e-p-i* plasma model can be verified, both for lowest order nonlinearity as well as for contribution of higher order effects. Such experiment will not only pay rich dividends to soliton physics but also simulate the astrophysical and cosmological environment where *e-p-i* plasma is believed to exist.

8.5   CONCLUSION

In the present investigation, the dressed solitons in an *e-p-i* plasma consisting of two temperature electron species, i.e., trapped and free have been studied. The associated profile as given in Eq. (19) is of *sech⁴*-type. This has been found to be significantly modified by the inclusion of higher order dispersive and nonlinear effects. It is observed that the effect of second order nonlinearity leads to an increase in the amplitude of dressed solitons. The velocity of dressed solitons is found to decrease with positron density and nonisothermality. A slight increase in width-amplitude product was observed with positron density and nonisothermal parameter $\beta<1$, while for $\beta>1$, an opposite trend has been observed. This investigation is relevant to study astrophysical and cosmic plasma environment where *e-p-i* plasma is believed to exist.

ACKNOWLEDGMENT

Authors are thankful to the UGC (via F. No 42-1065/2013) and CSIR (grant no 21(0880)/11/EMR-II) for providing financial assistance.

KEYWORDS

- **dressed solitons**
- **electron-positron-ion**
- **ion-acoustic solitons**
- **nonisothermal distribution**
- **reductive perturbation method**
- **renormalization technique**

REFERENCES

1. Alinejad, H., Sobhanian, S., & Mahmoodi, J. (2006). Nonlinear propagation of ion-acoustic waves in electron-positron-ion plasma with trapped electrons. *Physics of Plasmas, 13*, 012304.
2. Bharuthram, R. (1992). Arbitrary amplitude double layers in a multispecies electron positron plasma. *Astrophysical and Space Science, 189*, 213–222.
3. Burns, M. L. (1983). Positron-Electron Pairs in Astrophysics, M. L. Burns, A. K. Harding, & R. Ramaty (ed.) American Institute of Physics, New York, pp. 281–286.
4. Demiray, H. (1999). A modified reductive perturbation method as applied to nonlinear ion-acoustic waves. *Journal of Physics Society of Japan, 68*, 1833–1837.
5. Demiray, H. (2002). Contribution of higher order terms in nonlinear ion-acoustic waves: strongly dispersive case. *Journal of Physics Society of Japan, 71*, 1921–1930.
6. El-Labany, S. K., & Shaaban, S. M. (1995). Contribution of higher order nonlinearity to nonlinear ion-acoustic waves in a weakly relativistic warm plasma. Part 2. Nonisothermal case. *Journal of Plasma Physics, 53*, 245–252.

7. El-Labany, S. K., Moslem, W. M., El-Shewy, E. K., & Mowafy, A. E. (2005). Higher order solutions of dust ion-acoustic solitons in a warm dusty plasma with vortex like electron distribution. *Chaos Solitons and Fractals, 23*, 581–588.

8. El-Labany, S. K. (1992). Nonlinear ion-acoustic solitons in a warm plasma with adiabatic positive and negative ions and of nonisotermal electrons. *Astrophysical and Space Science, 191*, 185–194.

9. El-Taibany, W. F., & Moslem, W. M. (2005). Higher-order nonlinearity of electron-acoustic solitary waves with vortex-like electron distribution and electron beam. *Physics of Plasmas, 12*, 032307.

10. Esfandyari-Kalejahi, A., Kourakis, I., & Shukla, P. K. (2006). Electrostatic mode envelope excitations in e-p-i plasmas-application in warm positron plasmas with a small fraction of stationary ions. *Journal of Physics A: Mathematical and General, 39*, 13817–13830.

11. Esfandyari-Kalejahi, A., Kourakis, I., & Shukla, P. K. (2008). Ion-acoustic waves in a plasma consisting of adiabatic warm ions, nonisothermal electrons, and a weakly relativistic electron beam: Linear and higher-order nonlinear effects. *Physics of Plasmas 15*, 022303.

12. Gahn, C., Tsakiris, G. D., Prezler, G., Witte, K. J., Delfin, C., Wahlstrom, C. G., & Habs, D. (2000). Generating positrons with femtosecond-laser pulses. *Applied Physics Letters, 77*, 2662–2664.

13. Ghosh, S., & Bharuthram, R. (2008). Ion-acoustic solitons and double layers in electron-positron-ion plasmas with dust particulates. *Astrophysical and Space Science, 314*, 121–127.

14. Gill, T. S., Bains, A. S., Saini, N. S., & Bedi, C. (2010a). Ion-acoustic envelope excitations in electron–positron–ion plasma with nonthermal electrons. *Physics Letters A, 374*, 3210–3215.

15. Gill, T. S., Bala, P., & Kaur H (2008). Higher order solutions to ion-acoustic solitons in a weakly relativistic two-fluid plasma. *Physics of Plasmas, 15*, 122309.

16. Gill, T. S., Bala, P., & Kaur, H. (2010b). Higher order nonlinear effects on wave structures in a multispecies plasma with nonisothermal electrons. *Z. Natureforsch A, 65*, 315–328.

17. Gill, T. S., Bala, P., & Bains, A. S. (2015). Electrostatic wave structures and their stability analysis in nonextensive magnetized electron-positron-ion plasma. *Astrophysical and Space Science, 357*, 63

18. Gill, T. S., Kaur, H., & Saini, N. S. (2003). Ion-acoustic solitons in a plasma consisting of positive and negative ions with nonisothermal electrons. *Physics of Plasmas, 10*, 3927–3932.

19. Gill, T. S., Singh, A., Kaur, H., Saini, N. S., & Bala, P. (2007). Ion acoustic solitons in weakly relativistic plasma containing electron-positron and ion. *Physics Letters A, 361*, 364–367.

20. Goldreich, P., & Julian, W. H. (1969). Pulsar Electrodynamics. *Astrophysical Journal, 157*, 869–880.

21. Greaves, R. G., & Surko, C. M. (1995). An electron-positron beam plasma experiment. *Physical Review Letters, 75*, 3846–3849.

22. Guio, P., Borve, S., Daldorff, L. K., Lynov, J. P., Michelsen, P., Pecseli, H. L., Rasmussen, J. J., Saeki, K., & Trulsen, J. (2003). Phase space vortices in collisionless plasmas. *Nonlinear Processes in Geophysics, 10*, 75–86.

23. Holloway, J. P., & Dorning, J. J. (1991). Undamped plasma waves. *Physical Review A, 44,* 3856–3868.

24. Ichikawa, Y. H., Mitsuhashi, T., & Konno, K. (1976). Contribution of higher order terms in the Reductive Perturbation Theory. I. A case of weakly dispersive wave. *Journal of Physics Society of Japan, 41,* 1382–1386.

25. Jain, S. K., & Mishra, M. K. (2013). Arbitrary amplitude ion-acoustic solitons in two-electron temperature warm ion plasma. *Astrophysical and Space Science, 346,* 395–407.

26. Kodama, Y., & Taniuti, T. (1978). Higher order approximation in the Reductive Perturbation method. I. The weakly dispersive system. *Journal of Physical Society of Japan, 45,* 298–310.

27. Kodama, Y. (1978). Higher order approximation in the Reductive Perturbation method. II. The strongly dispersive system. *Journal of Physical Society of Japan, 45,* 311–314.

28. Kourakis, I., Verheest, F., & Cramer, N. (2007). Nonlinear perpendicular propagation of ordinary mode electromagnetic wave packets in pair plasmas and electron-positron-ion plasmas. *Physics of Plasmas, 14,* 022306.

29. Lai, C. S. (1979). Effects of higher order contribution and of ion temperature in ion acoustic solitary waves. *Canadian Journal of Physics, 57,* 490–495.

30. Lynov, J. P., Michelsen, P., Pecseli, H. L., & Rasmussen, J. J. (1980). Interaction between electron holes in a strongly magnetized plasma. *Physics Letters, 80,* 23–25.

31. Mahmood, S., Mushtaq, A.and Saleem, H. (2003). Ion acoustic solitary waves in homogeneous magnetized electron-positron-ion plasma. *New Journal of Physics, 5,* 28.

32. Mamun, A. A., Cairns, R. A., & Shukla, P. K. (1996). Effects of vortex like and nonthermal ion distributions on nonlinear dust acoustic waves. *Physics of Plasmas, 3,* 2610–2614.

33. Mamun, A. A. (1998). Nonlinear propagation of ion acoustic waves in a hot magnetized plasma with vortex like electron distribution. *Physics of Plasmas, 5,* 322–324.

34. Michel, F. C. (1982). Theory of pulsar magnetospheres. *Review in Modern Physics, 54,* 1–66.

35. Miller, H. R., & Witta, P. J. (1987). Active Galactic Nuclei. Springer-Verlag, Berlin, pp. 202.

36. Mondal, K. K. (2007). General theory of higher order corrections to solitary waves in multicomponent plasmas. *FIZIKA A, 16,* 11–24.

37. Moslem, W. M. (1999). Propagation of ion acoustic waves in a warm multicomponent plasma with an electron beam. *Journal of Plasma Physics, 61,* 177–189.

38. Moslem, W. M. (2000). Higher-order contributions to ion-acoustic solitary waves in a warm multicomponent plasma with an electron beam. *Journal of Plasma Physics, 63,* 139–155.

39. Moslem, W., Shukla, P. K., & Schlickeiser, R. (2007). Nonlinear excitations in electron-positron-ion plasmas in accretion disks of active galactic nuclei. *Physics of Plasmas, 14,* 102901.

40. Mushtaq, A., & Shah, H. A. (2005). Nonlinear Zakharov-Kuznetsov equation for obliquely propagating two-dimensional ion-acoustic solitary waves in a relativistic, rotating magnetized electron-positron-ion plasma. *Physics of Plasmas, 12,* 072306.

41. Nejoh, Y. (1996). The effect of ion temperature on large amplitude ion-acoustic waves in an electron-positron-ion plasma. *Physics of Plasmas, 3*, 1447–1451.

42. Popel, S. I., Vladimirov, S. V., & Shukla, P. K. (1995). Ion-acoustic solitons in electron-positrons-ion plasmas. *Physics of Plasmas, 2*, 716–719.

43. Ress, M. J. (1983). The Very Early Universe. G. W. Gibbons, S. W. Hawking, & S. Siklos (ed.), Cambridge Univ. Press, Cambridge, pp. 29.

44. Saeki, K., Michelsen, P., Pecseli, H. L., & Rasmussen, J. J. (1979). Formation and coalescence of electron solitary holes. *Physical Review Letters, 42*, 501–504.

45. Salahuddin, M. (2002). Ion acoustic envelope solitons in electron-positron-ion plasmas. *Physical Review E, 66*, 036407.

46. Schamel, H. (1972). Stationary solitary, snoidal and sinusoidal ion acoustic waves. *Plasma Physics, 14*, 905–924.

47. Schamel, H. (1973). A modified Korteweg-de Vries equation for ion acoustic waves due to resonant electrons. *Journal of Plasma Physics, 9*, 373–387.

48. Schamel, H. (1986). Electron holes, ion holes and double layers: electrostatic phase space structures in theory and experiment. *Physics Reports, 9*, 161–191.

49. Shukla, P. K., Mamun, A. A., & Stenflo, L. (2003). Vortices in a Strongly Magnetized Electron–Positron–Ion Plasma. *Physica Scripta, 68*, 295.

50. Stenflo, L., Sukla, P. K., & Yu. M. Y. (1985). Nonlinear propagation of electromagnetic waves in magnetized electron-positron plasmas. *Astrophysical and Space Science, 117*, 303–308.

51. Sugimoto, N., & Kakutani, T. (1977). Note on higher order terms in reductive perturbation theory. *Journal of Physical Society of Japan, 43*, 1469–1470.

52. Surko, C. M., Leventhal, M., Crane, W. S., Passner, A., Wysocki, F., Murphy, T. J., Strachan, J., & Rowan, W. L. (1986). Use of positrons to study transport in tokamak plasmas. *Review of Scientific Instruments, 57*, 1862–1867.

53. Tandberg-Hansen, E., & Emslie, A. G (1988). The Physics of Solar Flares. Cambridge Univ. Press, Cambridge, pp. 124.

54. Tinkle, M. D., Greaves, R. G., Surko, C. M., Spencer, R. L., & Mason, G. W. (1994). Low-order modes as diagnostics of spheroid non-neutral plasmas. *Physical Review Letters, 72*, 352–355.

55. Tiwari, R. S., Kaushik, A., & Mishra, M. K. (2007). Effects of positron density and temperature on ion acoustic dressed solitons in an electron-positron-ion plasma. *Physics Letters A, 365*, 335–340.

56. Tran, M. Q. (1979). Ion acoustic solitons in a plasma: A review of their experimental properties and related theories. *Physica Scripta, 20*, 317–327.

57. Tstovich, V., & Wharton, C. B. (1978). Laboratory electron-positron-ion plasma. A new research object. *Comments Plasma Physics and Controlled Fusion, 4*, 91–100.

58. Washimi, H., & Taniuti, T. (1966). Propagation of ion-acoustic solitary waves of small amplitude. *Physical. Review. Letters, 17*, 996–998.

59. Watanabe, S., & Jiang, B. (1993). Higher order solutions of an ion-acoustic solitary wave in plasma. *Physics of Fluids B, 5*, 409–414.

60. Yu, M. Y., Shukla, P. K., & Stenflo, L. (1986). Alfvén vortices in a strongly magnetized electron-positron plasma. *Astrophysical Journal, 309*, L63–66.

CHAPTER 9

ARBITRARY AMPLITUDE KINETIC ALFVÉN SOLITARY WAVES IN A PLASMA WITH TWO-ELECTRON TEMPERATURES

LATIKA KALITA

Department of Mathematics, Kamrup Polytechnic, Baihata Chariali, Kamrup, Assam, India, E-mail: latika84k@rediffmail.com

CONTENTS

ABSTRACT

To investigate the existence of kinetic Alfvén wave solitons, warm ions and two electron components, namely hot and cold are considered in a magnetized plasma. In this work, by using the Sagdeev pseudopotential method, an exact of analytical expression for arbitrary amplitude solitary kinetic Alfvén wave is derived. The Sagdeev potential (SP) $K(n_c)$ has been calculated numerically and the range of parameters for the existence of solitary waves and their effects on the plasma medium are studied in details and presented graphically for the different set of plasma parameter values.

9.1   INTRODUCTION

Hasegawa and Mima [16] and Yu and Shukla [42] were the first to investigate the existence of solitary kinetic Alfvén waves propagating in an oblique direction with respect to the ambient magnetic field in a magnetized plasma with $\alpha \equiv \beta/2Q \gg 1$, where $\beta = n_0 T/(B_0^2/2\mu_0)$ being the ratio of the thermal to magnetic pressure with n_0, T, B_0, and μ_0 the equilibrium plasma density, temperature, ambient magnetic field, and the permeability of free space, and $Q = m_e/m_i$ is the ratio of electron mass to ion mass, respectively. Alfvén waves have been very actively investigated and have found numerous recent major applications in laboratory, space, solar coronal plasma heating [33], astrophysical, tokamak plasma heating [7], and fusion plasmas [8, 17]. Their role in heating, acceleration of particles, and transport of magnetic energy in various space plasmas is well established. In the Earth's auroral plasmas, the concentration of oxygen ions can reach 30%–50%, often even as high as 90% or more [5, 6, 30]. Wu et al. [41] identified kinetic Alfvén solitary waves (KASWs) accompanied by both dip-type and hump-type density structures in the Freja observations. Louarn et al. [25] also examined the localized strong electromagnetic perturbations that were observed by the F4 experiment in the Freja satellite [21]. The acceleration of electrons and ions responsible for the aurora could be due to Alfvén waves [38]. Alfvén waves are used to heat ions through cyclotron resonance and electrons through Landau damping in a magnetically confined plasmas for fusion [16, 18]. When the amplitudes of Alfvén waves are sufficiently large, one cannot neglect nonlinearities. The latter contribute to the localization of waves and lead to different types of nonlinear structures, such as solitons [1, 10, 23, 24,

26, 32, 36], etc. Kinetic Alfvén waves are of importance in the study of coupling between the ionosphere and magnetosphere [30]. Sagdeev [31] used a mechanical analogy (pseudo-potential) approach to study the basic properties of the arbitrary/large amplitude ion acoustic solitary waves (IASWs). Alfvénic solitons are finite-amplitude waves of permanent form which owe their existence to a balance between nonlinear wave steepening and linear wave dispersion.

It is well-known that laboratory and space plasmas can contain distinct populations of hot and cold electrons [11, 27, 29]. In two-electron plasmas, electron-acoustic waves (EAWs) with wave frequency larger than the ion plasma frequency can be generated [12]. The dynamics of ion acoustic solitons in two-electron temperature plasma depends on the relative temperature and density ratio(s) between the two-electron components [13], which also affect the very conditions for existence of solitary excitations. At present decades, two-electron temperature plasmas appear to have received much new theoretical interest, as witnessed by an increasing number of studies on various plasma modes in such plasmas. The co-existence of cold and hot electrons has been revisited with respect to ion-acoustic and electron-acoustic pulses and double layers. Space observations have also attracted interest in the high-frequency electron modes in two electron temperature plasmas [35, 37], not overlooking related amplitude modulation studies [22, 34]. Two-electron temperature distributions are very common in space and laboratory plasmas [14, 28, 40]. Double layers can accelerate, decelerate, or reflect the plasma particles. A great deal of interest has been generated in understanding the formation of large amplitude ion-acoustic solitary waves and double layers in two-electron temperature plasmas. The large amplitude ion-acoustic solitary waves in a two-electron temperature plasma is investigated by Buti [4] and Devi and Kalita [9]. It is found that finite amplitude compressional and rarefactional ion-acoustic solitons exist in such a plasma. The theory of large amplitude ion-acoustic double layers in a stationary frame in an unmagnetized plasma with cold ions and Boltzmann distributed two temperature electrons is studied by Bharuthram and Shukla [3]. A large amplitude fluid theory of electrostatic ion-acoustic double layers [2] has been carried out and shown to be adaptable to stationary double layers in two-ion plasma with kinetically determined ions. In this chapter, we investigated the existence of kinetic Alfvén wave solitons in warm ions and two electron temperatures in a magnetized plasma. An exact of analytical expression for arbitrary amplitude solitary kinetic Alfvén wave is derived using the Sagdeev pseudopotential method. As mentioned earlier, the Sagdeev potential $K(n_c)$ has been calculated numerically and the range of parameters for the existence of solitary waves (SWs) and their effects on the plasma medium are studied for the different set of plasma parameter values.

9.2   BASIC EQUATIONS

We have considered the basic equations governing a plasma contaminated with two electron components, namely hot and cold electrons immersed in an uniform external magnetic field $\boldsymbol{B}_0 = B_0 \hat{z}$ directed along z-axis, assuming the wave vector to be (k_x, k_z). The electron distributions are considered as in Tagare [39]. The equations of motion for warm ions in $(x–z)$ plane are:

$$\frac{\partial n_i}{\partial t} + \frac{\partial}{\partial x}\left(n_i v_{ix}\right) + \frac{\partial}{\partial z}\left(n_i v_{iz}\right) = 0 \tag{1}$$

$$\frac{\partial v_{ix}}{\partial t} + v_{ix}\frac{\partial v_{ix}}{\partial x} + v_{iz}\frac{\partial v_{ix}}{\partial z} = \frac{\beta}{2}\left(-\frac{\partial \phi}{\partial x} - \frac{\alpha}{n_i}\frac{\partial n_i}{\partial x}\right) + v_{iy} \tag{2}$$

$$\frac{\partial v_{iy}}{\partial t} + v_{ix}\frac{\partial v_{iy}}{\partial x} + v_{iz}\frac{\partial v_{iy}}{\partial z} = -v_{ix} \tag{3}$$

$$\frac{\partial v_{iz}}{\partial t} + v_{ix}\frac{\partial v_{iz}}{\partial x} + v_{iz}\frac{\partial v_{iz}}{\partial z} = \frac{\beta}{2}\left(-\frac{\partial \psi}{\partial z} - \frac{\alpha}{n_i}\frac{\partial n_i}{\partial z}\right) \tag{4}$$

$$\frac{\partial n_e}{\partial t} + \frac{\partial}{\partial z}\left(n_e v_{ez}\right) = 0 \tag{5*}$$

$$\frac{\partial^4\left(\phi - \psi\right)}{\partial x^2 \partial z^2} = \frac{2}{\beta}\left[\frac{\partial^2}{\partial t^2}\left(n_e\right) + \frac{\partial^2}{\partial t \partial z}\left(n_i v_{iz}\right)\right] \tag{5}$$

$$n_e = n_h + n_c \tag{6}$$

where

$$n_h = v \exp \frac{\gamma\psi}{\mu + v\gamma} \text{ and } n_c = \mu \exp \frac{\psi}{\mu + v\gamma} \quad (7)$$

In the derivation of Eq. (5) we have used continuity equation for the electron (n_e), i.e., Eq. (5*). The ion polarization effect is not considered in our model. Therefore, the parallel ion inertia effect can be studied through our model.

We have normalized the ion, electron densities n_i, n_e by the equilibrium plasma density n_0, time (t) by ion gyro period Ω_i^{-1}, velocities by Alfvén velocity $v_A = cB_0/(4\pi n_0 m_i)^{1/2}$, space by $p_s = C_s/\Omega_i$, the potential by KT_{eff}/e, and magnetic field by B_0, temperature is normalized to effective temperature $T_{eff} = T_h T_c/(n_{0h}T_h + n_{0c}T_c)$. Here, $\alpha = T_i/T_{eff}$, $\gamma = T_c/T_h$, $\mu = n_{c0}/n_0$, $v = n_{h0}/n_0$.

9.3 DERIVATION OF THE SAGDEEV POTENTIAL (SP) EQUATION

We assume that the wave is propagating obliquely to the external magnetic field $B_0\hat{z}$ depending on the stationary independent variable,

$$\eta = xk_x + zk_z + \omega t \quad (8)$$

$$k_x^2 + k_z^2 = 1, \ \omega = \frac{wave\ velocity}{Alfvén\ velocity} = \frac{v}{v_A}$$

Using the new co-ordinate η moving with the wave, we get from Eq.(1), after integration and simplification,

$$k_x v_{ix} + k_z v_{iz} = \omega\left(1 - \frac{1}{n_i}\right) \quad (9)$$

In deducing these equations, we have applied the boundary conditions

$$v_{ix} = v_{iz} = 0, \ v_{ez} = 0, \ \phi = \psi = \partial n/\partial \xi = 0 \text{ at } n_i = 1 \text{ as } |\eta| \to \infty$$

Using the above transformation (8), Eqs. (2)–(5) can be simplified as

$$-\frac{\omega}{n_i}\frac{\partial v_{ix}}{\partial \eta} + \frac{\alpha}{n_i}k_x\frac{\partial n_i}{\partial \eta} = -\frac{\beta}{2}k_x\frac{\partial\phi}{\partial\eta} + v_{iy} \quad (10)$$

$$\frac{\omega}{n_i} \frac{\partial v_{iy}}{\partial \eta} = v \tag{11}$$

$$-\frac{\omega}{n_i} \frac{\partial v_{iz}}{\partial \eta} + \frac{\alpha k_z}{n_i} \frac{\partial n_i}{\partial \eta} = -\frac{\beta}{2} k_z \frac{\partial \psi}{\partial \eta} \tag{12}$$

$$k_x^2 k_z^2 \frac{\partial^4}{\partial \eta^4} (\phi - \psi) = \frac{2}{\beta} \left[\omega^2 \frac{\partial^2 n_e}{\partial \eta^2} - k_z \omega \frac{\partial^2}{\partial \eta^2} (n_i v_{iz}) \right] \tag{13}$$

Multiplying Eq. (10) by k_x and Eq. (12) by k_z and then adding

$$-\frac{\omega^2}{n_i^3} \frac{\partial n_i}{\partial \eta} + \frac{\alpha}{n_i} \frac{\partial n_i}{\partial \eta} = -\frac{\beta}{2} \left(k_x^2 \frac{\partial \phi}{\partial \eta} + k_z^2 \frac{\partial \overline{\psi}}{\partial \eta} \right) + k_x v_{iy} \tag{14}$$

Integrating Eq. (13) twice, subjected to the boundary conditions $v_{iz} = 0$, $\phi = \psi \rightarrow 0$, n_i, $n_e \rightarrow_0$ as $|\eta| \rightarrow \infty$ we get,

$$k_x^2 k_z^2 \left(\frac{\partial^2 \phi}{\partial \eta^2} - \frac{\partial^2 \psi}{\partial \eta^2} \right) = \frac{2}{\beta} \left[\omega^2 (n_e - 1) - k_z \omega n_i v_{iz} \right] \tag{15}$$

Differentiating Eq. (14) we get

$$\frac{\partial}{\partial \eta} \left[\left(\frac{\omega^2}{n_i^3} - \frac{\alpha}{n_i} \right) \frac{\partial n_i}{\partial \eta} \right] = \frac{\beta}{2} \left(k_x^2 \frac{\partial^2 \phi}{\partial \eta^2} + k_z^2 \frac{\partial^2 \psi}{\partial \eta^2} \right)$$
$$- \frac{n_i}{\omega} \left[\omega \left(1 - \frac{1}{n_i} \right) - k_z v_{iz} \right] \tag{16}$$

Taking logarithm, Eq. (7) can be written as,

$$\log(n_h) = \log \left(v \exp \frac{\gamma \psi}{\mu + v\gamma} \right), \quad \Rightarrow \quad \log(n_h) = \log v + \frac{\gamma \psi}{\mu + v\gamma}$$

Differentiating with respect to η we get, $\dfrac{dn_h}{d\eta} = \dfrac{n_h\gamma}{\mu+v\gamma}\dfrac{d\psi}{d\eta}$

$$\left.\begin{array}{l} \\ \dfrac{dn_c}{d\eta} = \dfrac{n_c}{\mu+v\gamma}\dfrac{d\psi}{d\xi} \end{array}\right\} \quad (17)$$

Similarly we get,

Differentiating Eq. (6) with respect to ξ and using Eq. (17) we get,

$$\frac{dn_e}{d\eta} = \frac{1}{\mu+v\gamma}\left[n_h\gamma + n_c\right]\frac{d\psi}{d\eta} \tag{18}$$

Combination of Eq. (6) gives,

$$n_h = \frac{v}{\mu^\gamma}n_c^\gamma \tag{19}$$

Therefore, Eq. (6) becomes

$$n_e = \frac{v}{\mu^\gamma}n_c^\gamma + n_c \tag{20}$$

Differentiating Eq. (20) with respect to η we get,

$$\frac{dn_e}{d\eta} = \left(1 + \frac{v\gamma}{\mu^\gamma}n_c^{\gamma-1}\right)\frac{dn_c}{d\eta} \tag{21}$$

Using Eq. (19) in Eq. (18) and using Eq. (21) we get,

$$\frac{dn_e}{d\eta} = \frac{1}{\mu+v\gamma}\left(\frac{v}{\mu^\gamma}n_c^\gamma\gamma + n_c\right)\frac{d\psi}{d\xi}$$

$$\frac{d\psi}{d\eta} = \frac{\mu+v\gamma}{n_c}\frac{dn_c}{d\eta} \tag{22}$$

Using Eq. (22) in Eq. (12) and using charge neutrality condition $n_i = n_e$, we get

$$\begin{aligned} v_z = {}& \frac{\beta}{2}\frac{k_z}{\omega}(\mu+v\gamma)\left[\left(n_c + \frac{v}{\gamma\mu^\gamma}n_c^\gamma\right) - \left(1 + \frac{v}{\gamma\mu^\gamma}\right)\right] \\ & + \frac{\alpha}{\omega}k_z\left[\left(n_c + \frac{v}{\mu^\gamma}n_c^\gamma\right) - \left(1 + \frac{v}{\mu^\gamma}\right)\right] \end{aligned} \tag{23}$$

Now putting the value of v_{iz} in Eq. (16) we get

$$\Rightarrow \frac{\beta}{2}\left(k_x^2 \frac{\partial^2 \varphi}{\partial \eta^2} + k_z^2 \frac{\partial^2 \psi}{\partial \eta^2} \right) =$$

$$\frac{\partial}{\partial \eta}\left[\left(\frac{\omega^2}{\left(n_c + \dfrac{v}{\mu^\gamma} n_c^\gamma\right)^3} - \frac{\alpha}{\left(n_c + \dfrac{v}{\mu^\gamma} n_c^\gamma\right)} \right) \times \left(1 + \frac{v\gamma}{\mu^\gamma} n_c^{\gamma-1}\right) \frac{dn_c}{d\eta} \right.$$

$$+ \frac{\left(n_c + \dfrac{v}{\mu^\gamma} n_c^\gamma\right)}{\omega}\left[\omega\left(1 - \frac{1}{\left(n_c + \dfrac{v}{\mu^\gamma} n_c^\gamma\right)} \right) \right]$$

$$- \frac{k_z}{\omega}\left(n_c + \frac{v}{\mu^\gamma} n_c^\gamma\right)\left[\frac{\beta}{2}\frac{k_z}{\omega}(\mu + v\gamma)\left\{\left(n_c + \frac{v}{\gamma\mu^\gamma} n_c^\gamma\right) - \right. \right.$$

$$\left. \left. \left. \left(1 + \frac{v}{\gamma\mu^\gamma}\right)\right\} + \frac{\alpha k_z}{\omega}\left\{\left(n_c + \frac{v}{\mu^\gamma} n_c^\gamma\right) - \left(1 + \frac{v}{\mu^\gamma}\right)\right\}\right] \right] \tag{24}$$

Again Eq. (15) we get

$$k_x^2 k_z^2 \left(\frac{\partial^2 \phi}{\partial \eta^2} - \frac{\partial^2 \psi}{\partial \eta^2} \right) = \frac{2}{\beta}\left[\omega^2\left(\left(n_c + \frac{v}{\mu^\gamma} n_c^\gamma\right) - 1 \right) - k_z \omega\left(n_c + \frac{v}{\mu^\gamma} n_c^\gamma\right) \right.$$

$$\times \left[\frac{\beta}{2}\frac{k_z}{\omega}(\mu + v\gamma)\right]\left[\left(n_c + \frac{v}{\gamma\mu^\gamma} n_c^\gamma\right) \right.$$

$$\left. \left. - \left(1 + \frac{v}{\gamma\mu^\gamma}\right)\right] + \frac{\alpha}{\omega}k_z\left[\left(n_c + \frac{v}{\mu^\gamma} n_c^\gamma\right) - \left(1 + \frac{v}{\mu^\gamma}\right)\right] \right] \tag{25}$$

Eliminating $\partial^2\phi/\partial\eta^2$ from Eqs. (24) and (25) we get,

$$\Rightarrow \frac{\beta}{2}k_z^2 \frac{\partial^2 \psi}{\partial \eta^2} = k_z^2 \frac{\partial}{\partial \eta}\left[\left(\frac{\omega^2}{\left(n_c + \frac{v}{\mu^\gamma}n_c^\gamma\right)^3} - \frac{\alpha}{\left(n_c + \frac{v}{\mu^\gamma}n_c^\gamma\right)}\right)\left(1 + \frac{v\gamma}{\mu^\gamma}n_c^{\gamma-1}\right)\frac{dn_c}{d\eta}\right.$$

$$-\omega^2\left(1 - \frac{k_z^2}{\omega^2}\right)\left[\left(n_c + \frac{v}{\mu^\gamma}n_c^\gamma\right) - 1\right] + k_z^2\alpha\left(n_c + \frac{v}{\mu^\gamma}n_c^\gamma\right)\left(1 - \frac{k_z^2}{\omega^2}\right)\left[\left(n_c + \frac{v}{\mu^\gamma}n_c^\gamma\right) - \left(1 + \frac{v}{\mu^\gamma}\right)\right]$$

$$\left.+ k_z^2\frac{\beta}{2}\left(1 - \frac{k_z^2}{\omega^2}\right)(\mu + v\gamma)\left(n_c + \frac{v}{\mu^\gamma}n_c^\gamma\right)\left[\left(n_c + \frac{v}{\gamma\mu^\gamma}n_c^\gamma\right) - \left(1 + \frac{v}{\gamma\mu^\gamma}\right)\right]\right]$$

Using Eq. (22) we get,

$$\Rightarrow \frac{\partial}{\partial \eta}\left[\left[\frac{\beta}{2}k_z^2\frac{(\mu + v\gamma)}{n_c} - k_z^2\right]\left[\left(\frac{\omega^2}{\left(n_c + \frac{v}{\mu^\gamma}n_c^\gamma\right)^3} - \frac{\alpha}{\left(n_c + \frac{v}{\mu^\gamma}n_c^\gamma\right)}\right)\left(1 + \frac{v\gamma}{\mu^\gamma}n_c^{\gamma-1}\right)\right]\frac{dn_c}{d\eta}\right]$$

$$= -\omega^2\left(1 - \frac{k_z^2}{\omega^2}\right)\left[\left(n_c + \frac{v}{\mu^\gamma}n_c^\gamma\right) - 1\right] + k_z^2\alpha\left(n_c + \frac{v}{\mu^\gamma}n_c^\gamma\right)\left(1 - \frac{k_z^2}{\omega^2}\right)\left[\left(n_c + \frac{v}{\mu^\gamma}n_c^\gamma\right) - \left(1 + \frac{v}{\mu^\gamma}\right)\right]$$

$$+ k_z^2\frac{\beta}{2}\left(1 - \frac{k_z^2}{\omega^2}\right)(\mu + v\gamma)\left(n_c + \frac{v}{\mu^\gamma}n_c^\gamma\right)\left[\left(n_c + \frac{v}{\gamma\mu^\gamma}n_c^\gamma\right) - \left(1 + \frac{v}{\gamma\mu^\gamma}\right)\right]$$

Multiplying both sides by the expression of the bracket of left hand side and after some algebra, we get an expression of the form,

$$\frac{1}{2}\left(\frac{dn_c}{d\eta}\right)^2 + K(n_c, \omega, k_z, \beta, \alpha, v, \gamma, \mu) = 0 \tag{26}$$

where,

$$K(n_c) = \lambda(n_c)\mu(n_c) \tag{27}$$

$$\lambda(n_c) = \frac{-1}{\left[\left[k_z^2\frac{\beta}{2}\frac{(\mu + v\gamma)}{n_c} - k_z^2\right]\left\{\frac{\omega^2}{\left(n_c + \frac{v}{\mu^\gamma}n_c^\gamma\right)^3} - \frac{\alpha}{\left(n_c + \frac{vn_c^\gamma}{\mu^\gamma}\right)}\right\}\left(1 + \frac{v\gamma}{\mu^\gamma}n_c^{\gamma-1}\right)\right]^2}$$

$$\mu(n_c) = k_z^4\alpha\frac{\beta}{2}\left(1 - \frac{k_z^2}{\omega^2}\right)(\mu + v\gamma)\left[\frac{1}{2}(n_c^2 - 1) + \frac{v^2}{\mu^{2\gamma}}\cdot\frac{1}{2\gamma}(n_c^{2\gamma} - 1) + \frac{2\gamma}{\mu^\gamma}\frac{1}{\gamma + 1}(n_c^{\gamma+1} - 1) - \left(1 + \frac{v}{\mu^\gamma}\right)\times\right.$$

$$\left\{\left(n_c + \frac{v}{\gamma\mu^\gamma}n_c^\gamma\right) - \left(1 + \frac{v}{\gamma\mu^\gamma}\right)\right\} - k_z^4\alpha\left(1 - \frac{k_z^2}{\omega^2}\right)\left[\omega^2\left\{Log\left(n_c + \frac{v}{\mu^\gamma}n_c^\gamma\right) - Log\left(1 + \frac{v}{\mu^\gamma}\right)\right\} - \frac{\alpha}{2}\times$$

$$\left\{\left(n_c + \frac{v}{\mu^\gamma}n_c^\gamma\right)^2 - \left(1 + \frac{v}{\mu^\gamma}\right)^2\right\} + \left(1 + \frac{v}{\mu^\gamma}\right)\omega^2\left\{\frac{1}{\left(n_c + \frac{v}{\mu^\gamma}n_c^\gamma\right)} - \frac{1}{\left(1 + \frac{v}{\mu^\gamma}\right)}\right\} + \alpha\left(1 + \frac{v}{\mu^\gamma}\right)\left\{\left(n_c + \frac{v}{\mu^\gamma}n_c^\gamma\right)\right.$$

$$\left.- \left(1 + \frac{v}{\mu^\gamma}\right)\right\}\right] - \omega^2\frac{\beta}{2}k_z^2\left(1 - \frac{k_z^2}{\omega^2}\right)(\mu + v\gamma)\left[\left\{\left(n_c + \frac{v}{\gamma\mu^\gamma}n_c^\gamma\right) - \left(1 + \frac{v}{\gamma\mu^\gamma}\right)\right\} - Log\,n_c\right] + \omega^2k_z^2\left(1 - \frac{k_z^2}{\omega^2}\right)$$

$$\times\left[-\omega^2\left\{\frac{1}{\left(n_c + \frac{v}{\mu^\gamma\gamma}n_c^\gamma\right)} - \frac{1}{\left(1 + \frac{v}{\gamma\mu^\gamma}\right)}\right\} - \alpha\left(1 + \frac{v}{\mu^\gamma}\right)\left\{\left(n_c + \frac{v}{\mu^\gamma}n_c^\gamma\right) - \left(1 + \frac{v}{\mu^\gamma}\right)\right\} + \frac{\omega^2}{2}\left\{\frac{1}{\left(n_c + \frac{v}{\mu^\gamma}n_c^\gamma\right)^2}\right.\right.$$

$$\left.\left.- \frac{1}{\left(1 + \frac{v}{\mu^\gamma}\right)^2}\right\} + \alpha\left\{Log\left(n_c + \frac{v}{\mu^\gamma}n_c^\gamma\right) - Log\left(1 + \frac{v}{\mu^\gamma}\right)\right\}\right] + k_z^4\frac{\beta^2}{4}\left(1 - \frac{k_z^2}{\omega^2}\right)(\mu + v\gamma)^2\left[\frac{1}{2\gamma}\mu^{-2\gamma}\left|\frac{v^2}{\gamma}\right.\right.$$

$$\times(n_c^{2\gamma-1}-1)+\gamma\mu^{2\gamma}(n_c^2-1)-2\mu^\gamma(v+\mu^\gamma\gamma)(n_c-1)+\frac{1}{\gamma}\{(-v+(n_c-1)\gamma\mu^\gamma)2\gamma n_c^\gamma+2v^2\}\}\right]$$

$$-k_z^4\frac{\beta}{2}\left(1 - \frac{k_z^2}{\omega^2}\right)(\mu+v\gamma)\left[\frac{1}{2\gamma(1+\gamma)}\left\{\frac{1}{(n_c\mu^\gamma+vn_c^\gamma)} - \frac{1}{(\mu^\gamma+v)}\right\}[\mu^{-2\gamma}\{-(1+\gamma)\alpha\gamma^3(n_c^{3\gamma}-1)\right.$$

$$-\alpha\mu^{3\gamma}\gamma(1+\gamma)(n_c^3-1)+2\alpha\mu^{2\gamma}(1+\gamma)(v+\mu^\gamma\gamma)(n_c^2-1)+4\alpha\mu^\gamma v(1+\gamma)(v+\mu^\gamma\gamma)(n_c^{1+\gamma}-1)$$

$$+2\alpha v^2(1+\gamma)(v+\mu^\gamma\gamma)(n_c^{2\gamma}-1)+2\mu^{2\gamma}\omega^2(1+\gamma)(v+\mu^\gamma\gamma)-2\mu^{3\gamma}\omega^2(\gamma^2-1)(n_c-1)$$

$$-(3+\gamma+2\gamma^2)\alpha\mu^\gamma v^2(n_c^{1+2\gamma}-1)-(2+\gamma+3\gamma^2)\alpha\mu^{2\gamma}v(n_c^{2+\gamma}-1)$$

$$+2\mu^{2\gamma}\omega^2\gamma(1+\gamma)\{(n_c\mu^\gamma+vn_c^\gamma)Log\,n_c\}\}]]$$

The boundary condition used in deriving Eq. (26) is $\frac{dn_c}{d\eta} = 0$, $n_c = 1$ at $\xi = \pm\infty$. Equation (26) can be interpreted as an energy integral of an oscillatory particle of an unit mass with velocity $\frac{dn_c}{d\eta}$ and position n_c in a potential well $K(n_c)$. That is, the above equation can be considered as a motion of a particle whose pseudoposition is n_c at pseudotime η with pseudovelocity $\frac{dn_c}{d\eta}$ in a pseudopotential well $K(n_c)$. That is why Sagdeev's potential is called pseudopotential and the Sagdeev potential is given by Eq. (27). The Sagdeev potential derived in this chapter shows that the effect of the electron continuity along the direction in the magnetic field in which the two temperature electron effect has been merged through Eq. (19).

9.4 MATHEMATICAL CONDITIONS FOR THE EXISTENCE OF SOLITON SOLUTION

The conditions for existence of SWs are as follows [Writing $K(n_c, \omega, k_z, \beta, \alpha, \nu, \gamma, \mu)$ as $K(n_c)$,],

1. $K(n_c) < 0$ between $n_c = 1$ and $n_c = N$, so that $dn_c/d\eta$ is real [Eq. (26)], here N gives the amplitude of the solitary wave. N can be both greater than 1 and less than 1. In the former case we have compressive solitary wave and in the latter case rarefactive solitary waves.

2. $K(n_c)$ must be a maximum at $n_c = 1$ which means, $\left.\dfrac{\partial K(n_c)}{\partial n_c}\right|_{n_c = 1} = 0$ and $\left.\dfrac{\partial K^2(n_c)}{\partial n_c^2}\right|_{n_c = 1} < 0$.

3. $K(n_c)$ should cross the 'n_c' axis from below near $n_c = N$; and $K(n_c) > 0$ for $n_c > N$.

9.5 RESULTS

We have numerically analyzed the effect of various sets of plasma parameters on the structures of kinetic Alfvén waves. In Figure 9.1, the Sagdeev

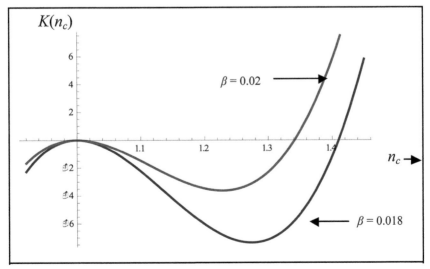

FIGURE 9.1   Exact Sagdeev potential $K(n_c)$ profile plotted against density n_c for two different values of $\beta = 0.018, 0.02$ and $\mu = 185$, $\nu = 0.09$, $\gamma = 0.9$, $\omega = 0.28$, $k_z = 0.009$, $\alpha = 0.0198$.

potential $K(n_c)$ verses density is plotted for different parameters by using Eq.(27). It can be seen that the amplitude of compressive potential profile increases with decrease of β. Compressive solitary waves are found to exist for ω = 0.28 (<1). It is found that the depth and amplitude of compressive pseudopotential decreases with increase of β(>0.02). Note that increasing of Sagdeev potential depth means that the width of the pulse decreases and vice verse. It may be noted that the compressive solitons are more affected by the ratio of the thermal to magnetic pressure (β) effects.

Figure 9.2 shows that the pseudopotential for different three values of the direction cosine (k_z). It is found that the depth and amplitude of compressive pseudopotential increases with the decrease of the direction cosine k_z. Also we noticed that an increase of the depth of the Sagdeev potential makes the solitary pulse narrower. Though k_z has a small value, it affects the pulse shape causing broadening of its profile. It may be mentioned that the compressive solitons cease to exist when the parameter k_z crosses a certain limit ($k_z > 0.007$), which of course depends on the other parameter.

In Figure 9.3, the effect of the Mach number (ω) on the formation of compressive soliton has been investigated. It is clear that compressive potential increases with the increase of Mach number ω. For $ω>0.3$ $K(n_c)$

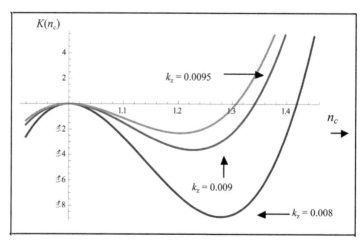

FIGURE 9.2 Exact Sagdeev potential $K(n_c)$ profile plotted against density n_c for three different values of $k_z = 0.008, 0.009, 0.0095$ and μ= 185, ν = 0.09, γ = 0.9, ω = 0.28, β = 0.02, α = 0.0198.

it does not satisfy the existence of solitary waves and hence soliton solution does not exist. We observed that the solitary pulses can propagate only in the region between $0.25 < \omega < 0.03$. Therefore, it is clear that the compressive solitons, depending on the plasma parameters can propagate in subsonic regimes in our plasma model.

Next we discuss the effect of α parameter on the nonlinear structures. The temperature (α) also plays an important role in the formation of nonlinear structures, we kept all other parameters fixed and varied α within the permitted range in space plasma which is depicted in Figure 9.4. It is noticed that the amplitude and depth of solitary wave increases with decrease of α. Solitary waves exists for lower values of $\alpha < 0.03$.

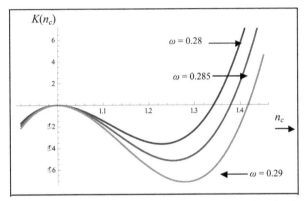

FIGURE 9.3 Exact Sagdeev potential $K(n_c)$ profile plotted against density n_c for three different values of $\omega = 0.28, 0.285, 0.29$ and $\mu = 185$, $\nu = 0.09$, $\gamma = 0.9$, $\beta = 0.02$, $k_z = 0.009$, $\alpha = 0.0198$.

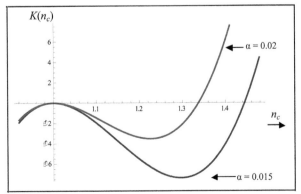

FIGURE 9.4 Exact Sagdeev potential $K(n_c)$ profile plotted against density n_c for two different values of $\alpha = 0.015, 0.02$ and $\omega = 0.28$, $\mu = 185$, $\nu = 0.09$, $\gamma = 0.9$, $\beta = 0.02$, $k_z = 0.009$.

9.6 DISCUSSIONS

Some recent observational [19–21] studies on data from Freja satellite showed that the low-frequency auroral electromagnetic fluctuations resulting in strong electric spikes, which can be interpreted as density pulses exhibiting kinetic Alfvén wave field characteristics. Due to the strong nonlinearity of the dispersive plasma medium, these may emerge out in the form of solitary kinetic Alfvén wave, etc. In our present work, the set of basic equations governing warm ions and two electron components, namely hot and cold have been reduced to a single equation known as the Sagdeev potential equation. A pseudopotential technique has been applied to study the arbitrary amplitude of kinetic Alfvén wave solitons. Furthermore, numerical calculations reveal that the present plasma system supports compressive subsonic solitons. The dependence of the solitary excitation characteristics on the different plasma parameters has been investigated. Our present theoretical studies could be of interest for explaining some of the recent satellite observations of the nonlinear wave structures in space and astrophysical scenarios where two temperature electrons are present.

ACKNOWLEDGMENTS

The author is grateful to Dr. Nirupama Devi, Ex Associate Professor, Department of Mathematics, Cotton University, Guwahati, Assam, for her constructive comments and suggestions, which helped the paper in its present form.

KEYWORDS

- kinetic Alfvén wave
- Sagdeev potential
- solitary waves

REFERENCES

1. Alinejad, H. (2012). Effect of nonthermal electrons on oblique electrostatic excitations in a magnetized electron-positron-ion plasma. *Phys. Plasmas, 19*, 052302.
2. Baboolal, S., Bharuthram, R., & Hellberg, M. A. (1988). Arbitrary-amplitude rarefactive ion-acoustic double layers in warm multi-fluid plasmas. *J. Plasma Phys., 40*, 163.
3. Bharuthram, R., & Shukla, P. K. (1986). Large amplitude ion-acoustic double layers in a double Maxwellian electron plasma. *Phys. Fluids, 29*, 3214.
4. Buti, B. (1980). Ion-acoustic holes in a two-electron-temperature plasma. *Phys. Lett. A, 76*, 251.
5. Chaston, C. C., Carlson, C. W., Peria, W. J., Ergun, R. E., & McFadden, J. P. (1999). FAST Observations of Inertial Alfven Waves in the Dayside Aurora. *Geophys. Res. Lett., 26*, 647, doi: 10.1029/1998GL900246.
6. Chaston, C. C., Peticolas, L. M., Bonnell, J. W., Carlson, C. W., Ergun, R. E., & McFadden, J. P.(2003). Width and brightness of auroral arcs driven by inertial Alfven waves. *J. Geophys. Res., 108*, 1091, doi: 10.1029/2001JA007537.
7. Chen, L. (2008). Alfvén waves: a journey between space and fusion plasmas. *Plasma Phys. Controlled Fusion, 50*, 124001.
8. Cramer, N. F. (2001). The Physics of Alfvén Waves, Wiley-VCH, Berlin.
9. Devi, N., & Kalita, L. (2012). Propagation of solitary waves in warm magnetized plasma with two temperature electrons. *Bulletin of Society for Mathematical Services and Standards, 1*(2), 46–52.
10. El-Awady, E. I., & Moslem, W. M. (2011). On a plasma having nonextensive electrons and positrons: Rogue and solitary wave propagation. *Phys. Plasmas, 18*, 082306.
11. Ferrante, G., Zarcone, M., Uryupina, D. S., & Uryupin, S. A. (2003). Radiation absorption and reflection by a plasma with cold and hot electrons. *Phys. Plasmas, 10*, 3344.
12. Gary, S. P., & Tokar, R. L. (1985). The electron-acoustic mode. *Phys. Fluids, 28*, 2439.
13. Goswami, B. N., & Buti, B. (1976). Ion acoustic solitary waves in a two-electron-temperature plasma. *Phys. Lett., A57*, 149.
14. Hairapetian, G., & Stenzel, R. L. (1990). Observation of a stationary, current-free double layer in a plasma .*Phys. Rev. Lett., 65*, 175.
15. Hasegawa, A. (1976). Particle acceleration by MHD surface wave and formation of aurora. *J. Geophys. Res., 81*, 5083, doi: 10.1029/JA081i028p05083.
16. Hasegawa, A., & Mima, K. (1976). Exact Solitary Alfvén Wave. *Phys. Rev. Lett., 37*, 690.
17. Hasegawa, A., & Uberoi, C. (1982). The Alfvén Wave (Technical Information Centre, U.S. Department of Energy, Washington, DC).
18. Horton, W., Zhu, P., Hoang, G., Aniel, T., Ottaviani, M., & Garbet, X. (2000). Electron transport in Tore Supra with fast wave electron heating. *Phys. Plasmas, 7*, 1494.
19. Huang, D. J. Wu, G.-Li, Wang, D.–Yu, & Falthammar, C. G. (1996). Solitary kinetic Alfvén waves in the two-fluid model. *Phys. Plasmas 3*, 2879.
20. Huang, G. L., Wang, D. Y., Wu, D. J., Feraudy, H. de, Queau, D. Le, Volwerk, M., & Holback, B. (1997). The eigenmode of solitary kinetic Alfvén waves observed by Freja satellite. *J. Geophys. Res., 102*, 7217, doi: 10.1029/96JA02607.
21. Huang, G.-L., Wang, D.-Y., Wu, D.-J., de Feraudy, H., Le Queau, D., Volwerk, M., & Holback, B. (1997). The eigenmode of solitary kinetic Alfvén waves observed by Freja satellite. *JGR, 102*(A4), 7217.
22. Kourakis, I. and. Shukla, P. K. (2004). Electron-acoustic plasma waves: Oblique modulation and envelope solitons. *Phys. Rev. E, 69*, 036411.

23. Kourakis, I., Sultana, S., & Hellberg, M. A. (2012). Dynamical characteristics of solitary waves, shocks and envelope modes in kappa-distributed non-thermal plasmas: an overview. *Plasma Phys. Controlled Fusion, 54*, 124001.

24. Lakhina, G. S., Kakad, A. P., Singh, S. V., & Verheest, F. (2008). Ion- and electron-acoustic solitons in two-electron temperature space plasmas. *Phys. Plasmas, 15*, 062903.

25. Louarn, P., Wahlund, J.-E., Chust, T., de Feraudy, H., Roux, A., Holback, B., Dovner, P. O., Eriksson, A. I., & Holmgren, G. (1994). Observation of kinetic Alfvén waves by the FREJA spacecraft. *Geophys. Rev. Lett., 21*, 1847.

26. Mamun, A. A., & Shukla, P. K. (2002). Obliquely propagating electron-acoustic solitary waves. *Phys. Plasmas, 9*, 1474.

27. Montgomery, D. S., Focia, R. J., Rose, H. A., Russell, D. A., Cobble, J. A., Fernandez, J.C., & Johnson, R. P. (2001). Observation of Stimulated Electron-Acoustic-Wave Scattering. *Phys. Rev. Lett., 87*, 155001.

28. Nishida, Y., & Nagasawa, T. (1986). Excitation of ion-acoustic rarefactive solitons in a two-electron-temperature plasma. *Phys. Fluids, 29*, 345.

29. Pottelette, R., Ergun, R. E., Treumann, R. A., Berthomier, M., Carlson, C. W., McFadden, J.P., & Roth, I. (1999). Modulated electron-acoustic waves in auroral density cavities: FAST observations. *Geophys. Res. Lett., 26*, 2629, doi: 10. 1029/1999GL900462.

30. Roychoudhury, R. (2002). Arbitrary-amplitude solitary kinetic Alfvén waves in a non-thermal plasma. *J. Plasma Phys., 67*, 199.

31. Sagdeev, R. Z. (1966). Cooperative Phenomena and Shock Waves in Collisionless Plasmas. Reviews of Plasma Physics, Consultants Bureau, New York, Vol. 4, pp. 23.

32. Saini, N. S., Kourakis, I., & Hellberg, M. A. (2009). Arbitrary amplitude ion-acoustic solitary excitations in the presence of excess superthermal electrons. *Phys. Plasmas, 16*, 062903.

33. Shukla, P. K., Bingham, R., Eliasson, B., Dieckmann, M. E., & Stenflo, L. (2006). Nonlinear aspects of the solar coronal heating. *Plasma Phys. Controlled Fusion, 48*, B249–B255.

34. Shukla, P. K., Hellberg, M.A., & Stenflo, L. (2003). Modulation of electron-acoustic waves. *J. Atmosph. Solar-Terrestrial Phys. 65*, 355.

35. Singh, S. V., & Lakhina, G. S. (2001). Generation of electron-acoustic waves in the magnetosphere. *Planet. Space Res., 49*, 107.

36. Singh, S. V., Devanandhan, S., Lakhina, G. S., & Bharuthram, R. (2013). Effect of ion temperature on ion-acoustic solitary waves in a magnetized plasma in presence of superthermal electrons. *Phys. Plasmas, 20*, 012306.

37. Singh, S. V., Reddy, R. V., & Lakhina, G. S. (2001). Broadband electrostatic noise due to nonlinear electron-acoustic waves. *Adv. Space Res, 28*, 1643.

38. Stasiewicz, K., Bellan, P., Chaston, C., Kletzing, C., Lysak, R., Maggs, J., Pokhotelov, O., Seyler, C., Shukla, P., Stenflo, L., Streltsov, A., & Wahlund, J.-E. (2000). Small Scale Alfvénic Structure in the Aurora. *Space Sci. Rev. 92*, 423.

39. Tagare, S. G. (2000). Ion-acoustic solitons and double layers in a two-electron temperature plasma with hot isothermal electrons and cold ions. *Phys. Plasmas, 7*(3).

40. Temerin, M., Cerny, K., Lotko, W., & Mozer, F. S. (1982). Observations of Double Layers and Solitary Waves in the Auroral Plasma. *Phys. Rev. Lett., 48*, 1175.

41. Wu, D. J., Huang, G.-L., Wang, D.-Y., & Falthammar, C.-G. (1996). Solitary kinetic Alfvén waves in the two-fluid model. *Phys. Plasmas, 3*(8), 2879.

42. Yu, M. Y., & Shukla, P. K. (1978). Finite amplitude solitary Alfven waves. *Phys. Fluids, 21*, 1457.

CHAPTER 10

NONLINEAR DUST KINETIC ALFVÉN WAVES IN A DUST–ION PLASMA WITH IONS FOLLOWING q-NONEXTENSIVE VELOCITY DISTRIBUTION

M. K. AHMED and O. P. SAH

Department of Physics, Birjhora Mahavidyalaya, Bongaigaon – 783380, Assam, India, E-mail: mnzur_27@rediffmail.com, opbngn@gmail.com

CONTENTS

ABSTRACT

Exact nonlinear dust kinetic Alfvén waves are investigated in collisionless, low but finite β plasma, whose constituents are negatively charged dust grains and non-extensive ions. It is found that the non-extensive dust-ion plasma model considered here supports only sub-Alfvénic compressive solitary dust KAWs and double layers. Owing to ions' non-extensivity, solitary dust KAWs may undergo amplitude fluctuations before they transform in to double layers. The implication of our results to Saturn's rings is pointed out.

10.1 INTRODUCTION

It is well-known that Alfvén wave is nondispersive in the ideal magnetohydrodynamic description. However, the wave can be dispersive for its oblique propagation to the direction ambient magnetic field in plasma when shear Alfvén waves are either affected by the ion gyroradius (for $Q \ll \beta \ll 1$; Q is the electron-to-ion mass ratio, and β is the thermal to magnetic pressure ratio), or by the electron inertia length (for $\beta \ll Q$). The resulting dispersive Alfvén waves are often known as kinetic Alfvén wave (hereafter KAW) and inertial Alfvén wave in the former and the latter cases, respectively. The dispersive character of KAWs when balanced with nonlinear steepening may lead to the formation of nonlinear structures like solitary KAWs and double layers. The data received from space satellites [1, 2] has revealed solitary like structures by strong electromagnetic spikes which could possibly be interpreted as KAWs with dip or hump type solitary structures. The study of nonlinear phenomena of KAWs in plasma has drawn much attention because of their relevance in explaining electromagnetic fluctuations and nonlinear structures observed in space, astrophysical, and laboratory plasmas.

Over the last few decades, there has been a growing interest in the understanding of various collective processes in dusty plasmas (plasmas with extremely massive and highly charged dust grains) because they are ubiquitous in laboratory, space, and astrophysical plasma environments, such as cometary tails, asteroid zones, planetary rings, interstellar medium, earth's environments, etc. [3–6]. The presence as well as dynamics of such massive charged dust grains in plasma not only modify the existing plasma wave spectra, but also introduce new wave modes [7–9]. A good number of authors have studied nonlinear wave structures in unmagnetized [10–12] as well as magnetized [13, 14] dusty plasmas. Investigations of nonlinear structures of dust KAWs in dusty plasmas have also been received much attention by several authors [15–17] in order to understand the formation of coherent structures in space and laboratory plasmas; and to make a diagnosis of dust in magnetized plasmas. Dust KAWs are Alfvén-like waves driven by the polarization drift of the dust and bending of the magnetic field lines. Chen and Yu (2000) investigated nonlinear dust KAWs in a low $\beta_d(\frac{m_i}{m_d} \ll \beta_d \ll 1/Z_d)$, where $\beta_d = 2\mu_0 n_{do}T/B_o{}^2$, m_i and m_d are the ion and dust masses, Z_d represents the number of charges residing on the dust grain surface, n_{do} is the equilibrium dust density, B_o is the external magnetic field, and T is the effective temperature) collisionless plasma. They studied the dynamics of negatively charged dust in the background of thermal electrons and ions obeying Boltzmann distributions. According to their analysis, solitons involving smooth density dips, and cusped density humps can coexist. Later, Mahmood et al. [16] showed that cusped density humps becomes smoother when complete dust nonlinearity and finite β effects are taken into account. Assuming equilibrium distribution to be Maxwellian for ions, Arshad et al. [17] studied fully nonlinear dust kinetic Alfvén waves in a collisionless, low but finite β two component dust–ion plasma. They found the existence of sub-Alfvénic solitons with density humps and density dips while at super-Alfvénic speed, only density humps were reported.

In recent years, there has been a great deal of interest to study the nonlinear behavior [12, 18–20] of plasma within a new statistical frame work based on q-entropic measure first conceived by Renyi [21] and subsequently proposed by Tsallis [22]. This statistics is thought to be a useful generalization of the conventional Boltzmann–Gibbs statistics and suitable for the statistical description of systems with long-range interaction such as plasma (Coulombian long-range interaction). In non-extensive statistical mechanics, the distribution function which maximizes the q-entropy is non-Maxwellian and is termed as q-non-extensive distribution [23]. Lima et al. [24] in their model have shown that q-non-extensive distribution can fit experimental data better than the standard Maxwellian distribution. Invoking Sagdeev pseudopotential method [25], the authors of this chapter propose to investigate the nonlinear structures of dust KAWs in low but finite β collisionless dust-ion plasma with ions exhibiting equilibrium q-non-extensive velocity distribution and dust temperature effect. It is found that this non-extensive dust-ion plasma model supports only sub-Alfvénic compressive solitons and double layers.

10.2 MATHEMATICAL FORMULATION

Two-component dusty plasma model corresponds to a state where most of the electrons from the ambient plasma are attached to the dust grain surface so that we may assume that $n_{eo} << Z_d n_{do}$, where n_{eo} (n_{do}) is the unperturbed electron (dust grain) number density and Z_d is the number of electrons residing on the dust grain surface. However, the depletion of electrons cannot be complete because the minimum value of the ratio between the electron and ion number densities turns out to be $(m_e/m_i)^{1/2}$ as the grain surface potential approaches zero, where $m_e(m_i)$ is the electron (ion) mass. Here, the dusty plasma may be regarded approximately as two-component plasma composed of negatively charged dust grains and ions. The latter shield the dust grains. The relevance of this model can be found in planetary ring systems, such as in Saturn's F-rings [5, 9, 28], and in comets (e.g., Halley's comet [13, 26]). This model is valid because for a situation $(n_{eo} << Z_d n_{do})$, we have $m_e/m_i << 1$, where n_{io} is the unperturbed ion number density. Hence, at equilibrium, we have $n_{io} \simeq Z_d n_{do}$.

The spatial scales of the problem under consideration as well as the inter-grain distance are assumed to be much larger than the grain size, so that the effects of dust charging in a magnetized plasma can be ignored and that the dust grains can be assumed to be of uniform mass and charge. We assume that $v_{Ad} > c_d$, where $c_d = (Z_d T/m_d)^{1/2}$ and $v_{Ad} = (B_o/\mu_o n_{do} m_d)^{1/2}$ are the dust acoustic and dust Alfvén speeds, respectively. In terms of β this assumption can also be written as $\beta_d \equiv 2\mu_o n_{do} T/B_o^2 < 1/Z_d$. Here, Z_d β_d can be of the same order as β. Furthermore, for very low-frequency fluctuations on the long dust timescale, we assume that $v_{ti} >> v_{Ad}$ (i.e., $\beta_d >> m_i/m_d$ and max $\{v_{di}, v_{dd}\} << kv_{Ad}$, where v_{ti} is the ion thermal speed and v_{dj} ($j = i, d$) is the dust collision frequency. This means that dust–ion and dust–dust collisions, which are much less frequent than ion-ion collisions, can be neglected. Thus, for the wave motion, the ions are fully relaxed and in local thermodynamic equilibrium whereas the dust fluid remains cold and ion skin-depth effects can be ignored [15]. Dusts being cold are strongly affected by magnetic field while ions are considered to follow q-non-extensive velocity distribution at equilibrium.

Let us suppose that perturbed number densities of non-extensive ions and inertial dusts of dust-ion plasma be respectively n_i and n_d. To model the effects of ions' non-extensivity, we refer to the following one-dimensional equilibrium q-distribution function for ion density [18, 23]

$$n_i(\psi) = n_{io} \left[1 - (q - 1) \frac{e\psi}{T_i} \right]^{\frac{1}{q-1} + \frac{1}{2}} \tag{1}$$

Here, the parameter q lies in the range $-1 < q < 1$ or $q > 1$ and stands for the strength of ions' non-extensivity. The distribution function (1) reduces to Maxwellian distribution in the extensive limiting case $q \rightarrow 1$.

The equation of state of an ideal ion gas at thermal pressure P and temperature T_i in the non-extensive kinetic theory is obtained as [24]

$$P = \frac{2}{3q-1} n_{io} T_i \tag{2}$$

Equation (2) can also be written in the following form

$$P = n_{io} T_{eff} \tag{3}$$

where $T_{eff} = \frac{2}{3q-1} T_i$ is the effective temperature for the non-extensive ions. In the limit $q \rightarrow 1$, the effective temperature reduces to the thermal temperature for Maxwellian distribution for ions. The effective temperature will change with the value of T_i provided q remains unchanged. We find from Eq. (2) that $q > 1/3$.

The ratio between the effective plasma thermal pressure and magnetic pressure is $\beta_{eff} = \beta_d T_{eff}/T_i$, where $\beta_d = \frac{2\mu_0 n_{do} T_i}{B_o{}^2}$. We assume that β_{eff} is small but much larger than the ion to dust mass ratio, i.e., $m_i/m_d << \beta_{eff} << 1/Z_d$. For low β plasma, the compressive component of the magnetic field perturbation can be ignored. This means that the z-component of the magnetic field perturbation is almost zero. This allows us to use two potential theory [27] for the electric field variables, i.e.,

$$E_x = -\frac{\partial \varphi}{\partial x} \quad , \quad E_z = -\frac{\partial \psi}{\partial z} \tag{4}$$

The set of equations governing the dynamics of dust KAWs [15] in dust-ion plasma valid for $m_i/m_d \ll \beta_{eff} \ll 1/Z_d$ are as follows

$$n_i = n_{io} \left[1 - (q-1)\frac{e\psi}{T_i} \right]^{\frac{1}{q-1}+\frac{1}{2}} \tag{5}$$

$$\frac{\partial n_i}{\partial t} + \frac{\partial}{\partial z}(n_i v_{iz}) = 0 \tag{6}$$

$$\frac{\partial n_d}{\partial t} + \frac{\partial}{\partial x}(n_d v_{dx}) + \frac{\partial}{\partial z}(n_d v_{dz}) = 0 \tag{7}$$

$$\frac{\partial v_{dz}}{\partial t} + v_{dx}\frac{\partial v_{dz}}{\partial x} + v_{dz}\frac{\partial v_{dz}}{\partial z} = \frac{Z_d e}{m_d}\frac{\partial \psi}{\partial z} - \frac{T_d}{m_d n_d}\frac{\partial n_d}{\partial z} \tag{8}$$

where v_{dx} is the dust perpendicular drift velocity given by

$$v_{dx} = \frac{m_d}{Z_d e B_o^2}\frac{\partial^2 \varphi}{\partial x \partial t} \tag{9}$$

For current density in the Z-direction, Ampere's law, and Faraday's law together yield

$$\frac{\partial^4}{\partial z^2 \partial x^2}(\varphi - \psi) = \mu_o \frac{\partial^2}{\partial z \partial t} j_z \tag{10}$$

where

$$\frac{\partial j_z}{\partial z} = -Z_d e \left[\frac{\partial n_d}{\partial t} + \frac{\partial}{\partial z}(n_d v_{dz}) \right] \tag{11}$$

In the above equations, we have used two potential representations for electric field variables, ion continuity equation, and quasi-neutrality condition $Z_d n_d \cong n_i$ wherever they were needed.

10.3   DERIVATION OF THE SAGDEEV POTENTIAL

In order to obtain one-dimensional localized stationary planar solutions, we define moving coordinate by $\eta = K_x x + K_z z - Mt$, where $M = V/C_d$ is the Mach number, V is the speed of nonlinear structure, and K_x and K_z are the direction cosines related by $K_x^2 + K_z^2 = 1$. Thus, on normalizing time to inverse of dust cyclotron frequency (Ω_d^{-1}), velocities to dust acoustic speed C_d, space coordinates to dust gyroradius (C_d/Ω_d), electric potentials to T_i/e, and particle densities to their respective equilibrium values, we get from Eqs. (5)–(11)

$$n_i = [\,1 - (q-1)\psi\,]^{\frac{1}{q-1}+\frac{1}{2}} \tag{12}$$

$$-M\partial_\eta n_d + K_x\partial_\eta(n_d v_{dx}) + K_z\partial_\eta(n_d v_{dz}) = 0 \tag{13}$$

$$v_{dx} = -K_x M\partial_\eta^2 \varphi \tag{14}$$

$$-M\partial_\eta v_{dz} + K_x v_{dx}\partial_\eta v_{dz} + K_z v_{dz}\partial_\eta v_{dz} = K_z\partial_\eta\psi - \sigma K_z\partial_\eta(\ln n_d) \tag{15}$$

where σ is the ratio of the dust to ion temperature.

$$2K_z^2 K_x^2 \partial_\eta^4 (\varphi - \psi) = -\beta_d Z_d M\big[M\partial_\eta^2 n_d - K_z\partial_\eta^2(n_d v_{dz})\big] \tag{16}$$

Integrating Eqs. (13) and (15) we get

$$K_x v_{dx} + K_z v_{dz} = M\left(1 - \frac{1}{N}\right) \tag{17}$$

$$M v_{dz} = K_z A \tag{18}$$

where

$$A = \frac{2}{3q-1}\left(N^{\frac{3q-1}{q+1}} - 1\right) + \sigma(N-1) \tag{19}$$

Eqs. (14) and (17) along with Eq. (19) give

$$-K_x^2\partial_\eta^2\varphi - \left(1 - \frac{1}{N}\right) + \frac{K_z^2}{M^2}A = 0 \tag{20}$$

The Eq. (16) leads to

$$2K_z^2 K_x^2 \partial_\eta^2 (\varphi - \psi) = -\beta_d Z_d [M^2(N-1) - K_z^2 NA] \quad (21)$$

In deriving Eqs. (17)–(21), we have replaced n_d by N and use the boundary conditions and $N = 1$ at $|\eta| = \infty$ for localised solutions to determine the constants of integration.

Combining Eqs. (20) and (21) and making use of Eq. (12), we have

$$\frac{K_x^2}{(q-1)} \partial_\eta^2 N^{\frac{2(q-1)}{q+1}} + \left(a - \frac{1}{N}\right)(N-1) + \frac{\beta}{2}\left(\frac{1}{a} - N\right)A = 0 \quad (22)$$

where $a = M_{Ad}^2/K_z^2$ is a function of speed and angle of propagation of the solitary dust KAWs, $M_{Ad} = V/v_{Ad}$ is the soliton speed in units of dust Alfvén speed and $\beta = \beta_d Z_d$.

Integrating Eq. (22) once we get the following after simplification

$$\frac{1}{2}\left(\frac{dN}{d\eta}\right)^2 + F(N) = 0 \quad (23)$$

where $F(N)$ is the Sagdeev potential which takes the following form

$$
\begin{aligned}
F(N) =\ & \frac{1}{K_x^2}\left(\frac{q+1}{2}\right) N^{\frac{-2(q-3)}{q+1}} \left[\left(\frac{q+1}{q-3}\right)\left(N^{\frac{q-3}{q+1}} - 1\right) - \frac{\beta}{2(3q-1)}\left(\frac{q+1}{3q-1}\right)\right. \\
& \left(N^{\frac{2(3q-1)}{q+1}} - 1\right) + \frac{\beta}{a(3q-1)}\left(\frac{q+1}{5q-3}\right)\left(N^{\frac{5q-3}{q+1}} - 1\right) + \\
& \left(a + \frac{\beta}{3q-1}\left(1 + \frac{\sigma(3q-1)}{2}\left(1 + \frac{1}{a}\right)\right)\right)\left(\frac{q+1}{3q-1}\right)\left(N^{\frac{3q-1}{q+1}} - 1\right) - \\
& \left(a + 1 + \frac{\beta}{a(3q-1)}\left(1 + \frac{\sigma(3q-1)}{2}\right)\right)\left(\frac{q+1}{2(q-1)}\right)\left(N^{\frac{2(q-1)}{q+1}} - 1\right) \\
& \left. - \frac{\beta\sigma}{2}\left(\frac{q+1}{4q}\right)\left(N^{\frac{4q}{q+1}} - 1\right)\right]
\end{aligned} \quad (24)
$$

We note that Sagdeev potential $F(N)$ admits singularity for possible values of $q = 3/5$ and 3.

10.4   LOCALIZED SOLUTIONS FOR DUST KAWS

We now examine the Sagdeev potential to determine conditions for localized solutions of dust KAWs.

The conditions for solitary wave solutions are

$$F(1) = \left(\frac{dF(N)}{dN}\right)_{N=1} = F(N_m) = 0 \tag{25}$$

$$F(N) < 0 \text{ for } N_m < N < 1 \text{ or } 1 < N < N_m$$

where N_m corresponds to extremum of density, i.e., amplitude of the solitary wave. The inequalities $N_m < N < 1$ and $1 < N < N_m$ represent the conditions for refractive and compressive type solitary structures, respectively.

In addition to above requirements, $F(N)$ has to satisfy the following conditions for double layers to exist.

$$F(N_{ms}) = \left(\frac{dF(N)}{dN}\right)_{N=N_{ms}} = 0 \tag{26}$$

$$F(N) < 0 \text{ for } N_{ms} < N < 1 \text{ or } 1 < N < N_{ms}$$

where N_{ms} is the amplitude of a double layer.

Since N_m is an extremum for N, the condition $F(N_m) = 0$ leads to the following relation between a and N_m.

$$a_\pm = \frac{-H \pm \sqrt{H^2 - 4GI}}{2G} \tag{27}$$

where

$G = A_5 - A_3;$

$H = [(A_1 - A_3) - \frac{\beta}{3q-1}(A_2 - A_5) - \frac{\beta\sigma}{2}(A_6 - A_5)];$

$I = \frac{\beta}{3q-1}(A_4 - A_3) + \frac{\beta\sigma}{2}(A_5 - A_3);$

$A_1 = \frac{q+1}{q-3}[N_m^{\frac{q-3}{q+1}} - 1], A_2 = \frac{q+1}{2(3q-1)}[N_m^{\frac{2(3q-1)}{q+1}} - 1], A_3 = \frac{q+1}{2(q-1)}[N_m^{\frac{2(q-1)}{q+1}} - 1],$

$A_4 = \frac{q+1}{(5q-3)}[N_m^{\frac{(5q-3)}{q+1}} - 1],$

$A_5 = \frac{q+1}{3q-1}[N_m^{\frac{3q-1}{q+1}} - 1]$ and $A_6 = \frac{q+1}{4q}[N_m^{\frac{4q}{q+1}} - 1]$

The Eq. (27) relates the speed and direction of nonlinear structure of dust KAWs to its amplitude. We have seen further from Eq. (27) that nonlinear structures are grouped into two branches, i.e., a_- (acoustic) and a_+ (Alfvénic). We will focus here on the case for a_+ which corresponds to the dust KAWs.

To see the behavior of the function $F(N)$ near its zeroes, we expand it to obtain

$$F(N \approx 1) = -\frac{1}{2}\frac{q+1}{2}\left(a - \frac{\beta}{2}\left(\frac{2}{q+1} + \sigma\right)\right)\left(\frac{1}{a} - 1\right)(N-1)^2/K_x^2 \tag{28}$$

$$F(N \approx N_m) = -\frac{q+1}{2}N_m^{-2\left(\frac{q-1}{q+1}\right)}(a + C(N_m))\left(a - \frac{1}{N_m}\right)$$
$$(N_m^2 - N)(N - N_m)/K_x^2 \tag{29}$$

where

$$C(N_m) = \frac{\beta}{3q-1}\frac{N_m - N_m^{\frac{4q}{q+1}}}{N_m - 1} - \frac{\beta\sigma N_m}{2} \tag{30}$$

It is apparent from Eq. (28) that $N = 1$ is a double root which is a necessary condition for the existence of the solitary waves.

Since $F(N)$ must be negative between $N = 1$ and $N = N_m$ for localized solutions to exist, we obtain the following conditions from Eqs.(28) and (29):

$$\left(a - \frac{\beta}{q+1} - \frac{\beta\sigma}{2}\right)\left(\frac{1}{a} - 1\right) > 0 \tag{31}$$

$$(a + C(N_m))\left(a - \frac{1}{N_m}\right) > 0 \tag{32}$$

The inequality Eq. (31) would be reduced to the following form if we consider the facts that $(1/3) < q$, $\beta < 1$ and $\sigma \leq 1$.

$$1 > a > \frac{\beta}{q+1} + \frac{\beta\sigma}{2} \tag{33}$$

From the inequalities (32) and (33), it can be shown further that

$$a < 1 < N_m \qquad (34)$$

From the above analysis, it is found that only sub-Alfvénic solitary dust KAWs with density humps (compressive) can exist in the non-extensive dust-ion plasma considered here subject to the conditions that the Eq. (27) for a_+ along with inequalities given by relations (31) and (32) being satisfied. The existence of double layers further requires that the following condition is satisfied.

$$a + C(N_{ms}) = 0 \qquad (35)$$

In order to identify the existence regions of solitary dust KAWs and to see how they are affected by different plasma parameters, viz., q, β and σ, we have plotted a vs. N (Figures 10.1 and 10.2) based on Eq. (27)

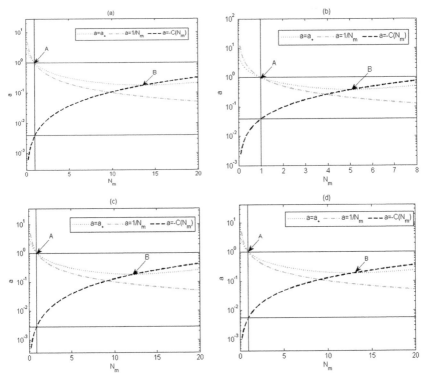

FIGURE 10.1 Existence regions of solitary dust KAWs in the subextensive case ($q > 1$) for (a) $q = 2$, $\beta = 0.01$, $\sigma = 0.1$, (b) $q = 2$, $\beta = 0.1$, $\sigma = 0.1$, (c) $q = 2.5$, $\beta = 0.01$, $\sigma = 0.1$, and (d) $q = 2$, $\beta = 0.01$, $\sigma = 0.5$.

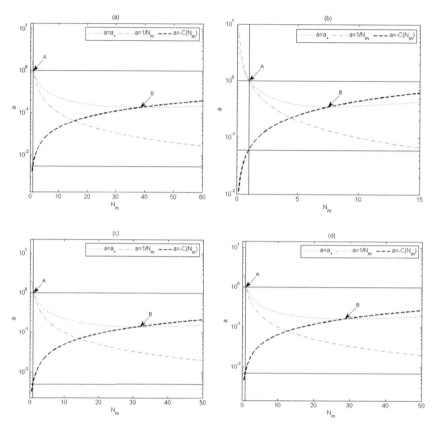

FIGURE 10.2 Existence regions of dust solitary KAWs in the superextensive case ($q < 1$) for (a) $q = 0.8$, $\beta = 0.01$, $\sigma = 0.1$, (b) $q = 0.8$, $\beta = 0.1$, $\sigma = 0.1$, (c) $q = 0.9$, $\beta = 0.01$, $\sigma = 0.1$, and (d) $q = 0.8$, $\beta = 0.01$, $\sigma = 0.5$.

and satisfying inequalities (31)–(33). In each figure, the portion AB of the dotted line denotes the existence region of dust solitary KAWs. From the coordinates of A and B in Figures 10.1–10.2, it is evident that the localized structures will exist only in the region $1 < N_m < N_{max}$, where N_{max} is the maximum limiting value of amplitude of solitary structures determined by q, β and σ. On the other hand, conditions for double layers to exist are satisfied at the point B only. We see from Figures 10.1 and 10.2 that an increase in any one of the parameters q, β, or σ reduces the maximum limiting value N_{max} thereby reducing the existence regions. On increasing plasma β from 0.01 to 0.1, it is evident from panels (a) and (b) of Figures 10.1–10.2 that the existence region is greatly reduced. The effect of q on the existence regions can be seen from panels (a) and (c) while that of σ can be seen from panels (a) and (d) of Figures 10.1–10.2.

Sagdeev potential function F is plotted against N in Figures 10.3–10.4 for various values of plasma parameters, viz. a, q, β and σ to show the formation of nonlinear structures of dust KAWs (both solitons and double layers) and to see how they are affected by these parameters. From Figures 10.3(a)–10.4(a), it is seen that the amplitude of dust solitary KAWs decreases with increase in soliton parallel velocity; and when this velocity attains a certain critical value determined by other plasma parameters, solitary waves transform into double layers. We see from panels (c) and (d) of Figures 10.3 and 10.4, respectively that the amplitude solitary waves increases with increase in both plasma β and dust temperature σ; and these

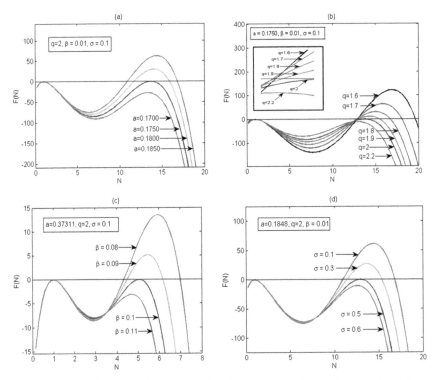

FIGURE 10.3 Plots of Sagdeev potential against N in the subextensive case (q>1) for k_x = 0.2 showing the effects of (a) a, (b) q, (c) β, and (d) σ on the formation of dust kinetic Alfvén solitons and double layers. The values of other parameters are as indicated in each figure.

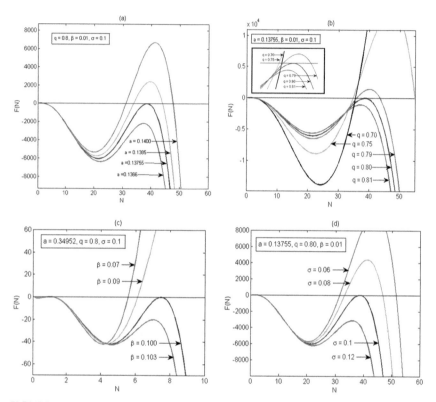

FIGURE 10.4 Plots of Sagdeev potential against N in the superextensive case ($k_x = 0.2$) showing the effects of (a) a, (b) q, (c) β, and (d) σ on the formation of dust kinetic Alfvén solitons and double layers. The values of other parameters are as indicated in each figure.

parameters on reaching certain critical values determined by other plasma parameters, double layers are formed. On increasing the value of q, it is seen from panel (b) of Figures 10.3 and 10.4 that the amplitude of solitary waves fluctuates before they transform into double layers. There exist critical values of q (for particular sets of other plasma parameters) at which double layers are seen to have formed. The effects of σ and q on soliton amplitude can also be seen from density profiles depicted in Figure 10.5.

As a possible application of our investigation, we consider for instance Saturn's rings for which typical dusty plasma parameters [28, 29] are $B_0 \approx 0.1 - 0.2\ G$, $T_i = 10\ eV$, $n_{d0} \approx 10\ cm^{-3}$, $Z_d \approx 10^3$, $\beta (= \frac{2\mu_0 n_i T_i}{B_o^2}) \sim 10^{-4}$, $r_d = 0.25\ \mu m$, where r_d is the dust particle radius. For these parameters, we find that

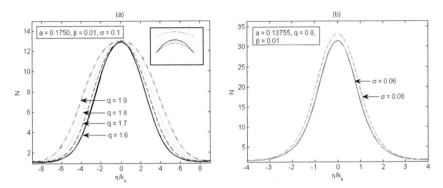

FIGURE 10.5 Density profiles of sub-Alfvénic compressive solitary dust KAWs showing the effects of (a) q and (b) σ. The values of other parameters are as indicated in each figure.

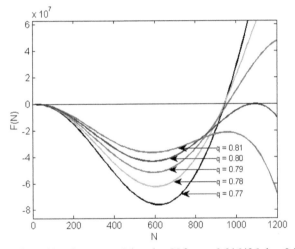

FIGURE 10.6 Plots of Sagdeev potential against N for $a = 0.016436$, $k_x = 0.2$, $\beta = 0.0001$, and $\sigma = 0$. The values of other parameters are as indicated in each figure.

$\dfrac{c_d{}^2}{v_{Ad}{}^2} \ll 1$. This means that in Saturn's rings the dust-Alfvén wave propagation is much more prominent than the long-wavelength dust acoustic wave propagation, and the waves involving perturbation of magnetic field in the formation of nonlinear structures cannot be ruled out. The Figure 10.6 depicts the formation of dust kinetic Alfvén solitons and double layers for the set of plasma parameters relevant for Saturn's ring along with other parameters as indicated therein. The amplitudes of the corresponding density profiles of dust kinetic Alfvén solitons and double layers for $n_{do} =$ 10 cm^{-3} turn out to be ~943.126×10^3 cm^{-3} (for $q = 0.78$) and ~1091×10^3 cm^{-3} (for $q = 0.80$), respectively. The q parameter because of its flexibility may be adjusted to fit the experimental data.

10.5 CONCLUSION

Sagdeev potential for arbitrary amplitude nonlinear structures of dust KAWs in a dust-ion plasma with non-extensive ions and dust temperature effect is derived through which the existence conditions of these structures have been analyzed theoretically and numerically. It is found that only sub-Alfvénic compressive solitary dust KAWs and double layers can exist in the non-extensive dust-ion plasma model considered here. Solitary dust KAWs are found to have an upper limit of wave amplitude, which decreases with increase in q, β and σ. Owing to ions' non-extensivity, the solitary dust KAWs undergo amplitude fluctuations before they transform into double layers, which is an interesting feature. On increasing both dust temperature and plasma, the amplitude of solitary dust KAWs is found to have increased. Considering the relevance of dust KAWs to space and astrophysical dusty plasmas, our study should be useful in understanding the formation and the properties of large amplitude localized electromagnetic fluctuations in such plasmas, particularly in planetary ring systems (namely Saturn's F-ring) and in cometary environments (namely Halley's comet), with a q-non-extensive ion velocity distribution.

KEYWORDS

- **double layers**
- **dust kinetic Alfvén waves**
- **dust-ion plasma**
- **q-non-extensive distribution**
- **Sagdeev potential**
- **solitons**

REFERENCES

1. Wahlund, J. E., Louran, P., Chust, T., Feraudy, H. de, Roux, A., Holback, B., Dovner, P. O., & Holmgren, G. (1994). On ion acoustic turbulence and the nonlinear evolution of kinetic Alfvén waves in aurora. *Geophys. Res. Lett., 21*, 1831.

2. Louran, P., Wahlund, J. E., Chust, T., Feraudy, H. de, Roux, A., Holback, B., Dovner, P. O., & Holmgren, G. (1994). Observation of kinetic Alfvén waves by the FREJA spacecraft. *Geophys. Res. Lett., 21*, 1847.

3. Horanyi, M., & Mendis, D. A. (1985). Trajectories of charged dust grains in the cometary environment. *Astrophys. J. 294*, 357.

4. Mendis, D.A., & Rosenberg, M. (1992). Some aspects of dust-plasma interactions in the cosmic environment. *IEEE Trans. Plasma Sci. 20*, 929.

5. Goertz, C. K. (1989). Dusty plasmas in the solar system. *Rev. Geophys. 27*, 271.

6. Verheest, F. (1996). Waves space plasmas and instabilities in dusty plasma. *Space Sci. Rev. 77*, 267.

7. Rao, N. N., Shukla, P. K., & Yu, M. Y. (1990). Dust -acoustic waves in dusty plasmas. *Planet. Space Sci. 38*, 543.

8. Rao, N. N. (1993). Low- frequency waves in magnetized dusty plasmas. *J. Plasma Phys. 49*, 375.

9. Shukla, P., & Silin, V. P. (1992). Dust ion-acoustic wave. *Phys. Scr. 45*, 508.

10. Mamun, A. A. (1999). Arbitrary amplitude dust-acoustic solitary structures in a three-component dusty plasma. *Astrophys. Space Sci., 268*, 443.

11. Roychoudhury, R., & Chatterjee, P. (1999). Arbitrary amplitude double layers in dusty plasma. *Phys. Plasmas 6*, 406.

12. Moslem, W. M., Sabry, R., El-Labany, S. K., & Shukla, P. K. (2011). Dust-acoustic rogue waves in a non-extensive plasma. *Phys. Rev. E 84*, 066402.

13. Ya Kotsarenko, N., Koshevaya, S. V., Stewart, G.A., & Maravilla, D. (1998). Electrostatic spatially limited solitons in a magnetized dusty plasma. *Planet. Space Sci. 46*, 429.

14. Mamun, A. A., & Shukla, P. K.(2003). Linear and nonlinear dust-hydromagnetic waves. *Phys. Plasmas 10*, 4341.

15. Chen, Y., Lu, W., & Yu, M. Y. (2000). Nonlinear dust kinetic Alfvén waves. *Phys. Rev. E 61*, 809.

16. Mahmood, M. A., Mirza, A. M., Sakanaka, P. H., & Murtaza, G. (2002). Fully nonlinear dust kinetic Alfvén waves. *Phys. Plasmas 9*, 3794.

17. Arshad, M. M., Mahmood, M.A., & Murtaza, G. (2003). Exact nonlinear dust kinetic Alfvén waves in a dust–ion plasma. *New Journal of Physics 5*, 1–11.

18. Tribeche, M., & Merriche, A. (2011). Non-extensive dust-acoustic solitary waves. *Phys. Plasmas 18*, 034502.

19. Liu, Y., Liu, S. Q., & Dai, B.(2011). Arbitrary amplitude kinetic Alfvén solitons in a plasma with a q-non-extensive electron velocity distribution. *Phys. Plasmas 18*, 092309.

20. Ahmed, M.K., & Sah, O. P. (2014). Effect of ion temperature on arbitrary amplitude Kinetic Alfvén solitons in a plasma with a q-non-extensive electron velocity distribution. *Astrophys. Space Sci. 353*, 145.

21. Rényi, A. (1955). On a new axiomatic theory of probability. *Acta Math. Acad. Sci. Hung. 6*, 285.

22. Tsallis, C. (1988). Possible generalization of Boltzmann-Gibbs statistics. *J. Stat. Phys. 52*, 479.

23. Silva, R., Plastino, A. R., & Lima, J. A. S. (1998). *A Maxwellian Path to the q-Nonextensive Velocity Distribution Function. Phys. Lett. A 249*, 401.

24. Lima, J. A. S., Silva, R. Jr., & Santos, J. (2000). Plasma oscillations and non-extensive statistics. *Phys. Rev. E 61*, 3260.

25. Sagdeev, R. Z. (1966). *Reviews of Plasma Physics.* Edited by M. A. Leontovich, Consultants Bureau (New York) Vol. 4, pp. 23.

26. de Angelis, U., Formisano, V., & Giordano, M. (1988). Ion plasma waves in dusty plasmas: Halley's comet. *J. Plasma Physics 40*, 399.

27. Kadomtsev, B. B. (1965). Plasma Turbulence. Academic, New York.

28. Mamun, A. A., Shukla, P. K., & Bingham, R. (2003a). Dust-Alfvén Mach Cones in Saturn's Dense Rings.*JETP Letters, 77*(10), 541–545.

29. Mamun, A. A., Shukla, P. K., & Bingham, R. (2003b). Comment on Mach Cones and Magnetic Forces in Saturn's Rings. *JETP Letters, 78*(2), 99–100.

30. Horanyi, M., & Mendis, D. A. (1986). The effects of electrostatic charging on the dust distribution at Halley's comet. *Astrophys. J. 307,* 800.

31. Mendis, D. A., & Rosenberg, M. (1994). Cosmic Dusty Plasma. *Ann. Rev. Astron. Astrophys. 32,* 419.

32. Rao, N. N. (1995). Magneto acoustic modes in a magnetized dusty plasma. *J. Plasma Phys. 53,* 317.

CHAPTER 11

STUDY ON GENERATION OF PULSED HIGH CURRENT WITH ALUMINUM ELECTROLYTIC CAPACITOR

PANKAJ DEB and ANURAG SHYAM

Energetics and Electromagnetics Division, Bhabha Atomic Research Centre, IDA Block B, 4th Cross Road, Autonagar, Visakhapatnam – 530012, India, E-mail: pankajdeb24@gmail.com

CONTENTS

ABSTRACT

Capacitor bank is the universal system being used as a pulsed power driver for generation of pulsed high current application. In pulsed power applications low-inductance and high-voltage capacitors are used to produce high-peak current (kA) and peak power in discharge conditions. Aluminum electrolytic capacitors have high-energy density and mainly used in power supply filters. This chapter describes the experimental analysis and behavior of an aluminum electrolytic capacitor when used to deliver pulsed high-peak current in an inductive load. With this aim we have made a 30 mF, 27 kJ capacitive storage pulsed power system with aluminum electrolytic capacitors in the laboratory. From the experiment we found that 30 mF, 27 kJ aluminum electrolytic capacitor bank delivers 35 kA peak current having a pulse duration of 850 μs when connected to a 4.5 μH inductive load. Solid-state switch (thyristor) is used to discharge the capacitor bank energy into the inductive load. This capacitor bank has energy/weight ratio of 1 kJ/kg. These capacitors can be used in many pulsed power applications like pulse forming system, electromagnetic launchers, rock fragmentation, and generation of pulsed high magnetic field inside a solenoid and also in space applications.

11.1   INTRODUCTION

Pulsed power has a wide variety of application, which includes nuclear fusion research, food processing, particle accelerators, medical treatment, defence sector, etc. [4]. High-voltage and low-inductance capacitors are mainly used in pulsed power application. These energy storage capacitors are used in the generation of high-peak current (kA) [8]. High-voltage energy storage capacitors have a paper dielectric and extended foil electrodes. They are impregnated with castor oil and are suitable for single-shot and repetitive rate applications [6]. Percentage voltage reversals of these high voltage capacitors are generally 80%. On the other hand, aluminum electrolytic capacitors have relatively larger capacitance per unit volume but aluminum electrolytic capacitors with non-solid electrolyte have different sizes, capacitance, and at low voltages. They have capacitance values 0.1 μF to 2.7 F with voltages values ranging from 4 V up to 630 V. They have a wide range of applications in SMPS based power supplies like DC–DC converters, rectified DC voltage, frequency converters as DC link capacitors for drives, inverters, etc. [1]. Aluminum electrolytic capacitors are polarized capacitors. They have to be operated with DC voltage with correct polarity and if correct polarity is not maintained, it will cause explosion. These capacitors, due to high energy density and low voltage, it became suitable to be used with unidirectional semiconductor switches. An aluminum electrolytic capacitor can be used in a limited space application. A capacitive storage pulsed power system consists of a power supply to charge the capacitor bank, a switch, and finally a load [5]. The stored energy in the capacitor bank is discharged into the load to produce a high current pulse.

Figure 11.1 represents the generator with capacitive energy storage and a closing switch. In this work, the aluminum electrolytic capacitors are arranged in series and parallel configuration to increase the effective voltage of the aluminum electrolytic capacitor bank, which can make the capacitor bank more compact. Behaviors of aluminum electrolytic capacitors as compared to high-voltage and low-inductance capacitors, which are commonly used in pulsed power applications, are experimentally studied. The thyristor switch is used as a discharge switch to generate pulsed high current with an inductive load to make the pulsed power system compact.

11.2 HIGH-ENERGY DENSITY CAPACITORS

High-energy density capacitors with metallized film having energy density 3 J/cc and stored energy of 260 kJ are now presently available [7]. We have used aluminum electrolytic capacitors (wet or non-solid type) capacitance 15 mF, 450 V (Model no. PG-6DI) having size (diameter 9 cm × length 22 cm) for 30 mF, 27 kJ capacitor bank system [2]. The energy density of this aluminum electrolytic capacitor model no. is 1 J/cc.

11.3 HIGH-CURRENT SWITCH

The thyristor switches as a current discharge switches can be used at their surge current ratings [3]. The high-voltage energy storage capacitor uses

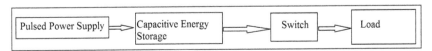

FIGURE 11.1 Block diagram of a pulsed power system.

gas discharge switches like spark gap, rail gap, thyratron, etc., for switching high currents to the inductive load due to their high voltage ratings and high voltage reversal. Thyristor (Model no. HST2000CT) has been used as a current discharge switch in the 30 mF, 27 kJ electrolytic capacitor bank and is reported in this chapter.

11.4   POWER SUPPLY

The 30 mF, 27 kJ capacitor bank is charged through a half-wave rectification technique through a rectifier diode of 2 kV peak inverse voltage rating. The input to the rectifier diode is provided by a 2 kV, 1 kVA single phase transformer. Auto variac is used in the input side of the single phase transformer to limit the inrush current. The power supply charges the capacitor bank in 4 minutes, and after that the power supply is decoupled from the capacitor bank by using a pneumatically controlled switch. After the trigger pulse to the thyristor switch it is ensured that the capacitor bank is completely discharged, and the residual energy of the capacitor bank was dumped into a ceramic carbon resistor having a high joule rating. Three ceramic carbon resistors of value 50 Ω, 50 kJ are connected in series for energy absorption of the 30 mF, 27 kJ capacitor bank. The advantage of this power supply is that it is low-cost, simple in construction, and reliable.

11.5   EXPERIMENT SETUP

The total dimension of the aluminum electrolytic capacitor bank is 66 cm × 40 cm × 45 cm, having 18 aluminum electrolytic capacitors [2]. One module contains six capacitors, which are connected in parallel. Three such modules are connected in a series configuration to increase the effective voltage of the capacitor bank to 1.35 kV. While charging the aluminum electrolytic capacitor bank, the individual electrolytic capacitor should not exceed the permissible rated voltage. Therefore, a shunt resistor is connected to each capacitor for equal distribution of charging voltage. The voltage of each aluminum electrolytic capacitor is monitored by a DC voltmeter. The thyristor switched capacitor bank with inductive load has a frequency dependent behavior. The equivalent series inductance of the circuit (Leq) constitutes of $L_{ESL} + L_{ckt} + L_l$, where L_{ESL} is the self inductance of capacitor, L_{ckt} is the residual inductance of circuit due to cables, connection, etc., and L_l is the load inductance.

The equivalent series resistance of the circuit (Req) constitutes of R_{ESR} + R_{ckt} + R_l, where R_{ESR} equivalent series internal resistance of capacitor, R_{ckt} is the residual resistance of circuit due to cables, connection, etc. and R_l is the load resistance. The aluminum electrolytic capacitor bank is discharged to a low inductive coil for generation of high-peak currents. Arrangement of thyristor switched aluminum electrolytic capacitor bank is shown in Figure 11.2. Experiments were carried out with thyristor switched 30 mF, 27 kJ aluminum electrolytic capacitor bank for the inductive load for pulsed current generation. The circuit diagram of aluminum electrolytic capacitor bank with control circuit is shown in Figure 11.3.

FIGURE 11.2   Arrangement of 30 mF, 27 kJ aluminum electrolytic capacitor bank.

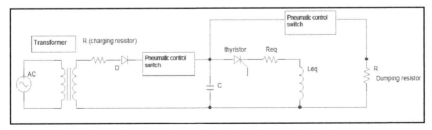

FIGURE 11.3   Circuit diagram of capacitor bank with control circuit.

The capacitor bank is charged to 970 V and is discharged into an inductive load value 4.5 μH. To determine the circuit inductance and resistance of the circuit obtained, experiment results is overlapped on the LCR simulated curve. The R_{eq} of the circuit estimated to be 17 mΩ. Simulated and experiment results obtained of the current traces are shown in Figure 11.4.

The peak current measured is 35 kA having a pulsed current duration of 850 μs. Peak discharge current is measured with a current monitor with attenuation ratio 1000:1. Figure 11.5 shows the oscilloscope trace peak discharge current of 30 mF, 27 kJ aluminum electrolytic bank.

11.6   CONCLUSION

We concluded that the peak current can be drawn from a 30 mF, 27 kJ aluminum electrolytic capacitor bank is 35 kA, pulse width 850 μs with

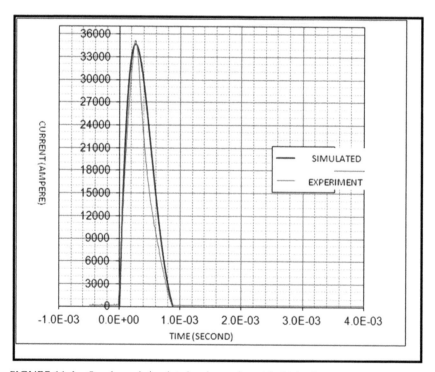

FIGURE 11.4   Overlapped simulated and experimental obtained current traces.

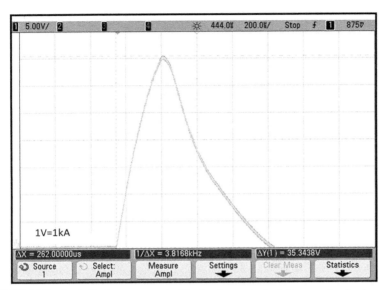

FIGURE 11.5   Peak discharge current with aluminum electrolytic capacitor bank.

4.5 μH inductive load. Thyristor switch can be used as a discharge switch at surge current rating. So we can use these capacitors with unidirectional semiconductor switch for discharging high pulsed current required in various pulsed power application. Safety has to be maintained during handling these capacitors as these are polarized capacitor. These capacitors cannot withstand negative voltage.

KEYWORDS

- aluminum electrolytic capacitor
- ceramic carbon resistor
- high current switches
- high energy density capacitors
- pulsed power system
- thyristor switch

REFERENCES

1. Aluminum electrolytic capacitor. https://en.wikipedia.org/wiki/Aluminum_electrolytic_capacitor (viewed on 19 August 2015).
2. Aluminum electrolytic capacitors. http://www.alconelectronics.com/aluminum-electrolytic-capacitors.php (viewed on 19 August 2015).
3. Heremans, G., Bockstal, L. V., & Herlach, F., (1989). Switching of a 0.5 MJ capacitor bank with thyristors at surge ratings. *Physica B: Condensed Matter, 155*(1–3), 48–50.
4. Akiyama, H., Sakugawa, T., Namihira, T., Takaki, K., Minamitani, Y., & Shimomura, N., (2007). Industrial applications of pulsed power technology. *IEEE Transactions on Dielectrics and Electrical Insulation, 14*(5).
5. Bluhm, H., (2006). *Pulsed Power Systems*. Springer, Verlag Berlin Heidelberg.
6. High Energy Capacitors Overview (2002). http://igor.chudov.com/manuals/Maxwell/high-energy-capacitors.pdf.
7. MacDonald, J. R., Schneider, M. A., Ennis, J. B., MacDougall, F. W., & Yang, X. H. (2009). High Energy Density Capacitors IEEE Electrical Insulation Conference, Montreal, QC, Canada.
8. Pulsed Power, https: //en.wikipedia.org/wiki/Pulsed_power (view on 18 July 2015).

CHAPTER 12

EFFECT OF RADIO FREQUENCY POWER ON MAGNETRON SPUTTERED TIO$_2$ THIN FILMS

R. SATHEESH,[1] S. SANKAR,[2] and K. G. GOPCHANDRAN[3]

[1]Assistant Professor, Department of Physics, S. V. R. NSS College, Vazhoor, T. P. Puram (P.O.), Kottayam, India

[2]Assistant Professor, Department of Physics, S. N. College, Chathannur, Karamkode (P.O.), Kollam, India

[3]Associate Professor, Department of Optoelectronics, University of Kerala, Kariavattom, Thiruvananthapuram, India, E-mail: satheeshr83@gmail.com

CONTENTS

ABSTRACT

Thin films of titanium dioxide (TiO$_2$) were synthesized using radio frequency (RF) magnetron sputtering. Structural and microstructural studies were performed on TiO_2 films deposited at different deposition powers. Films exhibited a transformation from amorphous to anatase phase and to anatase-rutile mixed phase with increase in deposition power. X-ray diffraction showed an increase in crystallite size with deposition power in support with scanning electron micrograph. Films exhibited good transmittance in the visible region and the spectra were oscillatory in nature. Band gap exhibited a decrease with increase in sputtering power. A band gap of 3.2 eV was obtained for anatase film, which coincides with its bulk value. The variation of refractive index with wavelength is used in the calculation of dispersion energy parameters in single oscillator model (i.e., Wemble and Didomenico model). The band gap of films calculated from dispersion energy parameter showed good agreement with the values calculated from transmittance spectra.

12.1 INTRODUCTION

Titanium dioxide (TiO_2) thin films exhibit excellent properties like high dielectric constant, high transmittance in the visible and near IR region, high refractive index, and wide optical band gap [1–3]. Bulk TiO_2 occurs in three crystalline polymorphs: anatase (tetragonal), rutile (tetragonal) and brookite (orthorhombic) [4]. Structure, phase composition, electrical, and optical properties of TiO_2 thin films are very sensitive to deposition conditions [5]. In catalysis, photocatalysis, and dye-sensitized solar cells, anatase has proven advantageous over the rutile phase [6–8]. Though rutile phase has been extensively investigated in the past, anatase is found to exhibit interesting properties recently [9, 10], which makes it a promising material for gas sensors, solar cells, and dielectrics in memory cell capacitors and semiconducting FET [11–15]. Also since TiO_2 films present good durability and a high refractive index; they are used as an anti-reflection coating, multilayer optical coatings, and optical wave-guides [18], accurate knowledge of the refractive indices and absorption coefficients in opaque and in band gap regions of semiconductor thin films is indispensable for the design and analysis of various optoelectronic devices. TiO_2 thin films can be fabricated using different methods, such as sputtering, chemical vapor deposition, ion beam assisted deposition, reactive evaporation, laser-assisted evaporation, sol-gel process, etc. [12]. However, magnetron sputtering seems to be the most favorable because the material can be supplied to grow a surface layer in correct proportions and with sufficient energy to ensure the formation of dense structure with easy control of deposition parameter [22].

In the present work, thin films of TiO_2 were prepared using RF planar magnetron sputtering on amorphous quartz substrates at different deposition power. Influence of sputtering power on thin film crystallization was carried out. A structural study reveals that the deposited film was polycrystalline and films showed a trend towards crystallinity with increase in sputtering power. Phase content and crystallite size of the deposited film showed dependence on sputtering power. Optical studies showed that all the films showed an average transmittance above 60% in the visible region and their transmittance decreased towards UV region. Films showed an increase in refractive index with increase in sputtering power. The calculated band gap (from transmission spectra) showed a decrease with increase in sputtering power. The variation of refractive index with wavelength is used in the discussion of refractive index dispersion in single oscillator model (i.e., Wemble and Didomenico model). From the calculated dispersion energy parameter, the band gap for anatase film and mixed phase film, that showed highest rutile content were calculated and is in good agreement with the values calculated from transmittance spectra.

12.2   EXPERIMENTAL SETUP

TiO_2 thin films were deposited on quartz substrate using RF planar magnetron sputtering. TiO_2 powder of 99.99% purity (Sigma Aldrich company) was powdered using an agate mortar and pressed into a dye of diameter $\sim 5 \times 10^{-2}$ m. The sputter deposition was performed using RF magnetron sputtering equipment (Hind Hivac. Planar magnetron sputtering model – 12. MSPT) with a frequency of 13.56 MHz and a variable RF power unit. The substrate–target distance was fixed at 3.5×10^{-2} m. Before sputtering, the chamber was evacuated to a pressure of $\sim 10^{-6}$ mbar and argon was introduced into the chamber at 40 lit/min. When the total pressure has reached 2.25×10^{-2} mbar, sputter deposition was performed for 50 minutes.

TiO_2 target was sputtered using RF powers ranging from 200 to 300 W in steps of 25 W. The crystal structures of the films were determined by X-ray diffractometer (XRD) using Philips PW 1710 diffractometer. The surface morphologies of the films were investigated using scanning electron microscopy (SEM) technique using the system Quanta 200, fitted with an energy dispersive spectrometer. Optical measurements were performed in wavelength range 200–900 nm using a double beam UV-visible spectrophotometer, JASCO – V550.

12.3   RESULTS AND DISCUSSION

12.3.1   X-RAY DIFFRACTION ANALYSIS

X-ray diffraction patterns of TiO_2 films deposited at different RF powers are shown in Figure 12.1. The lattice spacing (d), relative intensity of peaks (I/I_0) and phases identified are systematically presented for different sputtering power in Table 12.1. All the peaks in the diffraction pattern are indexed according to JCPDS data cards of TiO_2-rutile (21-1276) and TiO_2–anatase (21-1272). The XRD pattern reveal that TiO_2 film deposited at RF power of 200 W is amorphous. With increase in RF power to 225 W, weak diffraction peaks are observed around $2\theta = 25.3°$, 37.7°, 48.1°, and 55.1° which are assigned to the reflections due to (101), (004), (200), and (211) planes corresponding to anatase (TiO_2) structure implying tetragonal symmetry. TiO_2 film deposited at RF power of 250 W show intense diffraction peaks around $2\theta = 25.3°$, 37.8° and 55.1° and are assigned to reflections due to (101), (004), and (211) planes corresponding to anatase TiO_2 structure. No peaks corresponding to rutile phase were detected in this film. XRD pattern reveals a preferential growth along anatase (101) plane for TiO_2 films deposited at 225 and 250 W. With further increase in RF power to 275 W, diffraction peaks were observed around $2\theta = 25.3°$, 27.4° and 37.9° and are assigned to reflections due to anatase (101), rutile (110), and anatase (004) planes. Thus, the presence of mixed phase with a shift in the preferential orientation from anatase (101) plane to rutile (110) plane is identified. All planes belong to tetragonal symmetry. For TiO_2 film deposited at 300 W, diffraction peaks are observed around $2\theta = 25.3°$, 27.4°, 37.6°, 48°, and 56.6°; and are assigned to reflections due to anatase

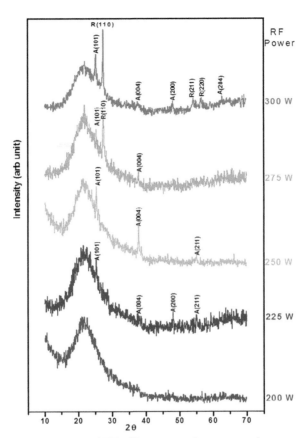

FIGURE 12.1   XRD patterns of TiO$_2$ films sputtered on quartz substrates at different RF powers.

(101), rutile (110), anatase (004), anatase (200), rutile (211), and rutile (220) planes implying tetragonal symmetry. More intense and sharper diffraction peak is observed along rutile (110) plane. With increasing sputtering power peaks become sharper and narrower, indicating an increase in crystallite size (Figure 12.1). The crystallite size of the preferentially oriented crystal plane is calculated using Debye–Scherrer formula [16]. The variation of crystallite size and full width at half maximum (FWHM) of anatase (101) diffraction peak on sputtering power is shown in Figure 12.2. The lattice constants, calculated from the interplanar seperation (d) and the corresponding miller indices values (hkl) are found to agree well with bulk values of tetragonal TiO$_2$ phase.

TABLE 12.1 The Variation in Lattice Constants a and c and Unit Cell Volume of TiO$_2$ Film with RF Power

RF Power (W)	Lattice Spacing, d (Å)	PHASE (hkl)	Relative Intensity (%)	a (Å) A-3.784 R – 4.593	c (Å) A-9.514 R-2.959	Volume (Å³) A –136.23 R – 62.4
225	3.5124	A (101)	100.00	3.7784	9.5297	A-136
	2.3832	A (004)	29.14			
	1.8893	A (200)	56.17			
	1.6667	A (211)	44.57			
250	3.51209	A (101)	100	3.7815	9.5046	A-135.9
	2.37615	A (004)	70.73			
	1.66588	A (211)	10.55			
275	3.5116	A (101)	65.76	3.7804	9.482	A-135
	3.2461	R (110)	100.00	4.5906		
	2.3705	A (004)	33.69			
300	3.5206	A (101)	58.85	3.7804	9.5456	A-136.4
	3.2456	R (110)	100.00	4.6019	2.975	R-63
	2.3864	A (004)	13.53			
	1.8920	A (200)	10.12			
	1.6909	R (211)	13.94			
	1.6236	R (220)	12.14			

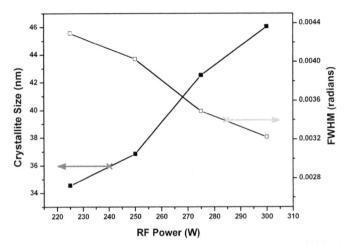

FIGURE 12.2 Dependence of crystallite size and FWHM of anatase (101) diffraction peak on sputtering power.

12.3.2 SCANNING ELECTRON MICROSCOPY

Figure 12.3 shows the scanning electron micrographs (SEM) of TiO_2 films deposited under different RF Powers. TiO_2 film deposited at 200 W exhibits nanocrystalline nature with a morphology consisting of closely packed grains. The films deposited at 250 W, which was shown to be pure anatase in XRD investigation, exhibits a morphology consisting of spheroids in nanoscale range, uniformly deposited all over the surface. But beyond RF power of 250 W, the morphology consists of a mixture of spheroid like particles and elongated ones. The morphological change observed in the SEM images increase with RF power, resembling the change observed in XRD investigation. It can also be seen that the average particle size increases with increase in RF power. The grain boundaries also become more well-defined with increase in RF power.

12.3.2.1 Optical Studies

The study of optical properties of TiO_2 thin films, particularly the absorption edge has proved to be very useful for elucidation of electronic structure

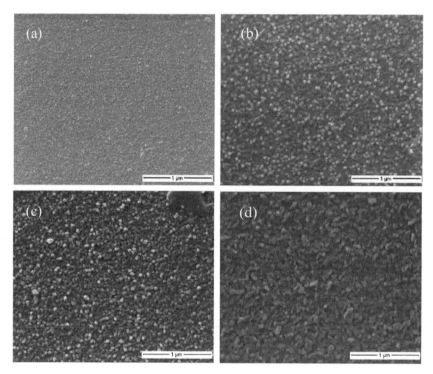

FIGURE 12.3   SEM images of TiO$_2$ thin films deposited at (a) 200 W, (b) 250 W, (c) 275 W, and (d) 300 W.

of these materials [17]. The spectral distribution curves of transmittance of TiO$_2$ films in the wavelength range 300–900 nm is shown for various RF power of deposition in Figure 12.4. It is observed that TiO$_2$ films shows good transparency (>60%) in the visible region and its transparency shows a sharp decrease in the UV region, because of the fundamental absorption edge [18].

After analysis of optical transmission spectra, it is clear that average optical transmittance of the films tends to decrease with increase in sputtering power. Similar reports of decrease in transmittance with increase in sputtering power can be seen in literature [19]. The decrease in average transmittance of film with increase in sputtering power could be accounted to the increase in film density and thickness [18]. With increase in sputtering power more and more particles per

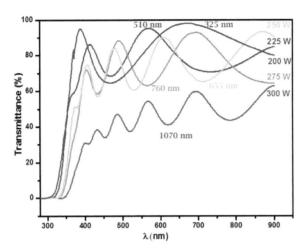

FIGURE 12.4   The spectral distribution of transmittance of films deposited at different RF power, in the wavelength range 300–900 nm.

unit time arrive at the substrate, which results in more energy being delivered to the growing film resulting in higher film density. From the transmission spectra, it is clear that the transmission edge shows a slight shift with increase in sputtering power. Shift in the transmission edge corresponds to a shift in the band gap. The transmission spectrum shows an oscillatory nature. These are interference fringes, which is the result of the interference of the light reflected between air-film and film-substrate interfaces [20]. As the RF power increases, the interference effects also become prominent showing the increase in thickness of the film. Using envelope method, if n_1 and n_2 are the refractive indices of two adjacent maxima/minima at wavelength λ_1 and λ_2, the thickness of film is found using the following expression. The thickness of film deposited at different RF powers is given in Figure 12.4. Increase in film thickness with sputtering power may be attributed with argon ions gain higher energy with an increasing RF power and hence dissipate more energy to TiO_2 targets thus promote higher sputtering rate at the target.

$$t = \frac{(\lambda_1 \lambda_2)}{2(\lambda_1 n_2 - \lambda_2 n_1)} \tag{1}$$

The structural transformation from anatase to rutile-anatase mixed phased films may be associated with increase in film thickness as reported by Kanjitvichyanukul et al. [21]. Index of refraction indicates the relative amount of light transmitted by a material. As the index of refraction increases, the transmitted light decreases. The amount of light reflected may be considered as inversely proportional to the amount transmitted, though much of the light may be dissipated [18]. The spectral distribution of refractive index n with wavelength is determined using Swanepoel method [20] in transmission spectrum.

The refractive index (n) of TiO_2 thin films is obtained using the following expression,

$$n = [M + (M^2 - S^2)^{\frac{1}{2}}]^{\frac{1}{2}} \tag{2}$$

where M is for weak and medium absorption region and is given by

$$M = 2S \frac{(T_M - T_m)}{T_M T_m} + \frac{(S^2 + 1)}{2} \tag{3}$$

where T_M is the maximum envelope of interference; T_m is the minimum envelope of interference; S is the refractive index of substrate which is equal to 1.54 for quartz and n, the refractive index of film. The spectral distribution of refractive index n verses wavelength λ of anatase-TiO_2 film and one of the mixed phase film with highest rutile content is illustrated in Figure 12.5. The refractive index of film is found to decrease with increase in incident photon wavelength. The difference in refractive index with deposition power may be associated with difference in thickness and also to the structural change (from anatase to mixed anatase – rutile phase).

12.3.2.2 Band Gap

The sharp decrease in the transparency of TiO_2 thin films in the UV region is caused by the fundamental light absorption. In the vicinity of the fundamental absorption, the absorption edge was calculated using the relation [17].

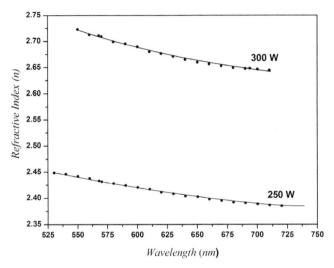

FIGURE 12.5 The spectral distribution of refractive index (n) verses wavelength (λ) of films deposited at two different RF power.

$$(\alpha h v) = A\left(h v - E_g\right)^m \tag{4}$$

where A is the edge width parameter representing the film quality, calculated from the linear part of the relation, E_g is the optical energy gap of the material and m determines the type of transition. The value of m is (1/2) for direct allowed, 2 for indirect allowed, (3/2) for direct forbidden and 3 for indirect forbidden transition [22]. The correct value of m is found by trying the different possible values and indirect allowed band transition was identified as the best fit for which m equals 2. The nature of plot indicates that above the absorption edge indirect allowed transition dominates the optical absorption for all the films studied. Thus,

$$(\alpha h v)^{1/2} = A\left(h v - E_g\right) \tag{5}$$

which is the value of energy gap E_g and is determined from the intercepts of the extrapolation to zero absorption to photon energy axis. The calculated band gap shows a decrease with increase in RF power. The decrease in band gap with increase in sputtering power can be explained in terms of quantum size effect [23]. According to this, we expect an elevated band gap for XRD amorphous/nanocrystalline material and a reduction in band gap for the large grain sized well-crystallized material. This is in complete agreement with XRD and SEM results. Also the obtained value of optical band gap for anatase films showed good agreement with their bulk values [E_g (anatase) = 3.20–3.3 eV]. Table 12.2 lists out the calculated optical band gap at different RF power.

12.3.2.3 Single Oscillator Model

Wemble and Didomenico used a single oscillator description of the frequency-dependent dielectric constant to define dispersion energy parameters E_d and E_0. The relation between refractive index (n), dispersion energy (E_d), and single oscillator strength (E_0), below the optical band gap is given by [24]

$$n^2 - 1 = \frac{E_d E_0}{\left(E_0^2 - (hv)^2\right)} \tag{6}$$

Thus,

$$\frac{1}{(n^2 - 1)} = \frac{\left(E_0 - (hv)^2\right)}{E_d E_0} \tag{7}$$

where E_d is the average strength of interband optical transition, and E_0 is the average of optical band gap (E_{opt}^{WD}). The variation of $1/(n^2-1)$ vs. $(hv)^2$

TABLE 12.2 Variation of Band Gap of TiO$_2$ Films with RF Power

RF power (W)	E_g^{opt} (eV)
200	3.33
225	3.3
250	3.2
275	3.13
300	3.11

for TiO_2 thin films deposited at two different RF powers is ploted in Figure 12.6. The plot yields a linear variation with intercept of extrapolation to zero photon energy equals E_0/E_d and its slope equals to $(E_0 E_d)^{-1}$.

The obtained values of E_0 and E_d together with values obtained by Abdel-Aziz et al. [17], Wemble and Didomenico [24], Toyoda [25], and Mandare and Hones [18] are given in Table 12.3. It is clear that the obtained values of E_0 and E_d for TiO_2 have some differences with those obtained by Wemble and Didomenico. The difference may be due to difference in structural properties of investigated oxide films as compared with single crystal form of TiO_2 studied by Wemble and Didomenico. The direct band gap values, E_g^1 were also calculated from Wemble and Didomenico dispersion parameter, E_0 using the relation [25],

$$E_g^1 = E_0 / 2 \tag{8}$$

The values of band gap obtained from transmission spectra are in good agreement with the band gap obtained from Wemble and Didomenico model. This shows the accuracy of the model.

12.4 CONCLUSION

Thin films of TiO_2 were prepared using RF planar magnetron sputtering on amorphous quartz substrates at different deposition power. Influence of sputtering power on thin film crystallization was carried out. Structural study reveals that the deposited film was polycrystalline and films showed

TABLE 12.3 Dispersion Energy Parameters Obtained in This Work Compared with Other Works

E_0 (eV)	E_d (eV)	Reference
6.38	27.7	Anatase TiO_2 (This work)
6.2085	34.24	Mixed TiO_2 (This work)
7.49	25.66	Abdel-Aziz
5.24	25.6	Wemble and Didomenico
5.09	22.9	Toyoda
5.24	25.7	Mandare and Hones

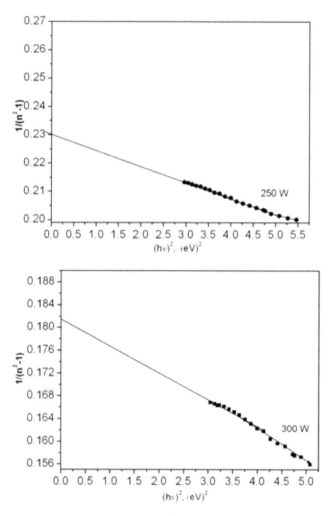

FIGURE 12.6   Variation of $1/(n^2-1)$ vs. $(hv)^2$ for TiO_2 thin films deposited at two different RF powers.

a trend towards crystallinity with increase in sputtering power. Crystallite size is found to increase with sputtering power. Phase content of the deposited film showed dependence on sputtering power. The morphological change observed in the SEM images increase with RF power, resembling the change observed in XRD investigation.

Optical studies showed that all the films showed an average transmittance above 60% in the visible region and their transmittance decreased towards UV region. Films showed an increase in refractive index with increase in sputtering power. The calculated band gap showed a decrease with increase in sputtering power. The variation of refractive index with wavelength is used in the discussion of refractive index dispersion in single oscillator model (i.e., Wemble and Didomenico model). From the calculated dispersion energy parameter, the band gap for anatase film and the mixed phase film showed that the highest rutile content were calculated, and is in good agreement with the values calculated from transmittance spectra.

KEYWORDS

- **microstructure**
- **phase transformation**
- **sputtering**
- **thin films**
- **vacuum deposition**

REFERENCES

1. Kadoshima, M., Hiratani, M., Shimamoto, Y., Torii, K., Miki, H., Kimura, S., & Nabatame, T. (2003). Rutile-type TiO_2 thin film for high-k gate insulator. *Thin Solid Films, 424*(2), 224–228.
2. Natsuhara, H., Matsumoto, K., Yoshida, N., Itoh, T., Nonomura, S., Fukawa, M., & Sato, K. (2006). TiO_2 thin films as protective material for transparent conducting oxides used in Si thin film solar cells. *Sol. Energy Mater. Solar Cells, 90*(17), 2867–2880.
3. Radecka, M., Zakrzewska, K., & Czternastek, H. (1993). The influence of thermal annealing on the structural, electrical and optical properties of TiO_2-x thin films. *Applied Surface Science, 65–66*, 227–234.
4. Sharma, A. K., Tareja, R. K., & Willer, U. (2003). Phase transformation in room temperature pulsed laser deposited TiO_2 thin films. *Applied Surface Science, 206*(1–4), 137–148.
5. Mardare, D., & Rusu, G. I. (2002). The influence of heat treatment on the optical properties of titanium oxide thin films. *Materials Letters, 56*(3), 210–214.

6. Hadjiivanov, K. I., & Klissurski, D. G. (1996). Surface chemistry of titania (anatase) and titania-supported catalysts. *Chemical Society Reviews, 25*(1), 61–69.
7. Linsebigler, A. L., Lu, G., & Yates, J. T. (1995). Photocatalysis on TiO2 Surfaces Principles, Mechanisms, and Selected Results. *Chemical Reviews, 95*(3), 735–758.
8. Regan, B. O., & Gratzel, M. (1991). Low-Cost, High Efficiency Solar Cell Based on Dye Sensitized Colloidal TiO_2 Film. *Nature, 353*(5), 737–739.
9. Stamate, M., Lazar, G., & Lazar, I. (2008). Anatase – Rutil TiO_2 Thin Films Deposited in a D.C. Magnetron Sputtering System. *Romanian Journal of Physics, 53*(1–2), 207–221.
10. Goodenough, J. B. (1971). Metallic Oxides. *Progress in Solid State Chemistry, 5,* 145–399.
11. Ferroni, M., Guidi, V., Martinelli, G., Faglia, G., Nelli, P., & Sberveglieri, G. (1996). Characterization of a nanosized TiO_2 gas sensor. *Nanostructured Materials, 7*(7), 709–718.
12. Tang, H., Prasad, K., Sanjines, R., & Levy, F. (1995). TiO_2 anatase thin films as gas sensors. *Sensors and Actuators B: Chemical, 26,* 71–75.
13. Nazeeruddin, M. K., Kay, A., Rodicio, I., Humphry Baker, R., Muller, E., Liska, P., Vlachopoulos, N., & Gratzel, M. (1993). Conversion of light to electricity by cis-X2bis(2,2'-bipyridyl-4,4'-dicarboxylate)ruthenium(II) charge-transfer sensitizers (X = Cl-, Br-, I-, CN-, and SCN-) on nanocrystalline titanium dioxide electrodes. *Journal of the American Chemical Society, 115*(14), 6382–6390.
14. Hart, J. N., Cheng, Y.-B. Simon, G. P., & Spiccia, L. (2008). Alternative materials and processing techniques for optimized nanostructures in dye-sensitized solar cells. *Journal of Nanoscience and Nanotechnology, 8*(5), 2230–2248.
15. Gilmer, D. C., Colombo, D. G., Taylor, C. J., Roberts, J., Haugstad, G., Campbell, S. A., Kim, H. S., Wilk, G. D., Gribelyuk, M. A., & Gladfelter, W. L. (1998). Low-Temperature CVD of Crystalline Titanium-Dioxide Films Using Tetranitratotitanium (IV). *Chemical Vapor Deposition, 4*(1), 9–11.
16. Cullity, B. D. (1978). Elements of X-ray Diffraction. Addison-Wesley Publishing Company, Inc., Massachusetts, pp. 99.
17. Abdul-Aziz, M. M., Yahia, I. S., Wahab, L. A., Fadel, M., & Afifi, M. A. (2006). Determination and analysis of dispersive optical constant of TiO_2 and Ti_2O_3 thin films. *Applied Surface Science, 252*(23), 8163–8170.
18. Mardare, D., & Hones, P. (1999). Optical dispersion analysis of TiO_2 thin films based on variable-angle spectroscopic ellipsometry measurements. *Materials Science and Engineering: B, 68*(1), 42–47.
19. Heo, C. H., Lee, S. B., & Boo, J. H. (2005). Deposition of TiO_2 thin films using RF magnetron sputtering method and study of their surface characteristics. *Thin Solid Films, 475*(1–2), 183–188.
20. Swanepoel, R. (1983). Determination of the thickness and optical constants of amorphous silicon. *Journal of Physics E: Scientific Instruments, 16*(12), 1214–1222.
21. Kanjitvichyanukul, P., Ananpattarachai, J., & Pongpom, S. (2005). Sol-Gel preparation and properties study of TiO_2 thin film for photocatalytic reduction of Chromium (VI) in photocatalysis process. *Science and Technology of Advanced Materials, 6,* 352–358.

22. Eufinger, K., Poelman, D., Poelman, H., Gryse, R. D., & Marin, G. B. (2007). Effect of microstructure and crystallinity on the photocatalytic activity of TiO_2 thin films deposited by dc magnetron sputtering. *Journal of Physics D: Applied Physics, 40*(17), 5232–5238.

23. Reddy, K. M., Manorama, S. V., & Reddy, A. R. (2002). Bandgap studies on anatase titanium dioxide nanoparticles. *Materials Chemistry and Physics, 78*, 239–245.

24. Wemple, S. H., & Didomenico, M. (1971). Behavior of the Electronic Dielectric Constant in Covalent and Ionic Materials. *Physical Review, B., 3*(4), 1338–1351.

25. Toyoda, T., Nakanishi, H., Endo, S., & Irie, T. (1985). Fundamental absorption edge in the semiconductor $CdInGaS_4$ at high temperatures. *Journal of Physics D: Applied Physics, 18*(4), 747–751.

PART II

SPACE AND ATMOSPHERIC PLASMA

CHAPTER 13

STABILITY OF KINETIC ALFVÉN WAVE (KAW) IN A COMETARY PLASMA WITH STREAMING ELECTRONS AND PROTONS

G. SREEKALA,[1] SIJO SEBASTIAN,[1] MANESH MICHAEL,[1] E. SAVITHRI DEVI,[1] C. P. ANILKUMAR,[2] and CHANDU VENUGOPAL[1]

[1]School of Pure and Applied Physics, Mahatma Gandhi University, Kottayam – 686560, Kerala, India

[2]Equatorial Geophysical Research Laboratory, Indian Institute of Geomagnetism, Krishnapuram, Tirunelveli, Tamil Nadu – 627011, India, E-mail: cvgmgphys@yahoo.co.in

CONTENTS

ABSTRACT

Alfvénic turbulence has been detected by Comet Halley at space-crafts Giotto and Vega and by Comet Giacobini-Zinner at International Cometary Explorer (ICE). We have studied the KAW instability driven by field aligned drifts of solar wind electrons (se), and protons (H) in a cometary pair–ion plasma consisting of cometary electrons (ce), positively and negatively charged oxygen ions, with each species being modeled by a drifting ring distribution [with $V_{dce} = V_{dO}^{+} = V_{dO}^{-} = 0$ and $V_{dH} = V_{dse}$].

In the low frequency regime, the dispersion relation is a polynomial equation of order four. We find that the growth rate increases with increasing drift velocities of hydrogen. The growth rate which increases with increasing hydrogen densities, decreases with increasing oxygen densities.

13.1 INTRODUCTION

The Alfvén and magneto acoustic waves, which are the basic low-frequency wave modes of magnetized plasmas, have been the subject of intense research. The main reason for the great interest in these waves is that they play important roles in the heating of, and the transport of energy in, laboratory [13, 18], space [23–26], and astrophysical plasmas [21, 22, 27–29]. The "Alfvén wave heating" scheme has been investigated theoretically and experimentally as a supplementary heating scheme for fusion plasma devices, and it has been invoked as a model for the heating of the solar and stellar coronae. The waves are believed to underlie the transport of magnetic energy in the solar and stellar winds, transfer angular momentum in interstellar molecular clouds during star formation, play roles in magnetic pulsations in the Earth's magnetosphere, and provide scattering mechanisms for the acceleration of cosmic rays in astrophysical shock waves [6].

The KAW was introduced by Hasegawa and co-workers [10, 11] and takes into account the effect of finite electron pressure and ion gyro-radius on the shear Alfvén wave.

The observations carried out in 1985/1986 by several space missions (ICE, VEGAs 1, 2, Suisei, Sakigake, and Giotto) on the environments of Comet Giacobini-Zinner and Comet Halley spurred intense cometary plasma wave research. Many features of the observed wave activity in comets have yet to be convincingly explained, meaning that there exists ample room for theoretical analyses aimed at the experimental cometary data.

Hydromagnetic wave activity was observed in distinct cometary environments. Specifically, Alfvénic turbulence was detected in the magnetic field [8, 19, 20] and in the electron distribution [16, 17] by Comet Giacobini-Zinner at the ICE spacecraft and by Comet Halley at Giotto and Vega spacecrafts.

KAWs can be excited in plasmas by many ways including temperature anisotropy, velocity shear, inhomogeneities in density and magnetic field, etc. [15]. When a comet approaches the Sun to within a few AU, the surface of the nucleus begins to evaporate; these evaporated molecules boil off and carry small solid particles with them, forming the comet's coma of gas and dust. There is a change of density inhomogeneity and temperature anisotropy in a comet when the solar wind sweeps past the comet. The drifting of the solar wind electrons and protons is the cause of a huge increase in free energy and which in turn drives instabilities.

In general, a cometary environment contains hydrogen and newborn heavier ions with relative densities depending on the distances from the nucleus [1]. However, the spacecraft Giotto observed negatively charged ions in the mass peaks of 7–19, 22–65, and 85–110 amu in the inner coma of Comet Halley [2], which indicates the presence of negatively charged oxygen ions unambiguously.

Chen et al. [3–5] studied the KAW instability driven by field aligned currents in high, low, and finite beta plasmas, where in they used a drifting Maxwellian distribution. However, in our case we use a ring type distribution obtained by subtracting two Maxwellian distributions with different temperatures.

We thus study the KAW instability in a five component plasma consisting of solar wind electrons and hydrogen ions which permeate through a cometary plasma consisting of cometary electrons, positively charged oxygen ions, and negatively charged oxygen ions. The growth rate which is aided by hydrogen ions and drift velocities, decreases with the addition of oxygen ions.

13.2 DISPERSION RELATION

We investigated the KAW instability driven by a field-aligned current, which is carried by the field-aligned drifts of both electrons and ions. As mentioned earlier we consider a five-component plasma consisting of streaming solar wind electrons and hydrogen ions and cometary electrons, positively charged oxygen ions and negatively charged oxygen ions. We model each component by a drifting ring distribution with zero drift for the cometary electrons and pair ions.

The drifting ring distribution is separable in velocity space [1]

$$F_s(v_\perp, v_\parallel) = f_{\perp s}(v_\perp) f_{\parallel s}(v_\parallel) \tag{1}$$

The parallel components are modeled by a drifting Maxwellian,

$$f_{\parallel s}(v_\parallel) = \frac{1}{\sqrt{\pi} V_{Ts}} \exp\left[-\left(\frac{v_\parallel - V_{ds}}{V_{Ts}} \right)^2 \right] \tag{2}$$

and the perpendicular ring is simulated by the subtraction of two Maxwellian distributions having different temperatures,

$$f_{\perp s}(v_\perp) = \frac{1}{\pi(a_s - b_s)V_{Ts}^2} \left[\exp\left(-\frac{v_\perp^2}{a_s V_{Ts}^2} \right) - \exp\left(-\frac{v_\perp^2}{b_s V_{Ts}^2} \right) \right] \text{ with } a_s > b_s. \tag{3}$$

The relevant dielectric elements using the distributions (2) and (3) have been derived earlier [7] and hence will not be repeated here.

In the low frequency regime, i.e., $\omega^2, k_\parallel^2 V_{Ts\parallel}^2, k_\parallel^2 V_{ds}^2 \ll \omega_{cs}^2$, we get the following final expressions for the dielectric tensor elements as

$$\varepsilon_{xx} = -\frac{c^2 k_\parallel^2}{\omega^2} + \frac{c^2}{V_A^2} \left(\chi_H^{(0)} \frac{\omega_*^2}{\omega^2} + A \right) \tag{4}$$

$$\varepsilon_{zz} = -\frac{c^2 k_\perp^2}{\omega^2} - \frac{\omega_{pse}^2}{\omega^2} - \frac{\omega_{pce}^2}{\omega^2} + 2i\sqrt{\pi} \frac{\omega_{pse}^2}{k_\parallel^2 V_{Tse}^2} \frac{\omega_*}{k_\parallel V_{Tse}} \exp\left[-\left(\frac{\omega_*}{k_\parallel V_{Tse}}\right)^2\right]$$

(5)

where $A = \dfrac{n_{O^+}}{n_H} \dfrac{m_H}{m_{O^+}} \chi_{O^+}^{(0)} + \dfrac{n_{O^-}}{n_H} \dfrac{m_H}{m_{O^-}} \chi_{O^-}^{(0)}$;

$$\chi_s^{(0)} = \frac{\Lambda_0\left(a_s L_{\perp s}\right) - \Lambda_0\left(b_s L_{\perp s}\right)}{\left(a_s - b_s\right) L_{\perp s}} \quad \text{and}$$

(6)

$$L_{\perp s} = \frac{k_\perp^2 V_{Ts}^2}{2\omega_{cs}^2} \quad \text{with} \quad s = H, O^+, O^-.$$

The kinetic dispersion equation is derived from the following form [14].

$$\det\begin{pmatrix} \varepsilon_{xx} - c^2 k_\parallel^2 \omega^{-2} & c^2 k_\parallel k_\perp \omega^{-2} \\ c^2 k_\parallel k_\perp \omega^{-2} & \varepsilon_{zz} - c^2 k_\perp^2 \omega^{-2} \end{pmatrix} = 0.$$

(7)

Expanding (7) and simplifying, we get the dispersion relation in final form

$$\frac{\omega^2}{k_\parallel^2 V_A^2} + \delta_e \frac{\omega_{*e}^2}{k_\parallel^2 V_A^2} - [\frac{\omega_*^2}{k_\parallel^2 V_A^2} + A \frac{\omega^2}{k_\parallel^2 V_A^2}][\delta_e + k_\perp^2 \lambda_e^2 \frac{\omega_{*e}^2}{k_\parallel^2 V_A^2} + \frac{\omega^2}{k_\parallel^2 V_A^2}] = 0 \quad (8)$$

where

$$\lambda_e = \frac{c^2}{\omega_{pe}^2}, \quad \delta_e = \frac{n_{ce}}{n_{se}}, \quad \omega_{*e} = \omega - k_\parallel V_{de} \quad \text{and} \quad \omega_* = \omega - k_\parallel V_{dH} \quad (9)$$

Putting $\omega = \omega_r + i\omega_i$ and $V_{de} = V_{dH}$ in Eq. (8); Real(D) = 0 implies a fourth order equation given as,

$$a_1 x^4 + a_2 x^3 + a_3 x^2 + a_4 x + a_5 = 0 \quad (10)$$

In Eq. (10)

$$a_1 = \left(1 + k_\perp^2 \lambda_e^2 + \delta_e\right)\left(\chi_H^{(0)} + A\right)$$

$$a_2 = 2\frac{V_{dH}}{V_A}\left[\left(1 + k_\perp^2 \lambda_e^2 + \delta_e\right)A + \chi_H^{(0)} + A\right]$$

$$a_3 = \left(\frac{V_{dH}}{V_A}\right)^2 \left[\left(1 + k_\perp^2 \lambda_e^2 + \delta_e\right)A + \chi_H^{(0)} + 5A\right] - \left(1 + \delta_e\right)$$

$$a_4 = 4A\left(\frac{V_{dH}}{V_A}\right)^3 - 2\left(\frac{V_{dH}}{V_A}\right) \quad \text{and}$$

$$a_5 = A\left(\frac{V_{dH}}{V_A}\right)^4 - \left(\frac{V_{dH}}{V_A}\right)^2.$$

$$(11)$$

Also $x = \dfrac{\omega_{r*}}{k_\parallel V_A} = \dfrac{\omega_r - k_\parallel V_{dH}}{k_\parallel V_A}$. Of the four roots, we take only the posi-

tive root satisfying the condition $\dfrac{\omega_{r*}}{k_\parallel V_A} < 1$ for numerical analyses.

An analytical expression for growth rate, obtained from the condition $\gamma = -\operatorname{Im} D/\left(\partial \operatorname{Re} D/\partial \omega_r\right)$, is given by

$$\frac{\gamma}{k_\parallel V_A} = \frac{2\sqrt{\pi}\,\dfrac{\omega_{r*}}{k_\parallel V_{Tse}}\dfrac{V_A^2}{V_{Tse}^2}\left[\chi_H^{(0)}\left(\dfrac{\omega_{r*}}{k_\parallel V_A}\right)^4\left(\dfrac{\omega_r}{k_\parallel V_A}\right)^2 + A\left(\dfrac{\omega_r}{k_\parallel V_A}\right)^4\left(\dfrac{\omega_{r*}}{k_\parallel V_A}\right)^2 - \left(\dfrac{\omega_{r*}}{k_\parallel V_A}\right)^2\left(\dfrac{\omega_r}{k_\parallel V_A}\right)^2\right]}{4a_1\left(\dfrac{\omega_{r*}}{k_\parallel V_A}\right)^3 + 3a_2\left(\dfrac{\omega_{r*}}{k_\parallel V_A}\right)^2 + 2a_3\left(\dfrac{\omega_{r*}}{k_\parallel V_A}\right) + a_4}$$

$$(12)$$

where values of the coefficients are given in Eq. (11).

13.3 DISCUSSION

From Eq. (8) it is clear that the dispersion relation for KAWs in a low β plasma obtained by Lysak and Lotko [14], may be recovered by neglecting contributions of both oxygen ions and cometary electrons and also by setting $V_d = 0$. We get

$$\frac{\omega^2}{k_\parallel^2 V_A^2} = \frac{1 + 3L_{\perp H}/4}{1 + k_\perp^2 \lambda_e^2} \qquad (13)$$

13.4   RESULTS

We now plot the growth rate of the KAW for the parameters observed in the coma of Comet Halley [1] namely hydrogen density $n_H = 4.95$ cm^{-3}, positively charged oxygen density $n_{O+} = 0.5$ cm^{-3}, negatively charged oxygen density $n_{O-} = 0.05$ cm^{-3}, hydrogen temperature $T_H = 9 \times 10^4$ K, solar electron temperature $T_{se} = 2 \times 10^5$ K, cometary electron temperature $T_{ce} = T_{se}/10$, and oxygen temperature $T_{O+} = T_{O-} = 1.16 \times 10^4$ K. The ring parameters 'a_s' and 'b_s' were held a constant for each species at $a_H = 1$, $a_{O+} = a_{O-} = 2.8$, $a_e = 1.8$ and $b_H = 1$, $b_{O+} = b_{O-} = 2.8$, $b_e = 1.2$, and the propagation angle $\theta = 70°$. The zero current condition was satisfied by setting $n_H V_{dH} = n_e V_{de}$.

Figure 13.1 is a plot of normalized growth rate $\gamma/k_\parallel V_A$ (V_A is the Alfvén velocity) versus $k_\perp \rho_{o+}$ (ρ_{o+} is the oxygen ion gyro-radius) as a function of

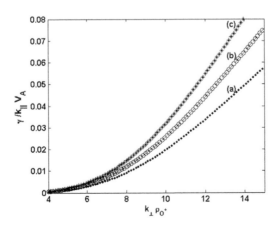

FIGURE 13.1   Variation of normalized growth rate $\gamma/k_\parallel V_A$ versus $k_\perp \rho_{o+}$ for various V_{dH}/V_A. Curve (a) is for $V_{dH}/V_A = 0.7$, curve (b) is for $V_{dH}/V_A = 0.9$, and curve (c) is for $V_{dH}/V_A = 1.1$.

the normalized drift velocities, $y = V_{dH}/V_A$. Curve (a) is for $y = 0.7$, curve (b) for $y = 0.9$, and curve (c) for $y = 1.1$; the other parameters are as given above. We find that the normalized growth rate increases with increasing normalized drift velocity. This indicates that the streaming components will drive the instability in a cometary plasma by increasing the free energy.

The influence of the hydrogen ions on the growth rate is studied next. Figure 13.2 is a plot of $\gamma/k_\parallel V_A$ versus $k_\perp \rho_{o+}$ for $y = 0.7$; the other parameters are as given above. Curve (a) is for $n_H = 3$ cm^{-3}, curve (b) is for $n_H = 5$ cm^{-3}, and curve (c) is for $n_H = 7$ cm^{-3}. We find that the growth rate increases with increasing hydrogen ion densities; this result is also consistent with the conclusion from Figure 13.1.

The effect of heavier ion is considered next. Figure 13.3 is thus a plot of $\gamma/k_\parallel V_A$ versus $k_\perp \rho_{o+}$ for $y = 0.7$; the other parameters are as given above. Curve (a) is for $n_{O+} = 0.1$ cm^{-3} curve, (b) is for $n_{O+} = 0.5$ cm^{-3}, and curve (c) is for $n_{O+} = 1.1$ cm^{-3}. From the figure we find that the growth rate decreases with increasing oxygen ion densities. This decrease in the growth rate could be due to the absorption of drift energy by the oxygen ions.

The effect of the ion temperatures is considered next. Figure 13.4 is thus a plot of $(\gamma/k_\parallel V_A)$ versus $k_\perp \rho_{o+}$ for different hydrogen ion temperatures T_H. Curve (a) is for $T_H = 4 \times 10^4$ K, curve (b) is for $T_H = 8 \times 10^4$ K, and

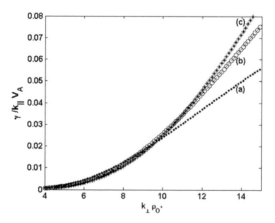

FIGURE 13.2 Variation of normalized growth rate $\gamma/k_\parallel V_A$ versus $k_\perp \rho_{o+}$ for different solar wind hydrogen ion densities. Curve (a) is for $n_H = 3$ cm^{-3}, curve (b) is for $n_H = 5$ cm^{-3}, and curve (c) is for $n_H = 7$ cm^{-3}.

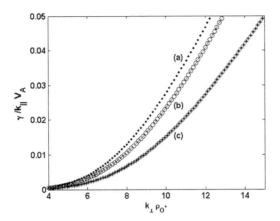

FIGURE 13.3 Variation of normalized growth rate $\gamma/k_{\parallel}V_A$ versus $k_{\perp}\rho_{o^+}$ of the wave for different oxygen ion densities. Curve (a) is for $n_{o^+} = 0.1$ cm^{-3}, curve (b) is for $n_{o^+} = 0.5$ cm^{-3}, and curve (c) is for $n_{o^+} = 1.1$ cm^{-3}.

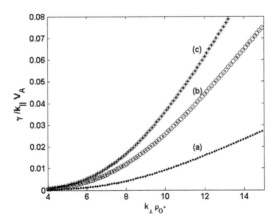

FIGURE 13.4 Variation of growth rate $\gamma/k_{\parallel}V_A$ versus $k_{\perp}\rho_{o^+}$ for different hydrogen ion temperatures T_H. Curve (a) is for $T_H = 4 \times 10^4$ K, curve (b) is for $T_H = 8 \times 10^4$ K, and curve (c) is for $T_H = 10 \times 10^4$ K.

curve (c) is for $T_H = 10 \times 10^4$ K. We find that the growth rate increases with increasing hydrogen temperatures.

Figure 13.5 is the variation of $\gamma/k_{\parallel}V_A$ versus $k_{\perp}\rho_{o^+}$ for $T_O{}^+ = 0.9 \times 10^4$ K (curve a), $T_O{}^+ = 1.5 \times 10^4$ K (curve b) and $T_O{}^+ = 2 \times 10^4$ K (curve c).

The figure shows that the growth rate decreases with increasing oxygen temperatures.

Finally, Figure 13.6 depicts the variation of $\gamma/k_{\parallel}V_A$ versus $k_{\perp}\rho_{o^+}$ for different temperatures of solar wind electrons. We can see that growth rate decreases with increasing electron temperatures.

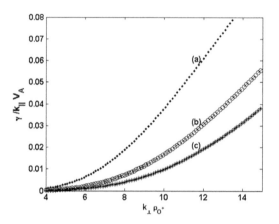

FIGURE 13.5 Variation of growth rate $\gamma/k_{\parallel}V_A$ versus $k_{\perp}\rho_{o^+}$ for different oxygen ion temperatures $T_O^+ = T_O^-$. Curve (a) is for $T_O^+ = 0.9 \times 10^4$ K, curve (b) is for $T_O^+ = 1.5 \times 10^4$ K, and curve (c) is for $T_O^+ = 2 \times 10^4$ K.

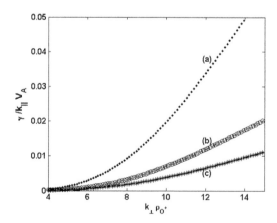

FIGURE 13.6 Variation of growth rate $\gamma/k_{\parallel}V_A$ versus $k_{\perp}\rho_{o^+}$ for different solar electron temperatures T_{se}. Curve (a) is for $T_{se} = 2 \times 10^5$ K, curve (b) is for $T_{se} = 4 \times 10^5$ K, and curve (c) is for $T_{se} = 6 \times 10^5$ K.

13.5   CONCLUSIONS

We have studied the stability of KAW in a five-component plasma, consisting of drifting electrons and hydrogen of solar origin. Cometary electrons and oxygen pair ions form the other constituents of the plasma. The wave growth which increases with increasing hydrogen ion densities and drift velocities decreases with increasing oxygen ion densities.

ACKNOWLEDGMENTS

Financial assistance from Kerala State Council for Science, Technology and Environment, Thiruvananthapuram, Kerala, India (JRFs), the University Grants Commission (EF) and Department of Science and Technology (FIST and PURSE Programs) is gratefully acknowledged.

KEYWORDS

- cometary plasma
- drift
- drifting Maxwellian
- growth rate
- instability
- kinetic Alfvén waves

REFERENCES

1. Brinca, A. L., & Tsurutani, B. T. (1987). Unusual characteristics of electromagnetic waves excited by cometary newborn ions with large perpendicular energies. *Astron. Astrophys., 187*, 311.
2. Chaizy, P., Reme, H., Sauvaud, J. A. et al. (1991). Negative ions in the coma of Comet Halley. *Nature, 349*, 393.
3. Chen, L., Wu, D. J., & Hua, Y. P. (2011), Kinetic Alfvén wave instability driven by a field-aligned current in high β plasmas. *Phys. Rev. E, 84*, 046406.

4. Chen, L., & Wu, D. J. (2012), Kinetic Alfvén wave instability driven by field-aligned currents in solar coronal loops. *Astrophys. J., 754*, 123.

5. Chen, L., Wu, D. J., & Huang, J. (2013). Kinetic Alfvén wave instability driven by field-aligned currents in a low β plasma. *J. Geo Phys. Res: Space Physics, 118*, 2951.

6. Crammer, F. N. (2001). The Physics of Alfvén waves. 1st edn., WILEY-VCH Verlag Berlin GmbH, Berlin (Federal Republic of Germany).

7. George, S., Savithri, D. E., & Venugopal, C. (2011). Kinetic Alfvén waves excited by cometary newborn ions with large perpendicular energies. *Plasma Sci. Tech., 13*, 135.

8. Gosling, J. T., Asbridge, J. R., Bame, S. J., Thomsen, M. F., & Zwickl, R. D. (1986). Large amplitude, low frequency plasma fluctuations at Comet Giacobini-Zinner. *Geophys. Res. Lett., 13*, 267.

9. Guang-Li, H., De-Yu, W., De-Jin, W., de Féraudy, H., Le Quéau, D., Volwerk, M., & Holback, B. (1997). The eigenmode of solitary kinetic Alfvén waves observed by Freja satellite. *J. Geophys. Res., 102*, 7217.

10. Hasegawa, A. (1976), Particle acceleration by MHD surface wave and formation of aurora. *J. Geophys. Res., 81*, 5083.

11. Hasegawa, A., & Chen, L. (1975), Kinetic process of plasma heating due to Alfvén wave excitation. *Phys. Rev. Let., 35*, 370.

12. Hasegawa, A., & Sato, T. (1989). Space Plasma Physics I – Stationary Processes. Springer Verlag, Berlin.

13. Huang, L., Quin, X. M., Ding, N., & Long, Y. (1991). Heating Finite Beta Tokamak-Plasmas by Alfvén Waves. *Chin. Phys. Lett., 8*, 232.

14. Lysak, R. L., & Lotko, W. (1996). On the kinetic dispersion relation for shear Alfvén waves. *J. Geophys. Res., 101*, 5085.

15. Naim, H., Bashir, M. F., & Murtaza, G. (2014). Drift kinetic Alfvén wave in temperature anisotropic plasma. *Phys. Plasmas, 21*, 032120.

16. Neubauer, P., Glessmeier, K. H., Pohl, M., Raeder, J., Acuna, M. H., Burlaga, L. F., Ness, N. F., Musmann, G., et al. (1986). First results from the Giotto magnetometer experiment at Comet Halley. *Nature, 321*, 352.

17. Reidler, W., Schwingenschuh, K., Yeroshenko, Y. G., Styashkin, V. A., & Russell, C. T. (1986). Magnetic field observations in Comet Halley's coma. *Nature, 321*, 288.

18. Ross, D. W., Chen, D. L., & Mahajan, S. M. (1982). Kinetic description of Alfvén wave heating. *Phys. Fluids., 25*, 652.

19. Tsurutani, B. T., & Smith, E. J. (1986a). Strong hydromagnetic turbulence associated with Comet Giacobini-Zinner. *Geophys. Res. Lett., 13*, 259.

20. Tsurutani, B. T., & Smith, E. J. (1986b). Hydromagnetic waves and instabilities associated with cometary ion pickup: ICE observations. *Geophys. Res. Lett., 13*, 263.

21. Voitenko, Y. M. (1998). Excitation of Kinetic Alfvén waves in a flaring loop. *Solar Phys. 182*, 411.

22. Voitenko, Y. M., & Goossens, M. (2000). Competition of damping mechanisms for the phase-mixed Alfvén waves in the solar corona. *Astron. Astrophys., 357*, 1086.

23. Wu, D. J. (2003a). Model of nonlinear kinetic Alfvén waves with dissipation and acceleration of energetic electrons. *Phys. Rev. E. 67*, 027402.

24. Wu, D. J. (2003b). Dissipative solitary kinetic Alfvén wave and electron acceleration. *Phys. Plasmas., 10,* 1364.

25. Wu, D. J., & Chao, K. J. (2003). Auroral electron acceleration by dissipative solitary kinetic Alfvén waves. *Phys. Plasmas 10,* 3787.

26. Wu, D. J., & Chao, K. J. (2004). Model of auroral electron acceleration by dissipative nonlinear inertial Alfvén wave. *J. Geophys. Res., 109,* 06211.

27. Wu, D. J., & Fang, C. (1999). Two-fluid motion of plasma in Alfvén waves and the heating of solar coronal loops. *Astrophys. J., 511,* 958.

28. Wu, D. J., & Fang, C. (2003). Coronal plume heating and kinetic dissipation of kinetic Alfvén waves. *Astrophys. J., 596,* 656.

29. Wu, D. J., & Fang, C. (2007). Sunspot chromospheric heating by kinetic Alfvén waves. *Astrophys. J., 659,* L181.

CHAPTER 14

EXTERNAL MAGNETIC FIELD EFFECT ON ABSORPTION OF SURFACE PLASMA WAVES BY METAL NANO-PARTICLES

DEEPIKA GOEL,[1] PRASHANT CHAUHAN,[1] ANSHU VARSHNEY,[1] D. B. SINGH,[2] and VIVEK SAJAL[1]

[1]*Department of Physics and Material Science and Engineering, Jaypee Institute of Information Technology, Noida – 201307, UP, India, E-mail: deepika7nov@yahoo.co.in; prashant.chauhan@jiit.ac.in; anshu.varshney@jiit.ac.in; vsajal@rediffmail.com*

[2]*Laser Science and Technology Center, Metcalfe House, Delhi – 110054, India, E-mail: dbsingh2@rediffmail.com*

CONTENTS

ABSTRACT

Configuration of metal surface embedded with metallic nanoparticles is used for the absorption of surface plasma waves (SPW) in the presence of external magnetic field (x-direction). The SPW (propagating in z-direction) excites resonant plasma oscillations in the particles incurring attenuation of the wave. For spherical metallic particles with plasma frequency ω_{pe}, the resonant plasma oscillations occur at $\omega^2 = \omega^2_{pe}/3$, where ω is the frequency of the SPW. At this frequency, energy is absorbed by the electrons inside nanoparticles and sharp increase in the absorption of SPW by the metallic particles, depending upon its size, interparticle separation, magnetic field strength and dielectric constant of the metal occurs. The change in absorption constant with external magnetic field is studied.

14.1 INTRODUCTION

Plasmonic nanostructures have been recently investigated as a possible way to improve absorption of light in solar cells and to modify the strength of optical interactions, resulting in overall increase in the power conversion efficiency [6]. Strong absorption and large enhancement in ablation have been reported when the wave frequency equals the frequency of surface charge oscillations of the nanoparticle [8, 12, 14]. The absorption of electromagnetic waves can be further modified by the surface plasma waves (SPW) [13, 10]. The surface plasma wave propagates over the metal surface and can be excited using an attenuated total reflection configuration when the wavenumber of the incident laser along the interface becomes equal to the SPW. The amplitude of SPW falls off exponentially with distance away from the interface in either medium [7, 11]. The SPW interacts with the metal nanoparticles embedded over the metal surface. Hwang et al. [4] attributed the reflectivity reduction to increase in effective surface area of the metal surface due to the presence of nanoparticles when the laser frequency is away from the surface plasma resonance. However, when the laser frequency equals the natural frequency of surface plasma oscillations, there is very significant absorption of surface plasma waves. Ahmad and Tripathi [1] have studied the absorption of laser normally incident on a metal surface embedded with nanoparticles. Kumar et al.

[9] theoretically observed enhanced absorption coefficient at resonant frequency ($\omega = \omega_{pe}/\sqrt{3}$) due to strong dissipation of the surface wave energy via collisions of the free electrons of the nanoparticle. The absorption of SPW at resonant frequency can be further modified by applying external magnetic field. Jazi et al. [5] analytically investigated the propagation of s-polarized electromagnetic waves through an inhomogeneous dissipative magnetised plasma slab having external magnetic field in parallel direction to plasma surface and observed that the absorption increases with increasing magnetic field. The effect of an external magnetic field on the reflection, absorption and transmission of uniform and non-uniform collisional plasma are also reported in literature [3].

In this study, our focus is on the effect of external magnetic field on absorption of SPW by metal nanoparticles embedded over the metal surface. The external magnetic field is applied in x-direction. The SPW (propagating in z-direction) excites resonant plasma oscillations in the nanoparticles incurring attenuation of the SPW, due to absorption of energy by the nanoparticles. We have also studied the change in absorption coefficient on varying external magnetic field, radius and interparticle separation between the nanoparticles to get the resonance in the desired range for applications in magneto plasma solar cells. This chapter has been organised into three sections where introduction is presented as Section 14.1. The propagation of SPW over the metal-vacuum interface in the presence of magnetic field and mathematical formalism for absorption of SPW by metal nanoparticles is established in Section 14.2. The discussion of the obtained results is given in Section 14.3 and finally conclusions in Section 14.4.

14.2   MATHEMATICAL MODEL

Consider the interface of metal and free-space at $x = 0$ embedded with metal nanoparticles having radius r_c with interparticle separation d. The metal occupies the half space ($x < 0$) and free space is ($x > 0$) as shown in Figure 14.1. External static magnetic field ($\bar{B}_s$) is applied in $\hat{x}$ direction.

In the presence of external magnetic field, the effective permittivity of metal ($\tilde{e}$) is a dielectric tensor given by

$$
\tilde{\varepsilon} =
\begin{pmatrix}
\varepsilon_L\left(1 - \dfrac{\omega_p^2}{\omega(\omega + \upsilon i)}\right) & 0 & 0 \\[3ex]
0 & \varepsilon_L\left(1 - \dfrac{\omega_p^2(\omega + i\upsilon)}{\omega((\omega + i\upsilon)^2 - \omega_c^2)}\right) & \dfrac{i\omega_c}{\omega}\dfrac{\varepsilon_L\omega_p^2}{((\omega + i\upsilon)^2 - \omega_c^2)} \\[3ex]
0 & \dfrac{-i\omega_c}{\omega}\dfrac{\varepsilon_L\omega_p^2}{((\omega + i\upsilon)^2 - \omega_c^2)} & \varepsilon_L\left(1 - \dfrac{\omega_p^2(\omega + i\upsilon)}{\omega((\omega + i\upsilon)^2 - \omega_c^2)}\right)
\end{pmatrix}
\tag{1}
$$

where ε_L is the lattice permittivity $\omega_p^2 = 4\pi n e^2 / \varepsilon_L m$ and $\omega_c = eB_s / m$ (B_s is the strength of external magnetic field) are the plasma and cyclotron frequency respectively. Here, e and m are the charge and effective mass of electron. n is the electron density at the metal surface and v is the electron ion collision frequency. Suppose the SPW propagates through this configuration with t, z variation as $e^{-i(\omega t - k_z z)}$. Maxwell's equations are used to study the dispersion relations of surface plasma waves in this configuration.

$$
\vec{\nabla} \times \vec{B} = \frac{1}{c^2}\frac{\partial}{\partial t}(\tilde{\varepsilon}\vec{E})
\tag{2}
$$

$$
\vec{\nabla} \times \vec{E} = -\frac{\partial \vec{B}}{\partial t}
\tag{3}
$$

On eliminating $\vec{B}$ from Eqs. (2) and (3), the wave equation can be found as

$$
\vec{\nabla} \times \vec{\nabla} \times \vec{E} = \frac{\omega^2}{c^2}(\tilde{\varepsilon}.\vec{E})
\tag{4}
$$

The electric field associated with the SPW, [Eq. (4)] can be obtained by satisfying $\nabla.\tilde{\varepsilon}E = 0$ in each region. On applying the boundary conditions, i.e., the normal component of $\vec{D}$ and the tangential component of $\vec{E}$ are continuous at the interface $x = 0$ and solving, we get the dispersion relation of SPW in the presence of external magnetic field [15], given by

$$
(\alpha_0 + \alpha_1 + \alpha_2)\alpha_1\alpha_2\varepsilon_1 + [\alpha_0(\alpha_1 + \alpha_2) + \alpha_1^2 + \alpha_1\alpha_2 + \alpha_2]\alpha_0\varepsilon_1\varepsilon_3 + \alpha_0 k_z^2\varepsilon_3(1 - \varepsilon_3) = 0
\tag{5}
$$

where $\alpha_0 = \left(k_z^2 - \dfrac{\omega^2}{c^2} \right)^{1/2}$,

$$\alpha_1 = \left(\left\{ k_z^2 \left(\frac{\varepsilon_1 + \varepsilon_3}{2\varepsilon_1} \right) + \frac{\omega^2}{c^2} \left(\frac{\varepsilon_2^2}{2\varepsilon_1} \right) \right\} + \left\{ \left(k_z^2 \left(\frac{\varepsilon_1 - \varepsilon_3}{2\varepsilon_1} \right) + \frac{\omega^2}{c^2} \left(\frac{\varepsilon_2^2}{2\varepsilon_1} \right) \right)^2 + k_z^2 \frac{\omega^2}{c^2} \left(\frac{\varepsilon_3}{\varepsilon_1} \right) \left(\frac{\varepsilon_2^2}{2\varepsilon_1} \right) \right\}^{1/2} \right)^{1/2}$$

$$\alpha_1 = \left(\left\{ k_z^2 \left(\frac{\varepsilon_1 + \varepsilon_3}{2\varepsilon_1} \right) + \frac{\omega^2}{c^2} \left(\frac{\varepsilon_2^2}{2\varepsilon_1} \right) \right\} - \left\{ \left(k_z^2 \left(\frac{\varepsilon_1 - \varepsilon_3}{2\varepsilon_1} \right) + \frac{\omega^2}{c^2} \left(\frac{\varepsilon_2^2}{2\varepsilon_1} \right) \right)^2 + k_z^2 \frac{\omega^2}{c^2} \left(\frac{\varepsilon_3}{\varepsilon_1} \right) \left(\frac{\varepsilon_2^2}{2\varepsilon_1} \right) \right\}^{1/2} \right)^{1/2}$$

$$\varepsilon_1 = \varepsilon_L \left(1 - \frac{\omega_p^2(\omega + i\upsilon)}{\omega((\omega + i\upsilon)^2 - \omega_c^2)} \right), \varepsilon_2 = \frac{i\omega_c}{\omega} \frac{\varepsilon_L \omega_p^2}{((\omega + i\upsilon)^2 - \omega_c^2)} \text{ and } \varepsilon_3 = \varepsilon_L \left(1 - \frac{\omega_p^2}{\omega(\omega + \upsilon i)} \right)$$

Under the influence of the SPW field $\left(E = \left[ik_z / \alpha_0 \, \hat{x} + \hat{z} \right] A e^{-\alpha_0 x} e^{-i(\omega t - k_z z)} \right)$ in the free space region, the electrons of the nanoparticles execute oscillations and the response is governed by equation of motion, given by

$$m \frac{d^2\vec{s}}{dt^2} + m\upsilon \frac{d\vec{s}}{dt} + m \frac{\omega_{pe}^2 \vec{s}}{3} = -e\vec{E}_s - e(\vec{V} \times \vec{B}_s) \tag{6}$$

where $\vec{s}$ and $\vec{V}$ are displacement and velocity of electrons, respectively. ω_p^2 is the plasma frequency of metal particle. It is same as ω_p^2 when metal particles and metal film are of same material. Taking $\partial/\partial t = -i\omega$ and solving Eq. (6) yields velocity components, given by

$$V_x = \frac{Ae\omega k}{\alpha_0 m \left(\omega^2 - \dfrac{\omega_{pe}^2}{3} + i\upsilon\omega \right)} e^{-i\omega t} \tag{6(a)}$$

$$V_z = \frac{-ie\omega A \left(\omega^2 - \dfrac{\omega_{pe}^2}{3} + \upsilon\omega i \right)}{m \left(\left(\omega^2 - \dfrac{\omega_{pe}^2}{3} + i\upsilon\omega \right)^2 - \omega^2 \omega_c^2 \right)} e^{-i\omega t} \tag{6(b)}$$

The energy absorbed by one electron per second in the influence of SPW field is

$$\varepsilon_{abs} = \frac{1}{2} \text{Re}[-eE^* \cdot V]$$

(7)

where $E^* = \left[-ik_z / \alpha_0 \, \hat{x} + \hat{z}\right] A e^{-\alpha_0 x} . e^{i(\omega t - k_z z)}$ is the complex conjugate of the electric field of SPW in free space. On substituting values from Eqs. 6(a) and 6(b) in Eq. (7) and solving,

$$\varepsilon_{abs} = \frac{e^2 A^2 v \omega^2 \left\{ \left(1 + \frac{k_z^2}{\alpha_0^2}\right)\left((\omega^2 - \frac{\omega_{pe}^2}{3})^2 + v^2 \omega^2\right) - \left(\frac{k_z^2}{\alpha_0^2} \omega^2 \omega_c^2\right)\right\}}{6\left((\omega^2 - \frac{\omega_{pe}^2}{3})^2 + v^2 \omega^2 - \omega^2 \omega_c^2\right)\left((\omega^2 - \frac{\omega_{pe}^2}{3})^2 + v^2 \omega^2\right)}$$

Suppose the electron density in metal nanoparticles is n_0. The power absorbed per second by the nanoparticles is given by

$$\varepsilon_{abs} = \frac{A^2 v \omega^2 \omega_{pe}^2 \left\{ \left(1 + \frac{k_z^2}{\alpha_0^2}\right)\left((\omega^2 - \frac{\omega_{pe}^2}{3})^2 + v^2 \omega^2\right) - \left(\frac{k_z^2}{\alpha_0^2} \omega^2 \omega_c^2\right)\right\} r_c^3}{6\left((\omega^2 - \frac{\omega_{pe}^2}{3})^2 + v^2 \omega^2 - \omega^2 \omega_c^2\right)\left((\omega^2 - \frac{\omega_{pe}^2}{3})^2 + v^2 \omega^2\right)}$$

(8)

The interparticle separation $d = N^{1/2} \geq r_c$, such that $N = \int n_p \delta(x - a) dx$, where N is the number of particles per unit area on the metal surface. Then, energy absorbed by n_p particles in distance dz is

$$dP = -\varepsilon_{abs} n_p dz = \frac{A^2 v \omega^2 \omega_{pe}^2 \left\{ \left(1 + \frac{k_z^2}{\alpha_0^2}\right)\left((\omega^2 - \frac{\omega_{pe}^2}{3})^2 + v^2 \omega^2\right) - \left(\frac{k_z^2}{\alpha_0^2} \omega^2 \omega_c^2\right)\right\} r_c}{6\left((\omega^2 - \frac{\omega_{pe}^2}{3})^2 + v^2 \omega^2 - \omega^2 \omega_c^2\right)\left((\omega^2 - \frac{\omega_{pe}^2}{3})^2 + v^2 \omega^2\right)} \left(\frac{r_c}{d}\right)^2 dz$$

(9)

Starting from the Poynting theorem, the energy flow for surface plasma wave over the metal surface is given by

$$P = \frac{A^2 c^2 k_z}{16\pi\alpha_2^2 \omega}\left(\alpha_2 - \frac{k_z^2}{\alpha_2}\right)$$ (10)

As the SPW propagates, the decay in energy of a beam propagating across a medium with metal nanoparticles is given by $P = P_0 e^{-k_{ip}z}$, $k_{i,p}$ is absorption constant. On differentiating this equation w.r.t z and dividing the two equations, we get

$$\int_{P_0}^{P}\frac{dP}{P} = \int_0^z k_{ip}\,dz + C$$ (11)

where P_0 is the power at $z = 0$, while P is the power of SPW after the absorption length z. Substituting values of dP and P from Eqs. (9) and (10), respectively in Eq. (11), we get the absorption constant k_{ip}, given as

$$k_{ip} = \frac{8\pi\alpha_0^2 v\omega^3\omega_{pe}^2\left\{\left(1+\frac{k_z^2}{\alpha_0^2}\right)\left((\omega^2 - \frac{\omega_{pe}^2}{3})^2 + v^2\omega^2\right) - \left(\frac{k_z^2}{\alpha_0^2}\omega^2\omega_c^2\right)\right\}r_c}{3k_z c^2\left((\omega^2 - \frac{\omega_{pe}^2}{3})^2 + v^2\omega^2 - \omega^2\omega_c^2\right)\left((\omega^2 - \frac{\omega_{pe}^2}{3})^2 + v^2\omega^2\right)(k_z^2/\alpha_0 - \alpha_0)}\left(\frac{r_c}{d}\right)^2$$ (12)

Eq. (12) is solved numerically for the parameters $\varepsilon_L = 4$, $\omega_{pe} = 4.079 \times 10^{15}$ rad/s, $d = 112$ nm, $r_c = 30$ nm, and $v = 7 \times 10^{12}$/sec.

14.3   RESULTS AND DISCUSSION

The absorption of SPW by metal nanoparticles is studied on varying external magnetic field strength, radius, interparticle separation between the nanoparticles, and dielectric constant of the metal. Figure 14.2 illustrates the change in normalized absorption constant ($k_{ip}c/\omega_{pe}$) verses normalized frequency (($\omega - \omega_{pe}/\sqrt{3})/\omega_{pe}$) on varying cyclotron frequency (ω_c/ω_p). The other parameters are $\omega_{pe} = 4.079 \times 10^{15}$ rad/s, $r_c = 30$ nm, $d = 112$ nm and $v = 7 \times 10^{12}$/sec. The resonance absorption constant increase from 0.11 to 0.16 as the cyclotron frequency is varied from 0.0 to 0.6. The SPW excites resonant plasma oscillations in the particles. The energy is absorbed by

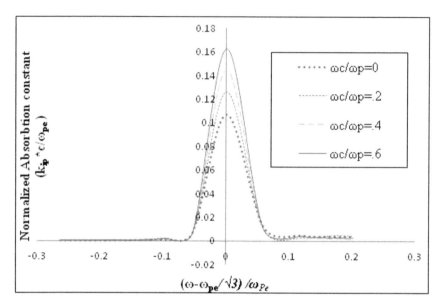

FIGURE 14.2   Plot of normalized absorption constant ($k_{ip}c/\omega_{pe}$) vs. normalized frequency $((\omega - \omega_{pe}/\sqrt{3})/\omega_{pe})$ on varying normalized cyclotron frequency (ω_c/ω_p) for nanoparticles of radius 30 nm.

the spherical metal nanoparticles at frequency $\omega^2 = \omega^2_{pe}/3$. At this frequency, absorption constant rises sharply which corresponds to the strong dissipation of the surface wave energy via collisions of the free electrons of the particle. Under the influence of external magnetic field, the absorption coefficient is dependent on the cyclotron frequency [Eq. (12)], which results in increased absorption of the waves.

Figure 14.3 unveils the effect of nanoparticle radius variation on the absorption constant of SPW at fixed ($\omega_c/\omega_{pe} = 0.2$) and distance between the particles $d = 112$ nm. The rest of parameters are same as used in Figure 14.3. Absorption constant is directly proportional to the cube of the nanoparticle radius as clear from Eq. (12). Rise in absorption constant is observed due to increase in number density of electrons with the size of nanoparticles. Figure 14.4 shows plot of normalized absorption constant verses normalized frequency on varying interparticle separation between the nanoparticle for ($\omega_c/\omega_{pe} = 0.2$) and $r_c = 30$ nm. Absorption constant being inversely proportional to the square of interparticle separation decreases with increase in separation between the particles. Figure 14.5 shows plot of normalized

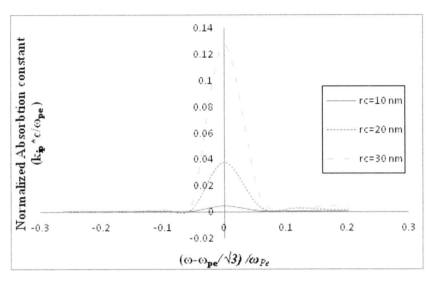

FIGURE 14.3 Plot of normalized absorption constant ($k_{tp}c/\omega_{pe}$) vs. normalized frequency (($\omega - \omega_{pe}/\sqrt{3})/\omega_{pe}$) on varying nanoparticle radius for $\varepsilon_L = 4$ and ($\omega_c/\omega_{pe} = 0.2$).

absorption constant versus normalized frequency for $\varepsilon_L = 4$. The rest of the parameters are the same as in Figure 14.4. It is observed from Figure 14.5 that the changing of the dielectric constant of the metal will induce extra absorption near the surface plasma resonance peak due to enhanced dipole moment between the nanoparticles. Cai et al. [2] reported the photocurrent photo voltage characteristics of the dye-sensitized ZnO based solar cell under the influence of an external magnetic field and observed sufficient increase in the photocurrent. Our study may be useful for improved absorption in solar cells by the proper adjustment of the parameters influencing the surface plasma resonance.

14.4 CONCLUSIONS

In conclusion, the study of the external magnetic field effect on absorption of SPW by metallic nano-particles adsorbed over the metal surface is facilitated by resonant plasma oscillations inside the particles. These oscillations occur for a specific frequency of the incident SPW for which plasma frequency of the nanoparticle becomes √3 times the incident laser frequency.

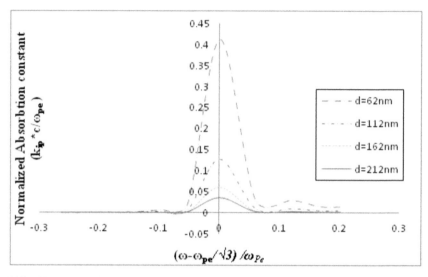

FIGURE 14.4 Effect of interparticle separation variation on normalized absorption constant ($k_{ip}c/\omega_{pe}$) of surface plasma waves for $\varepsilon_L = 4$, $\omega_c/\omega_{pe} = 0.2$, and $r_c = 30$ nm.

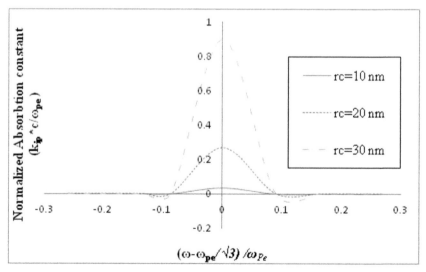

FIGURE 14.5 Variation of normalized absorption constant ($k_{ip}c/\omega_{pe}$) verses normalized frequency $((\omega - \omega_{pe}/\sqrt{3})/\omega_{pe})$ on varying radius of nanoparticles for $\varepsilon_L = 9$, and $\omega_c/\omega_{pe} = 0.2$.

The SPW propagating along z-axis is excited using Kretschmann geometry, having maximum amplitude at the metal-vacuum interface. At resonant frequency, the SPW energy is strongly dissipated via collisions of the free electrons of the particle resulting in increased absorption constant the wave. The resonant absorption of the surface plasma wave is influenced by the externally applied magnetic field, size of the metallic particles, interparticle separation between the nanoparticles and dielectric constant of the metal. Results revealed increased absorption coefficient with the increase in external magnetic field, size of particle, and decrease of the interparticle separation between the nanoparticles. By tuning the parameters, such as radius of the nanoparticles, external magnetic field strength, the interparticle separation between the nanoparticles and dielectric constant of the metal may be helpful in designing energy efficient solar cells. The present treatment is limited to low areal density of nanoparticles so that the SPW field structure is not drastically modified by the particles.

In conclusion, our results suggest that proper adjustment of the parameters influencing the absorption of surface plasma waves by nanoparticles leads to the generation of high efficiency solar cells.

KEYWORDS

- Kretschmann geometry
- magnetic field
- nanoparticles
- resonance absorption
- solar cells
- surface plasma waves

REFERENCES

1. Ahmad, A., & Tripathi, V. K. (2006). Nonlinear absorption of femtosecond laser on a metal surface embedded by metallic nanoparticles. *Applied Physics Letters, 89,* 153112.

2. Cai, F., Wang, J., Yuan, Z., & Duan, Y. (2012). Magnetic-field effect on dye-sensitized ZnO nanorods -based solar cells. *Journal of Power Sources, 216,* 269–72.

3. Gurel, C. S., & Oncu, E. (2010). Characteristics of electromagnetic wave propagation through a magnetised plasma slab with linearly varying electron density. *Progress in Electromagnetic Research B, 21,* 385–398.

4. Hwang, T. Y., Vorobyen, A. Y. & Guo, C. (2009). Ultrafast dynamics of femtosecond laser-induced nanostructure formation on metals. *Applied Physics Letters, 95,* 123111.

5. Jazi, B., Rahmani, Z., & Shokri, B. (2013). Reflection and absorption of electromagnetic wave propagation in a inhomogeneous dissipative magnetized plasma slab. *IEEE Transactions on Plasma Science, 41,* 2.

6. Kreibig, U., & Vollmer, M. (1995). Optical Properties of Metal Clusters. Springer-Verlag, Berlin.

7. Kretschmann, E., & Raether, H. (1968). *Radiative decay* of non-radiative *surface plasmons excited* by *light. Naturforsch, 23A,* 2135.

8. Kumar, A., & Verma, A. L. (2011). Nonlinear absorption of intense short pulse laser over a metal surface embedded with nanoparticles. *Laser and Particle Beams, 29,* 333–338.

9. Kumar, G., & Tripathi, V. K. (2007). Anomalous absorption of surface plasma wave by particles adsorbed on metal surface. *Applied Physics Letters, 91,* 161503.

10. Nancy Xu, X. H., Huang, S., Brownlow, W., Salaita, K., & Jeffers, R. B. (2004). Size and temperature dependence of surface plasmon absorption of gold nanoparticles induced by tris (2,2'-bipyridine) ruthenium. *Journal of Physical Chemistry B, 108,* 15543–15551.

11. Raether, H. (1988). Surface Plasmons on Smooth and Rouge Surfaces and on Gratings. Springer-Verlag, New York.

12. Rajeev, P. P., Ayyub, P., Bagchi, S., & Kumar, G. R. (2004). Nanostructures, local fields, and enhanced absorption in intense light–matter interaction. *Optics Letters, 29*(22), 2662–2664.

13. Schaadt, D. M., Feng, B., & Yu, E. T. (2005). Enhanced semiconductor optical absorption via surface plasmon excitation in metal nanoparticles. *Applied Physics Letters, 86,* 063106.

14. Vorobyev, Y., & Guo, C. (2005). Enhanced absorptance of gold following multipulse femtosecond laser ablation. *Physical Review B, 72,* 195,418–195,422.

15. Wallis, R. F., Brion, J. J., Burstein, E., & Hartstein, A. (1974). Theory of surface polaritons in anisotropic dielectric media with applications to surface magnetoplasmons in semiconductors. *Physical Review B, 8*(9), 3424.

CHAPTER 15

PLASMA WAVES BEYOND THE SOLAR SYSTEM

VIPIN K. YADAV and ANIL BHARDWAJ

Space Physics Laboratory (SPL), Vikram Sarabhai Space Centre (VSSC), Thiruvananthapuram – 695022, Kerala, India,
E-mail: vkyadavcsp@gmail.com, vipin_ky@vssc.gov.in

CONTENTS

ABSTRACT

Plasma waves are observed in planets with magnetospheres—Earth, Mercury, Jupiter, Saturn, Uranus, and Neptune—as well as in planets such as Venus and Mars, which are deprived of a global magnetic field [1, 2]. Plasma waves are detected in planetary satellites and comets. The plasma in the Sun itself supports plasma waves such as Alfvén waves [3], which are observed in solar corona along the solar magnetic field which travels up to interplanetary medium and are observed there. Plasma waves are predicted to exist in interstellar medium also where the Langmuir waves and Alfvén waves excited by the incoming anisotropic cosmic rays stream along the magnetic field lines. The Alfvén waves are also believed to be present in the dusty winds of cooled supergiant stars. An abrupt rise in temperature is observed with the distance from the surface of supergiant stars, which can be explained by the mechanical dissipation of these Alfvén waves. Plasma waves are believed to exist in many other natural plasma systems, such as pulsars, quasars, and galaxies. In this chapter, all these waves are discussed in the form of a mini review.

15.1 INTRODUCTION

Plasma waves are omnipresent and thus are a unique feature of space plasmas as they propagate energy across different space regions. They transport particles and accelerate them to attain high energies. They transmit information about the local plasma properties from regions not accessible for in situ measurements and are specific to the phenomena/instabilities as their properties depend upon the background plasma prevailing at that location.

The motivation behind the study of outer space is a direct consequence of the observation of plasma waves in the near Sun environment. The chapter is organized as follows: In Section 15.2, the plasma waves in interplanetary medium are given. The plasma waves in the interstellar medium are discussed in Section 15.3. Section 15.4 contains the prediction regarding the possibility of the presence of plasma waves in stars and in particular to massive Supergiants.

15.2 PLASMA WAVES IN THE INTERPLANETARY MEDIUM (IPM)

The IPM is the matter that fills the space between the Sun and the planets in the solar system. Thus, the IPM comprises of the solar plasma coming as solar wind, interplanetary dust as well as the highly energetic solar and cosmic rays. The typical IPM temperature is ~100 K and the IPM plasma density is variable (varies inversely to the square of the distance) with ~ 5 particles cm^{-3} near Earth and decreases on moving away from Sun. Since, the IPM plasma density is variable and can be influenced by the interplanetary magnetic field and solar events, such as the coronal mass ejections; it can go as high as 100 particles cm^{-3}. The IPM is highly conducting collisionless plasma with approximate equipartition between thermal and magnetic field energy densities. A rare and unique observational opportunity to study the physics of such plasmas comes when a spacecraft measures the fine-scale structure of the medium (< scale lengths of 0.01 AU). The astrophysical studies of these plasmas involve their wave and turbulence properties [4].

In the beginning, the general perception about the interplanetary medium was that it is almost a perfect vacuum with some residual dust particles. For in-situ measurements, there was no dedicated mission sent to study the local environment of the interplanetary medium. Therefore, most of the plasma and wave measurements were carried out by the missions which were either going to Venus, such as Mariner-2 and Mariner-5, or the missions launched exclusively to study the Sun such as Helios-1 and Helios-2.

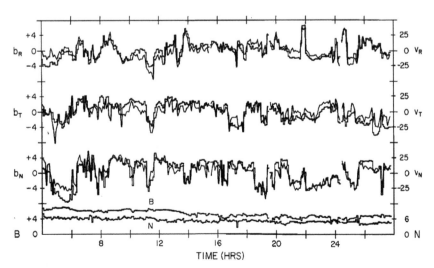

FIGURE 15.1   Mariner-5 data. The upper six curves are the bulk velocity (km/s) and magnetic field (nT). The lower two curves are the magnetic field strength and the proton density [4].

The initial data on interplanetary medium from spacecraft measurements was first obtained in early 1960s. The first attempt of simultaneous measurements of plasma velocity and the magnetic field in the interplanetary space was carried out by Mariner-2 in 1962. This spacecraft used positive ion spectrometer comprised of a cylindrical electrostatic analyzer and a flux-gate magnetometer when it was on its way to Venus. The main objective of this exercise was to find the properties of simultaneous variations in the magnetic field and the plasma velocity and to verify if the variations were generated by the hydromagnetic waves. The hydromagnetic waves are low frequency ion oscillations which take place in the presence of a magnetic field and have two main modes: Alfvén waves (along) and Magnetosonic waves (across) the magnetic field. However, the results were not conclusive [5].

The properties of the ionized gas and the magnetic field observed in the interplanetary space suggest that the IPM can support the propagation of hydromagnetic waves in it. The observations of Mariner-2, made by the magnetometer and the ion spectrometer, were compared with the variations that would be generated by hydromagnetic waves by simulating a first-order model [6]. Mariner-2 measurements had indicated that Alfvén waves are present in the interplanetary medium but the results were not conclusive either. Another such attempt was made in 1967 with Mariner-5 magnetic field measurements using low-field, vector helium, 3-axes magnetometer [7]. These magnetic measurements also indicated the existence of aperiodic Alfvén waves; propagating outward from the sun along the average magnetic field direction and dominating the fluctuations at least 30% time [7]. The magnetometer and the ion spectrometer measurements onboard Mariner-2 were analyzed again [4], shown in Figure 15.1, with following results:

1. Large-amplitude, non-sinusoidal Alfvén waves propagates outward from the Sun. These hydromagnetic waves generally have energy density comparable both to the unperturbed magnetic field energy density and to the thermal energy density.

2. These outwardly propagating Alfvén waves mainly occur in high-velocity solar wind streams and where the velocity decreases slowly with time. In low-velocity regions also, Alfvén waves are outwardly propagating but generally with smaller amplitudes. The large amplitude Alfvénic fluctuations are mostly found in the

compression regions of high-velocity streams where the solar wind velocity increases very fast with time.

3. The magnetoacoustic waves (MAW) were absent or rather not observed in this data. In case of occurrence, the MAW would have small average power of less than 10% than that of Alfvén waves.

The above-discussed results are the interpretation of a solar wind model. Most Alfvén waves in the IPM are supposed to be the undamped remnants of waves generated at or near the Sun. The high wave activity in high velocity and temperature streams can be a proof for the heating of streams by plasma (Alfvén) wave damping near the Sun. The extreme Alfvénic wave activity in the compression regions of high-velocity streams could be due to their amplification or the new generation of waves in these regions by two-stream interactions. The observed absence of magneto-sonic waves indicates their strong damping. These observations suggest the conversion of Alfvén waves to magnetosonic waves when they are convected away from Sun and by this process the energy from the small-scale field fluctuations is transferred to the solar wind plasma which is thermalized [4].

A theoretical study was carried out to study the propagation of arbitrarily large-amplitude, non-monochromatic, microscale Alfvén waves of any polarization in the IPM using Wentzel-Kramers-Brillouin (WKB) approximation [8]. Alfvén waves with periods of 2-3 hours and less are a major contributor to the overall level of interplanetary microscale fluctuation and are of particular importance in high-speed, solar wind streams. The origin of these Alfvén waves is believed to be close to the sun and then they propagate into the IPM with solar wind [9].

The Alfvénic character of the solar wind fluctuations, with period more than 1 hour, was also investigated from the magnetic field and plasma data of Helios 1 and 2 [10]. The results are shown in Figure 15.2 where complete correlation can be seen between the magnetic field components (B_x, B_y, and B_z) and corresponding plasma velocity components (v_x, v_y, and v_z) [10].

15.3 PLASMA WAVES IN THE INTERSTELLAR MEDIUM (ISM)

The ISM exists beyond our solar system where the IPM ends in the space between star systems in a galaxy which includes ionized gas, dust, and

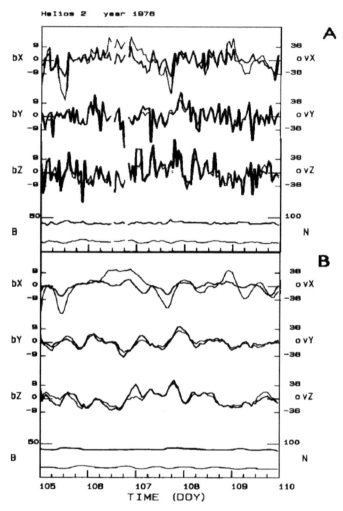

FIGURE 15.2 Helios 1 and 2 data: (A) the bold lines show the magnetic field (nT) variation and the light lines indicate the velocity (km/s). (B) Same data after filtration. The bottom two lines in (A) and (B) is the background magnetic field (nT) in bold and proton density (cm–3) in light line [10].

very high energy charged particles. The ISM is composed of multiple phases, distinguished by ionic, atomic, or molecular matter, the temperature, and the density of the matter. The ISM is highly turbulent with large pressure fluctuations due to energy ejection by stars and by instabilities in the disk of galaxy. The thermal pressures of multiple ISM phases are generally in equilibrium with one another. Magnetic fields and turbulent motions are also the sources, which provide pressure in the ISM; however, they are dynamically important than the thermal pressure [11].

The ISM is important in a way that this diffused medium forms the background in which molecular clouds form which acts as a seed for the cloud formation. The molecular cloud surfaces are almost at the same pressure as that of the background ISM. However, the bulk of this molecular cloud is self-gravitating and is at a high pressure. Hence, star formation takes place in this extremely dense region and the ISM gets refilled the material lost in the formation of a star from matter and energy through planetary nebulae, stellar winds, and supernovae [11].

The diffused matter in the ISM has four different states that are all at about the same pressure in a given region of the galactic disk [11, 12] as given below:

1. Cold neutral medium (CNM): In CNM, the existing matter is primarily in molecular form (mainly hydrogen) with a matter density of 10^6 molecules per cm^3. The typical temperature here is around 100 K only. It makes a fractional volume of maximum 5% of total ISM.

2. Warm neutral medium (WNM): In WNM, the interstellar matter is primarily ionized and the neutrals are mostly atomic in nature. By mass, only 1% of the ISM is in the form of dust and the rest is gas both ionized and neutral combined together. This warm medium generally has an extent between 300–400 pc and has temperature between $6000–10^4$ K and a mean density of about 0.3 cm^{-3}.

3. Warm ionized medium (WIM): It is primarily ionized and generated by hottest and massive O and B type stars. In the WIM, 89% is hydrogen atoms, 9% is helium, and remaining 2% is the heavier element atoms (i.e., metals after hydrogen and helium).

4. Hot ionized medium (HIM): The HIM is also known as the hot coronal gas (HCG). It is primarily generated by supernovae explosions. In the interstellar HIM/HCG, 89% of matter is hydrogen atoms, 9% is helium, and remaining 2% is the heavier element atoms (metals after H and He). The HIM/HCG has a plasma density between 0.0031–0.0047 cm^{-3}.

The important properties of the ISM are summarized in Table 15.1.

TABLE 15.1   The Physical and Plasma Parameters in the Interstellar Medium (Adapted from [13])

S. No.	Interstellar Region	Fractional Volume	Scale Height (pc)	Density (atom/cm³)	Temperature (K)	Magnetic Field (nT)*
1.	Cold neutral medium (CNM)	1–5%	100–300	20–50	50–100	0.1
2.	Warm neutral medium (WNM)	10–20%	300–400	2–5	6000–10,000	0.16
3.	Warm ionized medium (WIM)	20–50%	390–1000	0.2–0.5	8000	0.21
4.	Hot ionized medium (HIM)	30–70%	1000–3000	3–5×10^{-3}	10^6–10^7	0.5

Note: * denotes estimates.

Alfvén waves are predicted to exist in the interstellar medium. These hydromagnetic waves excited by the incoming anisotropic cosmic rays streaming along the magnetic field lines with a wide angle between the field and the direction of wave propagation [14, 15].

In any space plasma, the velocity distributions of electrons and ions consist of a dense cold component and a diffuse high-energy tail. If the high-energy protons (cosmic rays) are sufficiently intense and their pitch angle distributions sufficiently anisotropic, instability occurs for those waves propagating parallel to the magnetic field. If the spectrum of resonant protons is sufficiently hard, a reasonably large cone of propagating angles about the magnetic field can be unstable. Observed fluxes of trapped protons in the magnetosphere should destabilize the ion cyclotron wave at a lower intensity threshold than for at least one class of electrostatic waves. The interaction of relativistic cosmic rays with the interstellar plasma having a Maxwellian distribution gives rise to plasma oscillations due to non-linear transfer of energy to the ionized particles which finally result in the generation of electrostatic Langmuir waves [16].

Due to the stringent conditions of the warm intercloud medium, where very low particle density exists, the atomic and molecular radiative transitions efficiently cool the gas, so that an efficient heating mechanism is required in order to maintain the coronal gas temperature. One of the proposed processes of the ISM heating is by the Alfvén and magnetosonic waves [17–21].

15.4 PLASMA WAVES IN STARS

A star is a massive astronomical object of plasma held together by self-gravity which emits radiation due to its luminosity. The supergiants are among the most massive stars in the universe and can have mass 10 or more times than that of the solar mass ($M_\odot$) and the radius generally between 30–500 times the solar radii ($R_\odot$). These massive stars have their luminosities from 30,000 to over a million times than that of solar luminosity ($L_\odot$).

Plasma waves are observed in and around Sun [3] and are predicted to be present in the other massive stars such as supergiants due to the prevailing plasma conditions there. In general, in supergiants or stars, there is a huge movement of charged matter which produces a strong magnetic field in that region. This magnetic field when couples with the local plasma there gives rise to electromagnetic plasma waves such as Alfvén wave, a significant feature of solar plasma. Alfvén waves are believed to be the reason behind the high mass-loss rate in the cool supergiant stars and low terminal velocities of the winds. It is generally believed that Alfvén waves play the important role in driving stellar winds. In case of cool giant and supergiant stars, the driving mechanism is the outward directed flux of Alfvén waves which can produce a significant mass-loss even in those stars which do not have hot coronas or strong radiative fluxes [22].

15.5 SUMMARY

The plasma wave existence in the three outer space regions are summarized in the Table 15.2.

Plasma waves exist in IPM, ISM, and the supergiant stars. Alfvén waves, observed in the IPM, are most likely of the solar origin and believed

TABLE 15.2   Plasma Waves in Various Space Regions

S.No.	Outer space region	Observed/predicted plasma waves
1.	Interplanetary Medium (IPM)	Slow mode Alfvén waves
2.	Interstellar Medium (ISM)	Langmuir waves, Alfvén waves, Magnetosonic waves
3.	Supergiant stars	Alfvén waves

to reach IPM along with the solar wind. The Alfvén and magnetosonic waves, predicted to be present in the ISM, are believed to be produced by the cosmic ray protons interaction with the interstellar plasma as a result of the two-stream instability. In supergiant stars, the Alfvén wave mechanism is same as that in the Sun.

KEYWORDS

- Alfvén waves
- interplanetary medium
- interstellar medium
- Langmuir waves
- magnetosonic waves
- supergiants

REFERENCES

1. Kurth, W. S., & Gurnett, D. A. (1991). "Plasma Waves in Planetary Magnetospheres," *Journal of Geophysical Research, 96*(16), 18977–18991.
2. Yadav, V. K., Thampi, R. S., & Bhardwaj, A. (2013). "Plasma Waves in the Solar System," Proceedings of the 27th National Symposium on Plasma Science & Technology (Plasma-2012). Pudducherry, India; December 10–13, 2012, 454–458.
3. Yadav, V. K., Bhardwaj, A., & Thampi, R. S. (2014). "Plasma Waves in and around the Sun," Proceedings of the Regional Conference in Radio Science (RCRS 2014). Pune, India; January 2–5, 2014, 97–98.
4. Belcher, J. W., & Davis, Jr. L. (1971). "Large-Amplitude Alfvén Waves in the Interplanetary Medium 2," *Journal of Geophysical Research, 76* (16), 3534–3563.

5. Colman Jr. P. J. (1966). "Hydromagnetic Waves in the Interplanetary Plasma," *Physical Review Letters, 17* (4), 207–211.

6. Coleman, Jr. P. J. (1967). "Wave-Like Phenomena in the Interplanetary Plasma: Mariner 2," *Planetary Space Science, 15*, 953–973.

7. Belcher, J. W., Davis, Jr. L., & Smith, E. J. (1969). "Large-Amplitude Alfvén Waves in the Interplanetary Medium: Mariner 5," *Journal of Geophysical Research, 74*(9), 2302–2308.

8. Whang, Y. C. (1973). "Alfvén Waves in Spiral Interplanetary Field," *Journal of Geophysical Research, 78* (31), 7221–7228.

9. Belcher, J. W., & Burchsted, R. (1974). "Energy Density of Alfvén Waves Between 0.7 and 1.6 AU," *Journal of Geophysical Research, 79*(31), 4765–4768.

10. Bruno, R., Bavassano, B., & Villante, U. (1985). "Evidence of Long Period Alfvén Waves in the Inner Solar System," *Journal of Geophysical Research, 90* (A5), 4373–4377.

11. McKee, C. F., Hollenbach, D., & Wolfire, M. G. (1997). "Phases of the Interstellar Medium in the Milky Way," in The Dense Interstellar Medium in Galaxies, Springer-Verlag Berlin Heidelberg; Eds.: Susanne Pfalzner, Carsten Kramer, Christian Straubmeier, Andreas Heithausen, pp. 395–403.

12. Ferriere, K. M., Zweibel, G. E., & Shull, J. M. (1988). "Hydromagnetic wave Heating of the Low-Density Interstellar Medium," *The Astrophysical Journal, 332*, 984–994.

13. Ferriere, K. M. (2001). "The Interstellar Environment of our Galaxy," *Review of Modern Physics, 73*, 1031–1066.

14. Kulsrud, R., & Pearce, W. P. (1969). "The Effect of Wave-Particle Interactions on the Propagation of Cosmic Rays," *The Astrophysical Journal, 156*, 445–469.

15. Wentzel, D. G. (1968). "Hydromagnetic Waves Excited by Slowly Streaming Cosmic Rays," *Astrophysical Journal, 152*, 987–996.

16. Tsytovich, V. N. (1966). "The Isotropization of Cosmic Rays," *Soviet Astronomy*, AJ, *10*(3), 419–428.

17. Lerche, I., & Schlickeiser, R. (2001). "Linear Landau Damping and Wave Energy Dissipation in the Interstellar Medium," *Astronomy & Astrophysics, 366*, 1008–1015.

18. Lazar, M., Spanier, F., & Schlickeiser, R. (2003). "Linear Damping and Energy Dissipation of Shear Alfvén Waves in the Interstellar Medium," *Astronomy & Astrophysics, 410*, 415–424.

19. Spangler, S. R. (1991). "The Dissipation of Magnetohydrodynamic Turbulence Responsible for Interstellar Scintillation and the Heating of the Interstellar Medium," *The Astrophysical Journal, 376*, 540–555.

20. Spanier Felix (2004). "Heating of the ISM by Alfvén-Wave Damping," Proceedings of "The Magnetized Interstellar medium," Antalya, Turkey, September 8–12, 2003; Eds.: Bulent Uyanıker, Wolfgang Reich & Richard Wielebinski, pp. 115–119.

21. Spanier, F., & Schlickeiser, R. (2005). "Damping and Wave Energy Dissipation in the Interstellar Medium II. Fast Magnetosonic Waves," *Astronomy & Astrophysics, 436*, 9–16.

22. Vidotto, A. A., & Pereira, V. J. (2005). "Alfvén waves in dusty winds of cool supergiant stars," *AIP Conference Proceedings 784*, 629–634.

CHAPTER 16

JEANS INSTABILITY OF A SELF GRAVIATATING VISCOUS MOLECULAR CLOUD UNDER THE INFLUENCE OF FINITE ELECTRON INERTIA, HALL EFFECT, FINE DUST PARTICLES, AND ROTATION

D. L. SUTAR[1] and R. K. PENSIA[2]

[1]Research Scholar Mewar University Gangrar, Chittorgarh, 312901, Rajasthan, India, E-mail: Devilalsutar833@gmail.com

[2]Department of Physics, Government Girls College, Neemuch (M.P.), 458441, India

CONTENTS

ABSTRACT

In this chapter, we study the jeans instability of gaseous plasma in the presence of fine dust particles incorporating the effects of finite electron inertia, viscosity, rotation, thermal conductivity, electrical conductivity, and Hall current with the help of the linearized perturbed equations of the problem. A general dispersion relation is obtained. The general dispersion relation is discussed for a different mode of propagation and about a different axis of rotations. The stability of the system is discussed by applying Rought-Hurwitz criteria. We hope that our result of the present problem will help to understand the astrophysical problems (PACS-Numbers 94.30.cq, 52.25.xz, 94.05.Dd.).

16.1 INTRODUCTION

The gravitational instability of a plasma media plays a crucial role in understanding various astrophysical problems and is believed to be responsible for cloudiness of a number of astrophysical objects, for the formation of different phases of the molecular cloud in diffuse interstellar and intergalactic media and planetary nebulae and for the formation of prominences in the solar corona. The self-gravitational instability was discovered by Jeans [10] and he suggested that an infinite homogeneous self-gravitating fluid is unstable for all wavelength λ greater than critical Jeans wavelength $\lambda_j = \left(\frac{\pi c^2}{G\rho}\right)^{\frac{1}{2}}$, where the symbols have their usual meaning.

A detailed account of the gravitational instability, under the varying assumption of hydrodynamics and hydromagnetic, has been investigated by Chandrasekhar [3]. Herrengger [9] has considered the problem of self gravitating two component gaseous plasma under the influence of collisions and gyroviscosity. Several works investigating the validity of Jeans criterion of the instability of the gaseous plasma [1, 4, 5, 11, 20] have appeared in literature considering the presence of various parameters in astrophysical fluids. Recently Rosenberg [16, 17] has studied cross field dust acoustic instability in a dusty negative ions plasma. Prajapati et al. [15] have discussed the effects of arbitrary radiative heat-loss functions and Hall current on the self-gravitational instability of a homogeneous, viscous, rotating plasma. Uberoi [24] has studied the non-ideal electron inertial effects on the gravitational instability of a medium permeated by a magnetic field both in the presence and absence of rotation.

Another important force on interstellar gaseous plasma is magnetic fields. This is the force which provides pressure support and inhibits the contraction and fragmentation of interstellar clouds. The magnetic field interacts directly only with the ions, electrons and charged grains in the gas, collisions of the ions with the predominantly neutral gas in clouds are responsible for the indirect coupling of the magnetic fields to the bulk of the gas. Hennebelle and Perault [8] have studied a dynamical condensation process in a magnetized and thermally conducting medium and stressed their importance for the thermal condensation modes with magnetic fields. Falceta - Goncalves et al. [7] have discussed the role of magnetic field in the stability of ISM molecular clouds. De Colle and Raga [6] have investigated the effects of the magnetic field on the H_α emission from astrophysical jets.

In addition to this astrophysical commentary and tokamak edge plasmas also contain grains which are small particles. There are several situations where the interaction between the ionized and neutral gas components becomes important in cosmic physics. Scanlov and Segel [18] have investigated the problem on the onset of the Benard convection in hydromagnetics. Several authors have discussed this problem from different physical points of view [21, 25]. Recently Vaidyanathan et al. [23] have studied the Benard convection is ferromagnetic fluids. Sunil et al. [22] have discussed the effect of dust particles on the thermal convection in the ferromagnetic fluid saturating a porous medium by assuming isotropic viscosity. However, these works deal with self-gravitating gaseous plasma and dusty plasma under the influence of various parameters where finite electron inertia and fine dust particles effects are not taken into account.

Recently Merlino et al. [12] have performed an experiment to describe observation and interpretation of large amplitude dust acoustic waves, large amplitude dust acoustic waves into shocks and the spontaneous formation of stationary, stable dust structures in a moderately coupled dusty plasma. Chakrabati [2] has discussed the secondary instability of Jean's mode in a gravitating fluid with uniform rotation. Prajapati [13, 14] has studied the effect of polarization force on the Jeans instability of self-gravitating dusty plasma both in the presence and absence of magnetic field. Sharma and Chhajlani [19] have investigated the effect of spin-induced magnetization and electrical resistivity incorporating the viscosity on the Jeans instability of quantum magnetoplasma.

In all these studies of Jeans instability of a self-gravitating medium whether in the absence or presence of general rotation of a plasma endowed with fine dust particles and finite electron inertia has not been analyzed. It would, therefore, be of interest to examine the Jeans instability of self-gravitating gaseous plasma under the influence of finite electron inertia, rotation, fine dust particles, and Hall effect. In the present work, we have investigated the problem of Jeans instability of a self-gravitating of gaseous plasma in the presence of fine dust particles including the simultaneous effects of finite electron inertia, viscosity, electrical resistivity, thermal conductivity, rotations and Hall current. The present study can serve as a theoretical support to understand the astrophysical problems. This problem, to the best of our knowledge, has not been studied yet.

16.2 LINEARIZED PERTURBATION EQUATIONS

We consider an infinite homogeneous, viscous, Self-gravitating gaseous plasma composed of gas and the fine dust particles (suspended particles) mixture with a uniform vertical magnetic field, finite electron inertia, and the Hall effect. In the unperturbed state, the fluid is assumed to be at rest. The equilibrium quantities are assumed independent of space and time. Due to the action of perturbing field, a small amplitude perturbation induces an oscillatory motion. If the amplitude of these perturbations grows in time then the system is said to be unstable. The unstable mode will grow when energy transferred to the system exceeds the dissipation. The perturbations in density, velocity, pressure, magnetic field, temperature and the gravitational potential are given as $\delta\rho, v, \delta P, \hbar, \delta T$ and $\delta\phi$, respectively. The perturbation state is given by

$$\rho = \rho_0 + \delta\rho, \qquad P = P_0 + \delta P, \qquad H = H_0 + h,$$
$$T = T_0 + \delta T, \qquad U = U_0 + \delta U, \qquad V = v, \qquad u = u \qquad (1)$$

Suffix '0' is dropped from the equilibrium quantities. Our starting point will be a set of linearized perturbation equations

$$\frac{\delta \vec{v}}{\delta t} = -\frac{\vec{\nabla}\delta P}{\rho} + \vec{\nabla}\delta\phi + \frac{K_s N}{\rho}(\vec{u} - \vec{v}) + \vartheta\nabla^2\vec{v} + \frac{\vartheta}{K_1}\vec{v}$$

$$+ \frac{1}{4\pi\rho}(\vec{\nabla} \times \vec{h}) \times \vec{H} + 2(\vec{v} \times \vec{\Omega}) \qquad (1)$$

$$\varepsilon\frac{\partial\delta\rho}{\partial t} = -\rho\vec{\nabla}.\vec{v} \qquad (2)$$

$$\delta P = C^2 \delta P \qquad (3)$$

$$\nabla^2\delta\phi = -4\pi G\delta\rho \qquad (4)$$

$$\left(\tau\frac{\partial}{\partial t} + 1\right)\vec{u} = \vec{v} \qquad (5)$$

$$\lambda\nabla^2\delta T = \rho C_p\frac{\partial\delta t}{\partial t} - \frac{\partial\delta P}{\partial t} \qquad (6)$$

$$\frac{\delta P}{P} = \frac{\delta T}{T} + \frac{\delta\rho}{\rho} \qquad (7)$$

$$\frac{\partial\vec{h}}{\partial t} = (\vec{H}.\vec{\nabla})\vec{v} - (\vec{v}.\vec{\nabla})\vec{H} + \frac{C^2}{\omega_{pe}^2}\frac{\partial}{\partial t}\nabla^2\vec{b} - \frac{C}{4\pi Ne}\vec{\nabla} \times [(\vec{\nabla} \times \vec{h}) \times \vec{H}] \qquad (8)$$

where, $\vec{v}(v_x, v_y, v_z)$, $\vec{u}(u_x, u_y, u_z)$, N, ρ, P, ϕ, $\vec{H}(0, 0, H)$, $\vec{\Omega}(\Omega_x, 0, \Omega_z)$, T, G, ϑ, C_p, λ, ε, R, m, ρ_s, ω_{pe}, K_s ($6\pi\rho vr$), and $\vec{h}(h_x, h_y, h_z)$ denote, respectively, the gas velocity, particle velocity, number of particle, density, pressure, gravitational potential, magnetic field, rotation, temperature, gravitational constant, kinematic viscosity, specific heat at constant pressure, thermal conductivity, porosity, gas constant, mass per unit volume of the particles its density, plasma frequency of electron, the constant in the Stokes drag formula and the perturbation in magnetic field.

16.3 DISPERSION RELATION

We assume that all the perturbed quantities as

$$exp\{i(k_x x + k_z z + \omega t)\} \tag{9}$$

where k_x, k_z are the wavenumbers of perturbation along the x and z-axis so that $k_x^2 + k_z^2 = k^2$ and the ω is the frequency of harmonic disturbances. Using Eqs. (2)–(9) in Eq. (1), we obtain the following algebraic equations for the components.

$$M_1 v_x - \left(\frac{K_z^2 V^2 K^2 A_3}{A_2} + 2\Omega_z\right) v_y + \frac{ik_x}{k^2}\Omega_T^2 s = 0 \tag{10}$$

$$\left(\frac{K_z^2 K^2 V^2 A_3}{A_2} + 2\Omega_z\right) v_x + M_2 v_y - 2\Omega_z v_z = 0 \tag{11}$$

$$2\Omega_x v_y + d_1 v_z + \frac{ik_z}{k^2}\Omega_T^2 s = 0 \tag{12}$$

The divergence of Eq. (1) with the aid of Eqs. (2)–(9) gives

$$\frac{ik_x k^2 V^2 A_1}{A_2} v_x - \left\{\frac{iK_x K_z^2 V^2 K^2 A_3}{A_2} + 2i(k_x\Omega_z - \Omega_x k_z)\right\} v_y - M_3 s = 0 \tag{13}$$

where $s = \delta\rho/\rho$ is the condensation of the medium, $\gamma = C_p/C_v = C^2/C'^2$ is the ratio of the specific heat, $V = B/\sqrt{4\pi\rho}$ is the Alfvén velocity, $A = (K_s N)/\rho$ has the dimension of frequency, $\tau = m/K_s$ is the relaxation time, $\sigma = i\omega$ is the growth rate of perturbation, $\theta = \lambda/(\rho C_p)$ is the thermometric conductivity, and C and C' are the adiabatic and isothermal velocities of sound. Also we have assumed the following substitutions:

$$\beta = \tau A = \frac{\rho_s}{\rho}, \qquad \Omega_\vartheta = \vartheta\left(K^2 + \frac{1}{K_1}\right), \qquad A_1 = \sigma f,$$

$$f = \left(1 + \frac{C^2 K^2}{\omega_{pe}^2}\right) \qquad\qquad d_1 = \left(\sigma + \Omega_\vartheta + \frac{\beta\sigma}{\sigma\tau + 1}\right),$$

$$d_2 = \left(\frac{k_z^2 k^2 V^2 A_3}{A_2} + 2\Omega_z\right), \qquad d_3 = \left(\frac{ik_x k_z^2 k^2 V^2 A_3}{A_2} + 2iM_4\right),$$

$$\Omega_{j'}^2 = \left(C'^2 k^2 - 4\pi G\rho\right), \qquad \Omega_j^2 = \left(C^2 k^2 - 4\pi G\rho\right)$$

$$\Omega_T^2 = \left(\frac{\sigma\Omega_j^2 + \theta_k \Omega_{j'}^2}{\sigma + \theta_k}\right), \qquad M_1 = \left(d_1 + \frac{V^2 k^2}{A_1}\right),$$

$$M_2 = \left(d_1 + \frac{V^2 k_z^2}{A_1}\right), \qquad M_3 = \left(\sigma\varepsilon d_1 + \Omega_T^2\right),$$

$$M_4 = (k_x \Omega_z - \Omega_x k_z), \qquad A_2 = (A_1^2 + A_3^2 k_z^2 k^2),$$

$$\theta_k = \gamma\theta k^2 \qquad\qquad A_3 = \left(\frac{CH}{4\pi Ne}\right),$$

The nontrivial solution of the determinant of the matrix obtained from Eqs. (11)–(13) with (v_x, v_y, v_z), and s having various coefficients that should vanish gives the following dispersion relation.

$$\left(\sigma\varepsilon d_1 + \Omega_T^2\right)\left(M_1 M_2 d_1 + 4\Omega_x^2 M_1 + d_1 d_2^2\right) - \frac{k_x^2 V^2 A_1}{A_2}\Omega_T^2\left(M_2 d_1 + 4\Omega_x^2\right)$$

$$- 2\Omega_x k_z \Omega_T^2\left(\frac{iM_1 d_3}{k^2} + \frac{k_x A_1 d_2}{A_2}\right) + \frac{ik_x}{k^2} d_1 d_2 d_3 \Omega_T^2 = 0$$

$$(14)$$

The dispersion relation (14) shows the combined influence of fine dust particles thermal conductivity, finite electron inertia, magnetic field, viscosity and rotation on the self-gravitational instability of a homogeneous plasma. If we ignore the effect of finite electron inertia then Eq.

(14) reduces to Chhajlani and Vyas [6]. The present results are also similar to those of Chhajlani and Sanghvi [4] in the absence of rotation and finite electron inertia neglecting the contribution of finite Larmor radius (FLR) correction and Hall parameter in that case. In the absence of fine dust particles Eq. (14) gives similar results as are obtained by Prajapati et al. [11] excluding the effects of arbitrary radiative heat-loss functions in that case. Thus with these correlations, we find that the dispersion relation (14) is modified due to the combined effects of dust particles, finite electron inertia, rotation, magnetic field, viscosity and thermal conductivity. This dispersion relation will be able to predict the complete information about the acoustic wave, Alfvén wave and Jeans gravitational instability of the gaseous plasmas considered. The above dispersion relation is very lengthy and to analyze the effects of each parameter we now reduce the dispersion relation (14) for two modes of rotation and two modes of propagation.

16.4   ANALYSIS OF THE DISPERSION RELATION

Now we shall discuss the dispersion relation given by Eq. (14) for different cases of rotation and propagation as follows.

16.4.1   AXIS OF ROTATION ALONG THE MAGNETIC FIELD

In this case of rotational axis along the magnetic field we put $\Omega_x = 0$ and $\Omega_z = \Omega$ then relation (14) gives

$$
\left(\sigma \varepsilon d_1 + \Omega_T^2\right)\left\{M_1 M_2 d_1 + d_1 \left(\frac{k_z^2 k^2 V^2 A_3}{A_2} + 2\Omega\right)^2\right\} - \frac{k_x^2 V^2 A_1}{A_2}\Omega_T^2 d_1 M_2
$$
$$
+ \frac{ik_x}{k^2}d_1\left(\frac{k_z^2 k^2 V^2 A_3}{A_2} + 2\Omega\right)\left(\frac{ik_x k_z^2 k^2 V^2 A_3}{A_2} + 2ik_x\Omega\right)\Omega_T^2 = 0
$$

$$(15)$$

We find that when the axis of rotation along the magnetic field the dispersion relation is modified due to combined effects of finite electron inertia, thermal conductivity, viscosity, rotation, magnetic field, electrical resistivity, the presence of fine dust particles and Hall current. The dispersion relation (15) is further reduced for longitudinal mode propagation and transverse mode of propagation.

16.4.1.1 Longitudinal Mode of Propagation

For this case, we assume all the perturbations longitudinal to the direction of the magnetic field (i.e., $k_x = 0$, $k_z = k$). Thus the dispersion relation (15) reduces in the simple form to give

$$d_1 \left[M_1^2 + \left(2\Omega + \frac{k^4 V^2 A_3}{A_2} \right)^2 \right] (\sigma \varepsilon d_1 + \Omega_T^2) = 0$$

(16)

This dispersion expresses the combined influence of all parameters of the problem in the longitudinal mode of propagation when the axis of rotation is parallel to the magnetic field. The dispersion relation (16) has three different factors which show deferent modes of propagation incorporating different parameters as discussed below

The first factor of Eq. (16) gives, on substituting the value of d_1, the following second polynomial equation:

$$\tau \sigma^2 + \sigma \{ 1 + \tau (A + \Omega_g) \} + \Omega_g = 0 \qquad (17)$$

This dispersion relation (17) is a non-gravitating stable mode due to viscosity and influenced by the presence fine dust particles. This mode is independent of the magnetic field, finite electron inertia, finite electrical resistivity, rotation and the Hall current.

The second factor of Eq. (16) gives, on substituting the values of M_1, A_1 and A_3, the following eighth-degree polynomial equation.

$$\sigma^8 \tau^2 f^4 + \alpha_7 \sigma^7 + \alpha_6 \sigma^6 + \alpha_5 \sigma^5 + \alpha_4 \sigma^4 + \alpha_3 \sigma^3 + \alpha_2 \sigma^2 + \alpha_1 \sigma + \alpha_0 = 0$$

(18)

The dispersion relation (18) shows the Alfvén mode influenced by finite electron inertia, Hall current, the presence of fine dust particles, rotations, and thermal conductivity. The coefficients α_1, α_2, α_3, α_4, α_5, α_6, and α_7 are very length and constant term α_0 of the dispersion relation (17) is given below.

$$\alpha_0 = k^8\Omega_\vartheta^2 A_3^4 + 4\Omega^2 k^8 A_3^4 + 4\Omega k^8 V^2 A_3^3 + k^8 V^4 A_3^2$$

The third factor of Eq. (16) gives the following dispersion relation on re-substituting the values of d_1 and Ω_T^2:

$$\sigma^4\tau + \sigma^3\{1 + \tau(A + \Omega_\vartheta + \theta_k)\} + \sigma^2\left[(\Omega_\vartheta + \theta_k) + \tau\left\{\frac{\Omega_j^2}{\varepsilon} + \theta_k(A + \Omega_\vartheta)\right\}\right]$$

$$+ \sigma\left(\frac{\Omega_j^2}{\varepsilon} + \Omega_\vartheta\theta_k + \tau\theta_k\frac{\Omega_{j'}^2}{\varepsilon}\right) + \theta_k\frac{\Omega_{j'}^2}{\varepsilon} = 0 \tag{19}$$

The dispersion relation (19) is a gravitating mode influenced by viscosity, thermal conductivity, porosity and presence of fine dust particles. From the dispersion relation (19) we get the condition of instability for all Jeans length $\lambda > \lambda_j' = \left\{\left(\frac{\pi}{G\rho}\right)^{\frac{1}{2}}c'\right\}$. In the absence of thermal conductivity $\theta_k = 0$, the system change from isothermal behavior to adiabatic behavior.

Now we analyze the dynamical stability of the system represented by Eq. (19) by applying the Routh-Hurwitz criterion. If $\Omega_j^2 > 0$ and $\Omega_{j'}^2 > 0$ then all the coefficients of Eq. (19) are positive and the necessary condition for stability is satisfied. To obtain the sufficient condition, the principal diagonal minors of the Hurwitz matrix must be positive and we get.

$\Delta_1 = \{1 + \tau(A + \Omega_\vartheta + \theta_k)\} > 0$,

$$\Delta_2 = \left[(\Omega_\vartheta + \theta_k) + \tau\{A\theta_k + (\Omega_\vartheta + \theta_k)(A + \Omega_\vartheta + \theta_k)\} + \tau^2\theta_k\left(\frac{\Omega_j^2}{\varepsilon} - \frac{\Omega_{j'}^2}{\varepsilon}\right)\right.$$
$$\left. + \tau^2(\Omega_\vartheta + A)\left\{\frac{\Omega_j^2}{\varepsilon} + \theta_k(A + \Omega_\vartheta + \theta_k)\right\}\right] > 0,$$

$$\Delta_3 = \left[\left[\Omega_\vartheta\left\{\frac{\Omega_j^2}{\varepsilon} + \theta_k(\Omega_\vartheta + \theta_k)\right\} + \theta_k\left(\frac{\Omega_j^2}{\varepsilon} - \frac{\Omega_{j'}^2}{\varepsilon}\right) + \tau\left[\Omega_\vartheta\theta_k\{A\theta_k + (\Omega_\vartheta + \theta_k)(A + \Omega_\vartheta + \theta_k)\} + \right.\right.\right.$$
$$\left.\theta_k A\left(\frac{\Omega_j^2}{\varepsilon} - \frac{\Omega_{j'}^2}{\varepsilon}\right)\right] + \tau^2\left[\theta_k\left(\frac{\Omega_j^2}{\varepsilon} - \frac{\Omega_{j'}^2}{\varepsilon}\right) + \left(\frac{\Omega_j^2}{\varepsilon} + \tau\theta_k\frac{\Omega_{j'}^2}{\varepsilon}\right)\left\{(A + \Omega_\vartheta)\frac{\Omega_j^2}{\varepsilon} + \theta_k\left(\frac{\Omega_j^2}{\varepsilon} - \frac{\Omega_{j'}^2}{\varepsilon}\right)\right\} + (A + \right.$$
$$\left.\Omega_\vartheta)(A + \Omega_\vartheta + \theta_k + \Omega_\vartheta\theta_k)\frac{\Omega_j^2}{\varepsilon} + \theta_k\{A\theta_k + (A + \Omega_\vartheta)(A + \Omega_\vartheta + \theta_k)\}\frac{\Omega_j^2}{\varepsilon} + \Omega_\vartheta\theta_k(\Omega_\vartheta + \theta_k)(A + \right.$$
$$\left.\left.\left.\Omega_\vartheta + \theta_k)\right]\right]\right] > 0,$$

$$\Delta_4 = \theta_k \frac{\Omega_{j'}^2}{\varepsilon} \Delta_3 > 0,$$

All the $\Delta'S$ are positive, thereby, satisfying the Routh-Hurwitz criterion, according to which equation Eq. (19) will not admit any positive real root of $\sigma(= i\omega)$ or a complex root whose real part is positive, hence, it gives a stable mode independent the magnetic field. To analyze the role of viscosity, fine dust particles and thermal conductivity on the growth rate of an unstable mode, we choose the arbitrary values of these parameters in the present problem. We write the dispersion relation (19) in nondimensional form in term of self-gravitation as

$$\sigma^{*4}\tau^* + \sigma^{*3}\left[1 + \tau^*\left\{K_s^* + \vartheta^*\left(K^{*2} - \frac{1}{K_1^*}\right) + \lambda^*\right\}\right]$$
$$+ \sigma^{*2}\left[\vartheta^*\left(K^{*2} - \frac{1}{K_1^*}\right) + \lambda^* + \tau^*\left\{(K^{*2} - 1) + \lambda^*\left(K_s^* + \vartheta^*\left(K^{*2} - \frac{1}{K_1^*}\right)\right)\right\}\right]$$
$$+ \sigma^*\left[(K^{*2} - 1) + \lambda^*\vartheta^*\left(K^{*2} - \frac{1}{K_1^*}\right) + \tau^*\vartheta^*\left(K^{*2} - \frac{1}{K_1^*}\right)(K^{*2} - 1)\right]$$
$$+ \vartheta^*\left(K^{*2} - \frac{1}{K_1^*}\right)(K^{*2} - 1)$$
$$= 0 \tag{20}$$

where the various nondimensional parameters are defined as

$$\sigma^* = \frac{\sigma}{\sqrt{4\pi G\rho}}, K_s^* = \frac{K_s N}{\rho\sqrt{4\pi G\rho}}, \quad \lambda^* = \frac{\lambda}{\rho C_p\sqrt{4\pi G\rho}}, \quad K^* = \frac{KC}{\sqrt{4\pi G\rho}}, \quad \vartheta^* = \frac{\vartheta\sqrt{4\pi G\rho}}{C^2}$$
$$K_1^* = \frac{K_1\sqrt{4\pi G\rho}}{C^2}, \quad \tau^* = \tau\sqrt{4\pi G\rho} \tag{21}$$

In the present analysis, the expression for dispersion relation, and growth rate of instability are evaluated for the infinitely conducting medium. We have examined the effects of thermal conductivity, relaxation time and Stokes drag parameters on the growth rate of self-gravitational instability. The results are shown in Figures 16.1–16.3 which have depicted the non-dimensional growth rate versus non-dimensional wavenumber for various arbitrary values of the thermal conductivity (λ^*), relaxation time (τ^*), and Stokes drag parameters (K_s^*).

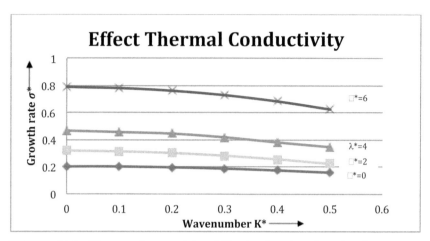

FIGURE 16.1 The growth rate of instability is plotted against the dimensionless wavenumber k^* with variation in the thermal-conductivity $\lambda^* = 0, 2, 4, 6$, with taking the values of k_s^*, ϑ, k_1^*, and τ^* as unity.

Figure 16.1 shows the growth rate of an unstable mode (positive real root of σ^*) against the wavenumber (k^*) with variation in the thermal conductivity (λ^*) parameter. We see that the growth rate of the instability increases with increases in the thermal conductivity (λ^*). The peak value of the growth rate is increased with increasing the thermal conductivity parameters. The present results are different to those of Prajapati et al. where the growth rate is unaffected by the presence of the thermal conductivity. This is the new finding, which shows that presence of fine dust particles and finite Larmor radius effect the peak value of growth rate of instability.

Similarly in Figure 16.2, we have depicted the growth rate (σ^*) of instability against wavenumber (k^*) for different value of relaxation time (τ^*) parameter. It is observed that the relaxation time parameter has a reverse effect on the growth rate, as compared to that of the thermal conductivity parameters. In other words, due to an increase in the relaxation time parameter, the growth rate of the instability decreases. Thus the relaxation time parameter has a damping effect on the growth rate of the system. Also the peak value of the growth rate decreases by increasing (τ^*).

In Figure 16.3 the effect of the Stoke drag parameter on the growth rate is shown by the depicting the curves between (σ^*) and (k^*) for the various values of k_s^*. From the curves, we see that Stokes drag parameter shows the

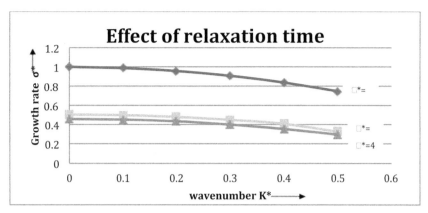

FIGURE 16.2 The growth rate of instability is plotted against the dimensionless wavenumber k^* with variation in the relaxation time $\tau^* = 0, 2, 4$, with taking the values of k_s^*, k_1^*, 9, and λ^* as unity.

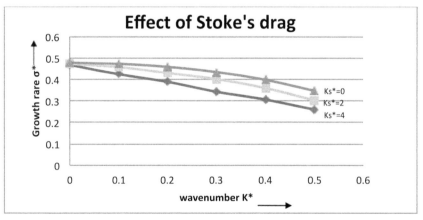

FIGURE 16.3 The growth rate of instability is plotted against the dimensionless wavenumber k^* with variation in the Stokes' drag constant $k_s^* = 0, 2, 4$, with taking the values of λ^*, k_1^*, 9^*, and τ^* as unity.

similar effect as shown by relaxation time parameter (τ^*). Thus, Stokes drag force has a stable influence on the self-gravitational instability of the system.

16.4.1.2 Transverse Mode of Propagation

In this case of a perturbation transverse direction of the magnetic field (i.e., $k_x = k$, $k_z = 0$), the dispersion relation (15) gives

$$d_1^2 \left[\sigma \varepsilon d_1^2 + d_1 \left(\Omega_T^2 + \frac{\sigma \varepsilon k^2 V^2}{A_1} \right) + 4\Omega^2 \varepsilon \sigma \right] = 0 \tag{22}$$

This dispersion relation is the product of two independent factors. These factors show the different mode of propagations incorporating different parameters as discussed below. The first factor of this dispersion relation is stable mode as discussed in the previous case and the second factor of the dispersion relations (22) after simplification written as

$$
\begin{aligned}
&\sigma^6 \tau^2 f + \sigma^5 \tau f \{ 2 + \tau (2A + 2\Omega_\theta + \theta_k) \} \\
&+ \sigma^4 \left[\tau^2 f \left\{ \frac{\Omega_{J'}^2}{\varepsilon} + A\theta_k + (A + \Omega_\theta)(A + \Omega_\theta + \theta_k) \right\} + 2\tau f \{ 1 + (A + 2\Omega_\theta + \theta_k) \} \right] \\
&+ \sigma^3 \left[\tau^2 f \left\{ (A + \Omega_\theta) \frac{\Omega_{J'}^2}{\varepsilon} + \Omega_\theta^2 + \theta_k \left(\frac{\Omega_{J'}^2}{\varepsilon} + A^2 \right) \right\} \right. \\
&\quad + \tau \left\{ \theta_k \left(2f \frac{\Omega_{J'}^2}{\varepsilon} + K^2 V^2 + 4\Omega^2 f \right) + (A + \Omega_\theta) \left(f \frac{\Omega_{J'}^2}{\varepsilon} + K^2 V^2 \right) \right. \\
&\quad \left. \left. + f\Omega_\theta \left(\frac{\Omega_{J'}^2}{\varepsilon} + \Omega_\theta \theta_k + \theta_k \right) \right\} + f(2\Omega_\theta + \theta_k) \right] \\
&+ \sigma^2 \left[\tau^2 (A + \Omega_\theta) \left(f\theta_k \frac{\Omega_{J'}^2}{\varepsilon} \right) \right. \\
&\quad + \tau \left\{ \theta_k \left(2f \frac{\Omega_{J'}^2}{\varepsilon} + K^2 V^2 + 4\Omega^2 f \right) + (A + \Omega_\theta) \left(f \frac{\Omega_{J'}^2}{\varepsilon} + K^2 V^2 + \Omega_\theta + \theta_k \right) \right. \\
&\quad \left. \left. + \theta_k f \left(\frac{\Omega_{J'}^2}{\varepsilon} + \Omega_\theta \theta_k \right) \right\} + \left(f \frac{\Omega_{J'}^2}{\varepsilon} + K^2 V^2 + \Omega_\theta + \theta_k \right) + f\Omega_\theta (\Omega_\theta + 2\theta_k) \right] \\
&+ \sigma \left[\tau \left\{ (A + \Omega_\theta)\theta_k \left(\frac{\Omega_{J'}^2}{\varepsilon} + K^2 V^2 \right) + \Omega_\theta \theta_k f \frac{\Omega_{J'}^2}{\varepsilon} \right\} + \theta_k \left(f \frac{\Omega_{J'}^2}{\varepsilon} + K^2 V^2 + 4\Omega^2 f \right) \right. \\
&\quad \left. + \Omega_\theta \left(f \frac{\Omega_{J'}^2}{\varepsilon} + K^2 V^2 + f\Omega_\theta \theta_k \right) \right] + \Omega_\theta \theta_k \left(f \frac{\Omega_{J'}^2}{\varepsilon} + K^2 V^2 \right) = 0 \tag{23}
\end{aligned}
$$

This is a sixth-degree polynomial equation and shows the combined influence of fine dust particles, viscosity, rotations, magnetic field, the thermal conductivity in the transverse mode of propagation when the axis of rotation is parallel to the direction of magnetic field. The modified Jeans condition of system is this case is given as $\lambda > \lambda_{j''} = \left\{ \pi \left(\frac{c'^2 + \frac{\varepsilon V^2}{f}}{G\rho} \right) \right\}^{\frac{1}{2}}$. We see that the expression of the critical Jeans wavelength is modified due to presence of porosity (ε) of the media, magnetic field and finite electron inertia. But it is unaffected by the Hall current, viscosity and the presence of fine dust particles in the present problem.

16.5   AXIS OF ROTATION PERPENDICULAR TO THE MAGNETIC FIELD

In the case of a rotation axis perpendicular to the magnetic field, we put $\Omega_x = \Omega$ and $\Omega_z = 0$ in the dispersion relation (14) and this gives,

$$
\begin{aligned}
(\sigma \varepsilon d_1 + \Omega_T^2) & \left\{ M_1 M_2 d_1 + 4\Omega^2 M_1 + d_1 \left(\frac{K_z^2 K^2 V^2 A_3}{A_2} \right)^2 \right\} - \frac{K_x^2 V^2 A_1}{A_2} \Omega_T^2 (d_1 M_2 + 4\Omega^2) \\
& - 2\Omega k_z \Omega_T^2 \left\{ \frac{iM_1}{K^2} \left(\frac{iK_x K_z^2 K^2 V^2 A_3}{A_2} - 2i\Omega k_z \right) + \frac{K_x A_1}{A_2} \left(\frac{K_z^2 K^2 V^2 A_3}{A_2} \right) \right\} \\
& + \frac{iK_x}{K^2} d_1 \left(\frac{K_z^2 K^2 V^2 A_3}{A_2} \right) \left(\frac{iK_x K_z^2 K^2 V^2 A_3}{A_2} - 2i\Omega k_z \right) \Omega_T^2 = 0 \qquad (24)
\end{aligned}
$$

This dispersion relation (24) represents the general dispersion relation of the problem when the axis of rotation is perpendicular to the magnetic field. The dispersion relation (24) is further reduced for longitudinal mode and transverse mode of propagation.

16.5.1.1   Longitudinal Mode of Propagation

For this case, we assume that all the perturbations longitudinal to the direction of the magnetic field (i.e., $k_x = 0$, $k_z = k$). Thus, the dispersion relation (24) reduces to the simple form to give

$$
d_1 \left[M_1 (\sigma \varepsilon d_1 + \Omega_T^2) M_1 + 4\Omega^2 \sigma + \frac{K^8 V^4 A_3^2}{A_2^2} (\sigma \varepsilon d_1 + \Omega_T^2) \right] = 0
\qquad (25)
$$

This dispersion relation is the product of two independent factors. These factors show the different mode of propagation incorporating different parameters as discussed below. The first factor of this dispersion

relation is stable mode as discussed in the previous case and the second
factor of the dispersion relations (25) after simplification is given as,

$$\sigma^{12}\tau^3 f^4 + \beta_{11}\sigma^{11} + \beta_{10}\sigma^{10} + \beta_9\sigma^9 + \beta_8\sigma^8 + \beta_7\sigma^7 + \beta_6\sigma^6$$
$$+ \beta_5\sigma^5 + \beta_4\sigma^4 + \beta_3\sigma^3 + \beta_2\sigma^2 + \beta_1\sigma + \beta_0 = 0 \qquad (26)$$

This is a twelfth-degree polynomial equation and shows the combined
influence of fine dust particles, viscosity, rotations, magnetic field, thermal
conductivity, electron inertia, porosity and Hall-effect in the longitudinal
mode of propagation when the axis of rotation is perpendicular to the mag-
netic field. Where the coefficients β_1, β_2, β_3, β_4, β_5, β_6, β_7, β_8, β_9, β_{10}, and β_{11}
are very lengthy and constant terms β_0 of the dispersion (26) is given below

$$\beta_0 = k^8 A_3^4 \Omega_\vartheta^2 \theta_k \frac{\Omega_{j'}^2}{\varepsilon} + k^8 V^4 A_3^2 \theta_k \frac{\Omega_{j'}^2}{\varepsilon}$$

16.5.1.2   Transverse Mode of Propagation

For this case, we assume all the perturbations transverse to the direction
of the magnetic field (i.e., $k_x = k$, $k_z = 0$). Thus, the dispersion relation (22)
reduces in the simple form as given below:

$$d_1\left(d_1^2 + 4\Omega^2\right)\left(M_1\varepsilon\sigma + \Omega_T^2\right) = 0 \qquad (27)$$

We find that in the transverse mode of propagation when the axis of
rotation is perpendicular to the magnetic field the dispersion relation is
modified due to the presence of fine dust particles, finite electron inertia,
Hall parameter, the porosity of the mediums, rotation, thermal conductivity, magnetic field, and viscosity. The dispersion relation (27) has three
independent factors which represent a different mode of propagation. The
first factor of the dispersion relation (27) is stable mode as discussed in the
previous case and the second factor of the dispersion relation (27) after
simplification is given as

$$\sigma^4\tau^2 + 2\sigma^3\tau\{1 + \tau(A + \Omega_\vartheta)\} + \sigma^2\left[\{1 + \tau(A + \Omega_\vartheta)\}^2 + \tau\left(2\Omega_\vartheta + \tau4\Omega^2\right)\right]$$
$$+ \sigma\left[2\Omega_\vartheta\{1 + \tau(A + \Omega_\vartheta)\} + \tau8\Omega^2\right] + \Omega_\vartheta^2 + 4\Omega^2 = 0$$
$$(28)$$

The above dispersion relation is non-gravitating mode influenced by rotation, the presence of fine dust particles and viscosity. This mode is independent of the magnetic field, finite electron inertia Hall current, and porosity of the system.

The third factor of the dispersion relation (27), gives on substituting the values of M_1 and Ω_T^2 the following fourth-degree polynomial equation:

$$
\sigma^4 \tau f + \sigma^3 f\{1 + \tau(A + \theta_k + \Omega_\vartheta)\} + \sigma^2 \left[f(\theta_k + \Omega_\vartheta) + \tau \left(f\frac{\Omega_J^2}{\varepsilon} + K^2 V^2 + f\Omega_\vartheta (A + \theta_k) \right) \right]
$$

$$
+ \sigma \left[\left(\frac{\Omega_J^2}{\varepsilon} + K^2 V^2 + f\Omega_\vartheta \theta_k \right) + \tau \Omega_\vartheta \left(f\frac{\Omega_J^2}{\varepsilon} + K^2 V^2 \right) \right] + \Omega_\vartheta \left(f\frac{\Omega_J^2}{\varepsilon} + K^2 V^2 \right)
$$

$$
= 0
$$

(29)

The dispersion relation (29) is a gravitating mode influenced by a magnetic field, finite electron inertia, porosity, presence of fine dust particles, viscosity, and thermal conductivity. This mode is independent of rotation and Hall current. The condition of instability is same as given by $\lambda > \lambda_{j,}$, we write the dispersion relation (29) in nondimensional form in term of self-gravitation as

$$
\sigma^{*4} \tau^* f^* + \sigma^{*3} f^* \left[1 + \tau^* \left\{ K_s^* + \vartheta^* \left(K^{*2} - \frac{1}{K_1^*} \right) + \lambda^* \right\} \right]
$$

$$
+ \sigma^{*2} \left[f^* \left\{ \vartheta^* \left(K^{*2} - \frac{1}{K_1^*} \right) + \lambda^* \right\} \right.
$$

$$
+ \tau^* \left\{ f^* (K^{*2} - 1)\frac{1}{\varepsilon} + K^{*2} V^{*2} + f^* \lambda^* \left(K_s^* + \vartheta^* \left(K^{*2} - \frac{1}{K_1^*} \right) \right) \right\} \right]
$$

$$
+ \sigma^* \left[(K^{*2} - 1)\frac{1}{\varepsilon} + K^{*2} V^{*2} + f^* \lambda^* \vartheta^* \left(K^{*2} - \frac{1}{K_1^*} \right) \right.
$$

$$
+ \tau^* \lambda^* \left\{ f^* (K^{*2} - 1)\frac{1}{\varepsilon} + K^{*2} V^{*2} \right\} \right] + \lambda^* \left(f^* (K^{*2} - 1)\frac{1}{\varepsilon} + K^{*2} V^{*2} \right) = 0
$$

(30)

where the various nondimensional parameters are defined as

$$
\sigma^* = \frac{\sigma}{\sqrt{4\pi G\rho}}, K_s^* = \frac{K_s N}{\rho\sqrt{4\pi G\rho}}, \quad \lambda^* = \frac{\lambda}{\rho C_p \sqrt{4\pi G\rho}}, \quad K^* = \frac{KC}{\sqrt{4\pi G\rho}}, \quad \vartheta^* = \frac{\vartheta\sqrt{4\pi G\rho}}{C^2},
$$

$$
f^* = f\sqrt{4\pi G\rho}, \quad K_1^* = \frac{K_1\sqrt{4\pi G\rho}}{C^2}, \quad \tau^* = \tau\sqrt{4\pi G\rho}, \quad V^* = \frac{V\sqrt{4\pi G\rho}}{C} \quad (31)
$$

From Figure 16.4, it is noted that the effect of thermal conductivity (λ^*) parameter on the growth rate of instability is same as shown in Figure 16.1 in the case of longitudinal propagation with the axis of rotation parallel to the direction of the magnetic field. Thus the effect of thermal conductivity (λ^*) parameter is found to destabilize the system in both the longitudinal and transverse modes of propagation.

From Figure 16.5 shows the variation of growth rate (σ^*) of instability against wavenumber (k^*) for different value of electron inertia (f^*) parameter. It is obvious that with an increase in electron inertia parameter these are an increase in the growth rate of instability in the system. The peak value of the growth rate gets increased due to an increase in the electron inertia parameter and it is the minimum for the non-electron inertial case, i.e., $f^* = 0$. Thus, the electron inertia has also a destabilizing influence in the transverse mode.

From Figure 16.6, it is clear that the effect of relaxation (τ^*) parameter on the growth rate of instability is same as shown in Figure 16.2 in the case of longitudinal propagation with the axis of rotation parallel to the direction of the magnetic field. Thus the effect of the relaxation time (τ^*) parameter is found to stabilize the system in both the longitudinal and transverse modes of propagation.

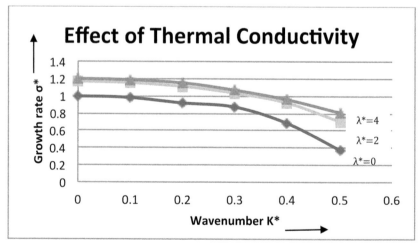

FIGURE 16.4 The growth rate of instability is plotted against the dimensionless wavenumber k^* with variation in the thermal-conductivity $\lambda^* = 0, 2, 4$, with taking the values of k_s^*, 9^*, V^*, f^*, k_1^*, and τ^* as unity.

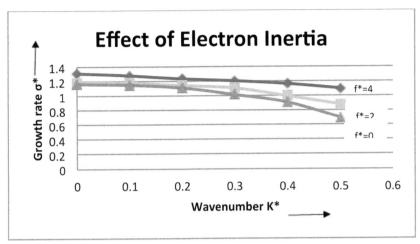

FIGURE 16.5 The growth rate of instability is plotted against the dimensionless wavenumber k^* with variation in the electron inertia $f^* = 0, 2, 4$, with taking the values of k_s^*, ϑ^*, λ^*, V^*, k_1^*, and τ^* as unity.

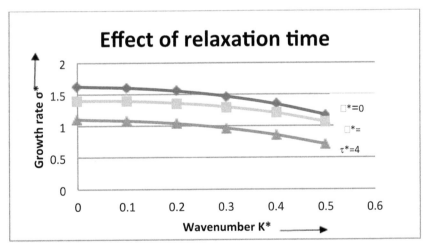

FIGURE 16.6 The growth rate of instability is plotted against the dimensionless wavenumber k^* with variation in the relaxation time $\tau^* = 0, 2, 4$, with taking the values of k_s^*, V^*, f^*, k_1^*, ϑ^*, and λ^* as unity.

16.6 CONCLUSIONS

We have studied the Jeans instability of a self gravitating system under the influence of finite electron inertia, Hall-effect, thermal conductivity, porosity, viscosity, rotation and presence of fine dust particles using normal mode analysis. The analytical expression of the general dispersion relation is obtained with the help of linearized perturbation equations. The findings of the present investigation are the following:

1. The general dispersion relation is modified due to the presence of these parameters. The Jeans condition of instability remains valid but the expression of the critical Jeans wavenumber is modified.

2. The thermal conductivity has a destabilizing influence on the growth rate of the system, in both the longitudinal and transverse modes of propagations. The thermal conductivity simply modifies the Jeans condition and it replaces adiabatic speed of sound by the isothermal one, in all the cases.

3. The viscosity parameter has a stabilizing effect on the system in both the longitudinal and transverse modes of propagation.

4. In the case of longitudinal propagation with the axis of rotation parallel to the direction of the magnetic field, the gravitating mode is influenced by viscosity, thermal conductivity, porosity and presence of fine dust particles but not affected by finite electron inertia, rotation, Hall current and magnetic field. The parameters of the magnetic field, finite electron inertia, fine dust particles, and porosity do not change the Jeans condition in this case. The dynamical stability of the system, in this case, is analyzed by applying the Routh-Hurwitz criterion. In this case, the curves depict the effects of thermal conductivity (λ^*) parameter, relaxation time (τ^*) parameter, and Stokes drag parameter (K_s^*) on the growth rate of instability.

5. In the transverse mode of propagation with the axis of rotation parallel to the direction of magnetic field, the self gravitating Alfvén mode is influenced by finite electron inertia, fine dust particles, viscosity, thermal conductivity and porosity of the medium. The Jeans condition of instability is modified by finite electron inertia, thermal conductivity, magnetic field and porosity of the medium but not affected by viscosity, rotation and the presence of fine dust particles. In this case, curves depict the effects of thermal conductivity (λ^*) parameter, relaxation time (τ^*) parameter, and finite electron inertia (f^*) parameters on the growth rate of instability.

6. From the curve, it is found that the thermal conductivity relaxation time shows mutually reverse effects on the growth rate of the instability. In other words, the thermal conductivity has a destabilizing influence, while the relaxation time has a stabilizing role on the growth rate of the system. The finite electron inertia has a destabilizing influence on the growth rate of the instability. Also, it decreasing the peak value of the growth rate means that the system becomes more and more unstable for higher values of the finite electron inertia.

KEYWORDS

- finite electron inertia
- Hall-effect
- magnetic field
- permeability
- porosity
- rotation
- suspended particles
- thermal-conductivity

REFERENCES

1. Bashir, M. F., Jamil, M., Murtaza, G., Solimullah, M., & Shah, H. A. (2012). Stability analysis of self-gravitational electrostatic drift waves for a streaming nonuniform quantum dusty magneto plasma. *Physics of Plasma, 19*, 43701–43707.
2. Chakrabati, N. (2011). Secondary instability of Jeans mode in a gravitating fluid with uniform rotation. *Physics of Plasma, 18*, 1–5.
3. Chandrashekhar, S. (1961). *Hydrodynamic and Hydro-Magnetic Stability*. Clarendon Press, Oxford, pp. 588–598.
4. Chhajlani, R. K., & Sanghvi, R. K. (1985). Finite armor radius and Hall current effects on magneto-gravitational instability of a plasma in the presence of suspended particles. *Astrophysics and Space Science, 124*, 33–42.
5. Chhajlani, R. K., & Vyas, M. K. (1988). Effect of thermal conductivity on the gravitational instability of a magnetized rotating plasma through a porous medium in the presence of suspended particles. *Astrophysics and Space Science, 145*, 223–240.

6. De Colle, F., & Raga, A. C. (2004). Effect of the magnetic field on the H alpha emission from jets. *Astrophys Space Sci., 293*(1–2), 173–180.

7. Falceta-Goncalves, D., Jatenco-Pereira, V., & De Juli, M. (2003). Dusty molecular cloud collapse in the presence of Alfvén waves. *Astrophysics, 597*, 970–974.

8. Hennebelle, P., &Perault, M.(2000). Dynamical condensation in a magnetized and thermally bistable flow. *Astron. Astrophys., 359*, 1124–1138.

9. Herrengger, F. J. (1972). Effects of collisions and Gyroviscosity on gravitational instability in a two-component plasma. *J. Plasma Physics, 8*, 393–400.

10. Jeans, J. H. (1902). The stability of spherical nebula. *Phil. Trans. Roy. Soc. London, 199*, 1–53.

11. Lima, J. A. S., Silva, R., & Santos, J. (2002). Jeans gravitational instability and nonextensive kinetic theory. *Astronomy &Astrophysics, 396*, 309–313.

12. Merlino, R. L., Heinrich, J. R., Kim, S. H., & Meyer, J. K. (2012). Dusty plasmas: experiments on nonlinear dust acoustic waves, shocks and structures. *Plasma Phys. Control, Fusion, 54*(2), 1–10.

13. Prajapati, R. P. (2011). Effect of polarization force on the Jeans instability of self gravitating dusty plasma. *Physics Letters A, 375*, 2624–2628.

14. Prajapati, R. P. (2013). Effect of magnetic field and radiative condensation on the Jeans instability of dusty plasma with polarization force. *Physics Latter A, 377*, 291–296.

15. Prajapati, R. P., Pensia, R. K., & Kaothekar, S. (2010). Self gravitating instability of rotating viscous Hall plasma with arbitrary radiative heat loss functions and electron inertia. *Astrophys Space Science, 327*, 139–154.

16. Rosenberg, M. (2009). On dust acoustic instability in a negative ion plasma. *Phys. Scr., 79*, 1–6.

17. Rosenberg, M. (2010). Cross-field dust acoustic instability in a dusty negative ion plasma. *Phys. Sci., 81*, 1–8.

18. Scanlon, J. W., &Segal, L. A. (1973). Some effects of suspended particles on the onset of Benard convection. *Phys. Fluids, 16*, 1573–1578.

19. Sharma, P., & Chhajlani, R. K. (2014). The effect of spin induced magnetization on Jeans instability of viscous and resistive quantum plasma. *Physics of Plasma, 21*, 1–10.

20. Sheikh, S., Khan, A., & Bhatia, P. K. (2007). Thermally conducting partially ionized plasma in a variable magnetic field. *Contrib. Plasma Phys., 47*(3), 147–156.

21. Siddheshwar, P. G., & Abraham, A. (2003). Effect of time-periodic boundary temperatures/body force on Rayleigh-Benard convection in a ferromagnetic fluid. *Acta Mech., 161*, 131–150.

22. Sunil, Sharma, D., & Sharma, R. C. (2005). Effect of dust particles on thermal convection in ferromagnetic fluid saturating a porous medium. *J. Magn. Magn. Metar, 288*, 83–95.

23. Vaidyanathan, G., Sekar, R., Hemalatha, R., Vasantha Kumari, R., & Sendhilnathan, S. (2005). Soret-driven ferrothermohaline convection. *J. Magn. Magn. Mater, 288*, 460–469.

24. Uberoi, C. (2009). Electron inertia effects on the transverse gravitational instability. *J. Plasma Fusion Res., 8*, 823–825.

25. Zebib, A. (1996). Thermal convection in a magnetic fluid. *J. Fluid Mech., 321*, 121–136.

PART III

NUCLEAR FUSION

CHAPTER 17

STRUCTURAL FABRICATION: STUDY OF INFRASTRUCTURE FACILITIES REQUIRED TO CONVERT CONCEPT TO REALITY

GAUTAM R. VADOLIA

Institute for Plasma Research, Near Indira Bridge, Bhat, Gandhinagar – 382428, Gujarat, India, E-mail: gautamv@ipr.res.in

CONTENTS

ABSTRACT

For the developing country like India, nuclear energy is important to ensure long-term energy security. Utilization of domestic reserve of thorium is important; and hence, research in fusion reactor is also important. Multilayered fusion reactor components require lot of research prior to actual component fabrication, which is required to be installed into the fusion reactor. This chapter deals with the study of infrastructure facilities required for such fabrications. Some of the facilities like HIP and Laser welding are discussed in detail.

17.1   INTRODUCTION

India is committed to making nuclear power—the backbone of its long-term power generation scheme. The key to nuclear power (both fission and fusion) is their high energy—flux density, which enables the society to use this source of power to perform all known industrial and manufacturing activities using very little fuel. Moreover, both nuclear fission and fusion are clean sources of energy and do not encumber the country's infrastructure unduly [10].

A paper on fusion research programme in India gives detailed explanation on importance of fusion research and justification of investment in the same. As an example, a 500 MW-thermal fusion reactor can be used to produce 5500 kg of U^{233} and 20 kg of T (tritium) per year from a thorium–lithium blanket, which in turn can support two 800 MWe LWRs (light water reactors) [7].

India joined the International Thermonuclear Experimental Reactor (ITER) consortium on 5-Dec-2005 with an aim to accelerate the gap-closure between indigenous technology and that is required to build a demonstration fusion power reactor (DEMO) [7].

Many technologies like functional material fabrication and characterization, liquid metal technologies, remote handling, etc., may be required to build a DEMO. Regarding structural fabrication of fusion reactor like DEMO, there are two distinct research requirements.

1.   Research in structural materials (including research on material production processes, fatigue, erosion, and nuclear transmutation); and

2. Research in manufacturing technology (including joining processes).

For research and development in fusion structural materials, laboratory scale facilities like vacuum induction melting (VIM), spark plasma sintering (SPS), and vacuum plasma spraying (VPS) may be required [2], but these are not discussed in this chapter.

Several components in ITER require advanced fabrication techniques and technology like large forging, cast with hot isostatic press (HIP), solid HIP, powder HIP, etc.. For ITER multilayer plasma facing walls [1], in addition to the conventional fabrication methods and joining techniques, electron beam (EB), TIG, laser, friction welding, brazing, etc., are used in the nuclear industry for plates, bars, tubes, etc.

This chapter is limited to manufacturing technology for structural materials relevant to nuclear applications. An effort is made to survey novel joining/manufacturing facilities required for such materials, facilities available within India and outside (with focus on few areas like HIP joining and laser welding) for fabrication of few components like divertor and blanket.

Nuclear Advanced Manufacturing Research Centre (UK, officially launched in 2012), which was established with an aim to give manufacturers a competitive advantage by reducing cost, risk, and lead times, is the main focus of this chapter.

17.2 NEED OF A MANUFACTURING RESEARCH CENTRE

Even for fission reactors, the process of generating the engineering design is often iterative. Specifications for the ordered equipment are generated by the designers and converted into engineering drawings that are handed over to the manufacturers, who frequently suggest changes to the design to improve its manufacturability.

In the absence of defined structural design criteria for fusion reactor components, the process of generating the design is likely to be more iterative. Component validation in laboratory environment (and subsequently in relevant environment) and prototype demonstration in relevant environment are much essential. Industries are generally not tuned to this type of research and development. Industries are generally tuned to follow codes like American Society of Mechanical Engineers (ASME—Section III for nuclear components). Such facility shall be helpful in establishing structural design criteria for fusion reactor components.

We generally specify pre-qualifying clause of experience with manufacturing equipment, but for fusion reactor components having typical set of requirements, it is unlikely get an experienced vendor (e.g., delamination of multilayer structure).

In India, the Bangalore based, Central Manufacturing Technology Institute (CMTI) is a research and development organization focusing on providing technology solutions to the manufacturing sector. CMTI (over the last five decades) has developed special purpose machines, inspection systems, and test rigs for qualification of testing products, tooling, and complex machined parts (for public and private sectors). Similarly, the Central Mechanical Engineering Research Institute (CMERI), at Durgapur, West Bengal, is for research in robotics and mechatronics, advanced manufacturing technology, and rapid prototyping and tooling.

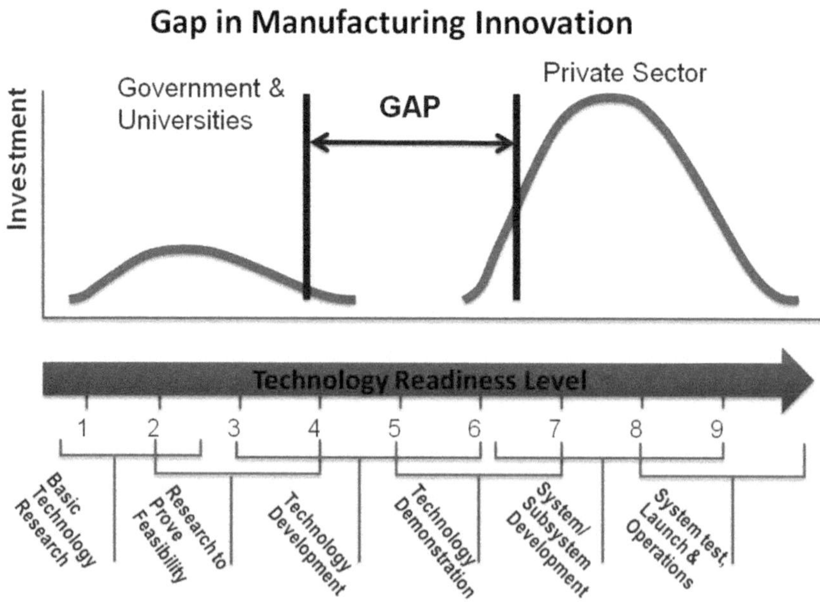

FIGURE 17.1   Gap in manufacturing innovation [15].

Commercialising Innovation

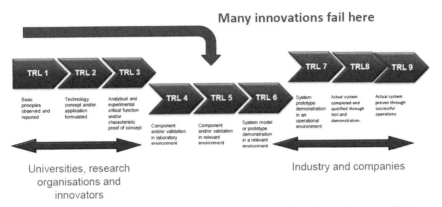

FIGURE 17.2 Technology readiness levels and places where innovations are likely to fail [16].

In India, though there are leading private organizations like Larsen & Turbo (L&T), Walchandnagar Industries, and Avasarala Technologies, there is no specific manufacturing research center for nuclear components. For some of the SST1 (Indian Steady State Superconducting Tokamak) components fabrication, Institute for Plasma Research (IPR) had to take help from Bharat Heavy Electricals (BHEL) Limited, which is an institution dedicated for mass production of electrical items rather than research and fabrication of nuclear components. However, the multilayer fusion reactor components, like divertor and shield blankets, which require different fabrication methodology, are unlikely to be taken up by BHEL for fabrication.

Manufacturing research center can help in this regard and bridge the gap between government and private sector. Refs. [15] and [16] gives an idea where innovations are likely to fail (Figures 17.1 and 17.2).

17.3   RESEARCH INVESTMENT SCENARIO

Investments made in manufacturing sector by countries like US, UK, etc. are mentioned in the following subsections.

17.3.1   UK

UK invested £200 million in 2010 in a network of Catapult Centers designed to close the gap between universities and industries and creating infrastructures, which would rapidly translate research into productivity. In 2013/14, the UK high value manufacturing CATAPULTs generated a private sector income of £65 million.

A report published in November 2014 by the entrepreneur Hermann Hauser for the Department of Business, Industry and Skills, marks the success of this investment in stimulating business development and economic growth, and calls for greater investment in UK research and development, leading to 30 CATAPULT centers across the UK by 2030 (see, www.gov.uk/government/publications/catapult-centres-hauserreview-recommendations) (Table 17.1).

One of these centers (see Table 17.1), the Nuclear Advanced Manufacturing Research Centre (Nuclear AMRC), aims to enhance the capabilities and competitiveness of the UK civil nuclear manufacturing industries and help British manufacturing companies compete for nuclear contracts worldwide. (https://hvm.catapult.org.uk/nuclear-amrc)

The core research areas include:

- machine tool optimization and process development;
- robotic machining;
- large-scale welding and cladding using robotics and adaptive control;
- production-scale demonstrators for innovative technologies and processes;
- non-destructive evaluation;
- large-scale metrology; and
- virtual simulation and design for manufacturing and assembly.

17.3.2   AUSTRALIA

The Australian Advanced Manufacturing Research Centre (AusAMRC) is an initiative led by Swinburne University of Technology and Boeing Company to research and develop new aerospace and other industry sector components, materials, and manufacturing technologies.

TABLE 17.1 CATAPULT Centers

High Value Manufacturing Catapult (HVM) has 7 physical centres across the UK (source: Presentation by Dick Elsy CEO January 2013)

Centre	Founding University
Advanced Forming Research Centre (AFRC)	University of Strathclyde
Advanced Manufacturing Research Centre (AMRC)	Sheffield University, University of Manchester
Centre for Process Innovation (CPI)	Wilton and Sedgefield
Manufacturing Technology Centre (MTC)	Loughborough University, Nottingham university, Birmingham University
Nuclear Advanced Manufacturing Research Centre (Nuclear AMRC)	Sheffield University
National Composite Centre (NCC)	University of Bristol
Warwick Manufacturing Group (WMG)	Warwick University

Industrial Partners: Many industrial partners like Areva, Tata steel, Rolls Royce, Westinghouse etc.

Return on Investment: Every £1 of Government funding generates £3.90 (According to a presentation on "High Value Manufacturing Catapult," by Paul John (Business Director), HVM Catapult receives strong support from industry – 40% income from Industry) [19].

AusAMRC works closely with the internationally acclaimed Advanced Manufacturing Research Centre (AMRC) in the United Kingdom.

17.3.3   US

According to June 2014 Wall Street Journal report, President Barack Obama has requested $1 billion from Congress to help fund a nationwide network of research institutes—the National Network for Manufacturing Innovation—that would work with companies and universities to develop manufacturing technology. The plan envisions at least 15 such institutes are funded by government and private-sector. So far, four such institutes have been announced or setup, including one in Youngstown, Ohio, which focuses on 3-D printing.

17.4   DISCUSSION ON SOME KEY FACILITIES

Efforts by various countries in fabrication of test blanket module first wall fabrication were surveyed. Use of hot isostatic pressing–diffusion welding (HIP–DW) is reported. The same are listed in Refs. [3–6, 9]. In the Ref. [9], it is mentioned that "a full-scale first-wall mockup of a breeding blanket was successfully developed using HIP fabrication with methods simplified to facilitate large-scale manufacturing procedures."

17.5   HIP UNIT (HOT ISOSTATIC PRESS) AT INTERNATIONAL LEVEL

HIP is a process to densify powders or cast and sintered parts in a furnace at high pressure (100–200 MPa) at temperatures from 900 to 1250°C. A few parameters for dissimilar metal joining for ITER Blanket first wall are mentioned below [14].

Three full-scale FW panel prototypes with HIP Be (beryllium) tiles have been completed as mentioned in the Table 17.2.

The HIP schematic and photograph of typical HIP system is shown in Table 17.3.

TABLE 17.2 The Process Parameter for Hipping of CuCrZr

Metals	Parameter	Remark
316L SS/CuCrZr joining	1040 C, 140 MPa, 2 hr.	Post-HIP solution annealing HT with fast cooling
CuCrZr/Beryllium joining	580 C, 140 MPa, 2 hr.	

TABLE 17.3 The HIP Schematic and Photograph of HIP System

HIP Schematic	HIP Photograph
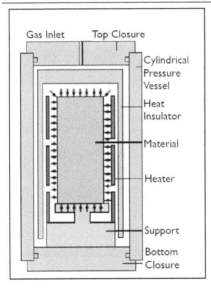	
	This 1.63 meter diameter HIP unit in Camas (Washington) is an example of the type of large, sophisticated equipment needed to process parts for the ITER project in a cost effective manner.
Image from internet	Photograph from www.bodycote.co.jp/eg/pdf/20080722171456.pdf

HIP Unit (Hot Isostatic Press) at International level (few examples of large work zone size facilities)

USA: 1.63 meter diameter HIP unit in Camas (Washington) to process parts for the ITER project

Sweden: 1.3 m diameter HIP facility at Bodycote, Surahammar

Japan: 2.05 m dia × 4.2 m length installed in 2010 [12]

UK: 0.45 m dia × 1.3 m length at Nuclear Advance Manufacturing Research Center (Nuclear AMRC)

India: 0.27 m dia. × 0.99 m height

TABLE 17.3 (Continued)

Why India should have bigger facility?

1. Prototype manufacturing is essential for components like blanket. For example, ITER blankets prototype manufacturing started 20 years ago and still not finalized. HIP fabrication route selected by the European Domestic Agency (EUDA) for the manufacture of FW panels [14].

2. Sizes available in India may not be suitable for large components such as divertor and shield blanket for next DEMO reactor to be built.

3. For ODS steels, TIG, EB and Laser welding are not suitable: ODS particle coarsens above ~1250°C [13].

4. Use of hipping reported in TBM first wall fabrication (India, Korea, China, Japan, EU). 1.5 m × 0.6 m × 0.16 m width first wall mockup successfully fabricated by Japan by Hipping using plates and rectangular tubes of F82H. However, author has not seen any claim for full scale actual job fabrication by Hipping.

Research areas:

1. Analysis of material properties of hipped parts compared with wrought or cast equivalent.

2. Generation of data to support code cases for adopting of powder metallurgy manufacturing within the civil nuclear sector.

3. Research into the consolidation of novel structures, such as dissimilar metal joints and development of procedures.

17.6 LASER WELDING AT INTERNATIONAL AND NATIONAL LEVEL

As per 2010 estimate, approx. 650 multi-kilowatt fiber lasers are used in production applications.

**USA:	INDIA
• 20 kW Fiber Laser – US Army (Test Facility)	Two decades of experience at RRCAT
• 15 kW IPG Fiber Laser – Edison Welding Institute (EWI), Columbus	Approx. 20 laser systems at DAE units, India cutting of stainless steel sheets up to 14 mm and welding up to depth of 2 mm were established.
• 10 kW CW YLR – US Naval Centre, Indiana	
• 6 kW Fiber Laser – LOS ALAMOS National Lab	**250 W** average power, 5 kW peak power, Nd-YAG
• 5 kW YLR – Gas Technology Institute	**500 W** average power 10 kW peak power, Nd-YAG

**GERMANY:

• 20 kW YLR – German Federal Institute for Material research and testing, BAM

• 10 kW Fiber Laser – Vietz Pipeline Laser System, GmBH

• 10 kW Fiber Laser – IMG, Rostock GmBH

• 4 kW Fiber Laser – Fraunhofer Institute for Laser Tech. ILT

**CANADA:

• 10 kW Fiber Laser – CSPTQ

• 5 kW Fiber Laser integrated with Arc welding for Hybrid laser arc at National Research council.

• 3 kW Laser system with metal powder and wire feeding for additive manufacturing.

**UK:

The University of Manchester Laser Processing Research Centre (LPRC). The facilities are coupled to 6 axis Kuka Robot

• 16 kW IPG fiber laser welding/cladding/ cutting system (Qty 01 each) with Precitec head suitable for purpose

• 8 kW Fiber Laser – Cranfield University

• 2 kW IPG Fiber Laser – Nottingham Uni, TWI

** JAPAN

• 10 kW, 6 kW and 4 kW fiber Laser at JWRI/Applied Laser technology Institute (JAEA)

• 20 kW at Sakurai Kogyo Co. Ltd

** Sweden

30 kW at IPG, 15 kW at Luleå

YLR is Continuous Wave (CW) Fiber Laser 1070 nm

880 W CW Nd-YAG Laser (4.4% electrical to laser conversion) for laser rapid manufacturing for fabrication of component of dissimilar metals and directly from CAD model (commissioned at MSD BARC)

2 kW cw Nd-YAG Laser at WRI Trichy

6 kW (Fiber coupled diode laser with 200 to 6 kW at ARCI)

2.2 kW (approx.) CO_2 Laser at Magod Laser

Tenders floated

2 kW Fiber Laser System DRDO (2012)

5 kW Fiber Laser System DRDO May 2014

10 kW Fiber/Disc/Diode Laser by IGCAR 2013

550 W Nd-YAG Laser Welding System IGCAR 2014

1 kW by Dayalbagh Educational Institute, Agra 2014

100 Watt CW Fiber laser system. *Qty 04 nos.* July 2013 by Laser Science and Technology Centre, Delhi

For useful laser based techniques in the field of nuclear engineering, high power (approx. 10 kW or higher power) Solid state laser of type fiber/disc, India may have to depend upon laser power source manufacturers like IPG Photonics, TRUMPF Inc., etc.

Apart from the joining processes like hipping, laser welding, and EB welding, the literature survey reveals that additive manufacturing (AM) is utilized for aerospace components and robotics components. However, limited use is reported in nuclear components. The same is not discussed over here, but use of this technology cannot be completely ruled out [18].

AM technologies refer to a group of technologies that build physical objects from computer aided design (CAD) data. Contrary to conventional subtractive manufacturing technologies, e.g., cutting, lathing, turning, milling, and machining, a part is created by the consecutive addition of liquids, sheet or powdered materials in ultra-thin layers. It is like building a loaf of bread from individual slices.

According to November 2014 report GE has announced it would spend $32 million to build a new research and education center focused on additive technologies in Pennsylvania [11].

17.7   REFERENCE FACILITIES AT NUCLEAR ADVANCED MANUFACTURING RESEARCH CENTRE AT UK (WWW:NAMRC.CO.UK)

Facility/Make	Photograph	Specification
VTL TurnMill Dorries Contumat		Max swing: 5 m, Max height: 3.145 m Max part weight 100 tonne (50 t crane limit) Table power: 160 kW spindle power: 37 kW
Horizontal Boring Soraluce FX 12000		X: 12 m, y (vertical): 5.3 m Z (cross): 1.9 m Spindle motor power 70 kW
Horizontal Boring/Mill-Turn Starrag HEC 1800 PTM		Vertical cylinder: 3.3 m dia. × 2.5 m Horizontal cylinder: 2.5 m dia. × 2.35 m Turning power: 55 kW @ 40% duty

Facility/Make	Photograph	Specification
5 axis Mill Turn Mori Seiki NT6600		Centre distance: 6.51 m Swing over bed: 1.07 m Power 30 kW
Deep Hole Drilling TBT ML 700		Drilling 5–110 mm diameter up to eight meter depth. (5–18 mm via gun drilling) Max work piece size: 0.4 m dia. × 8 m Max work piece weight: 8000 kg
Turning Irregular Parts Mazak Orbitec 20		Unique orbital headstock design capable of generating turned features of up to 508 mm diameter, while keeping the workpiece stationary. Horizontal travel: 0.6 m Vertical travel 0.6 m, Spindle power: 30 kW
Robotic Machining Fanuc 200i hexapod robot		Positional repeatability: ± 100 micron Lateral feedrate: 90 m/min Power: 8 kW
Submerged Arc Welding Miller 3 × DC1250 and 1 × AC/DC 1250		Column and boom: 5 × 5 m, Max work piece weight: 15 tonne Manipulator table Diameter: 3 m

Facility/Make	Photograph	Specification
GTAW Cell		GTAW Cell Column & boom 6 × 4 m Tilt & turn table: 7.5 tonne max weight Diameter: 1.5 m
Narrow groove welding Arc machines Inc. Model 52 narrow groove welding head with AMI model 415 WDR Power source	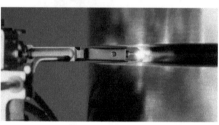	Torch AVC stroke: 101.6 mm; Travel speed: 5–500 mm/min; Tungsten diameter: 2.4, 3.2 or 4 mm; Filler wire diameter: 0.5–1.1 mm
Tube sheet welding Arc machines Inc., Model 96 and Model 6 weld heads with Model 227 power source.	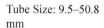	Tube Size: 9.5–50.8 mm Weld current 3-225 A DC @ 240 VAC input
Lincoln five wire submerged arc welding system		Column & boom: 6 × 6 m Max work piece weight: 70 tonne Power source: 8 x Lincoln Powerwave AC/DC 1000 SD
Electron Beam Welding Pro-Beam K25 Pro-Beam K 2000		Chamber Size: 1.6 × 1.2 × 1.2 m (K25 Model) Beam power: 40 kW X-Y table movement (max): 0.7 × 0.3 m Chamber size: 8.7 × 5.2 × 4.6 m (K2000 model)

Facility/Make	Photograph	Specification
		Beam power: 30 kW
		Max Work piece Load: 100 tonne
Virtalis ActiveCube Wall size: 3.2 × 2.45 m Floor size: 3.2 × 3.2 m Virtalis PowerWall Screen size: 4.5 × 2.8 m		The **ActiveCube** features 3D projection on three walls and floor, providing a fully immersive virtual environment for up to five people. **Power wall:** Large single wall with 3D back projection for presenting virtual environments to groups of up to 25 people.

17.8 CONCLUSION

Considering the facilities available at the institute and national level, from the survey conducted in available literature, and studying the strategies adopted by countries like UK for research in nuclear manufacturing, author draws following conclusion.

1. At institute level, many characterization facilities like scanning electron microscope (SEM), X-ray diffraction (XRD), high temperature X-ray diffraction (HTXRD), ellipsometry (instrument for contact free determination of thickness and optical constants of films of all kinds), atomic force microscopy (AFM), FTIR (a type of infrared spectroscopy), universal testing machine, creep testing, hardness (micro and macro), creep and thermo-mechanical fatigue (TMF), thermo-mechanical simulator (Gleeble 3800), etc., are available at various divisions of the Institute for Plasma Research.

   Institute has great potential in the development of computerized manufacturing environment and technology by expanding existing facilities by adding a few computerized controlled machines to train the personnel.

2. A nationwide thought process has to be initiated in manufacturing technology research. It is preferable to have a group of machines (in consultation with leading industries, leading government organizations and universities) dedicated to particular fusion reactor component. India may adopt the same methodology to attach each key facility to a university for better management and generation of employable manpower.

3. Balancing the research funding between material science and manufacturing technology may be essential.

With a key facility like a forge shop (Joint venture of Nuclear Power corporation of India and Larsen & Turbo (L&T), already available at L&T, Hazira, Surat (Gujarat, India), and with presence of Bharat Forge in this field, India can concentrate on reference facilities at Nuclear Advanced Manufacturing Research Centre at UK.

Although huge investment is required in creating a Nuclear Advanced Manufacturing Research Centre (e.g., £25 million (approx. $38 million) [17] were required for such a center in UK), the investment looks small compared to the cost of fusion reactor components. For example, the cost of a single component of ITER called cryostat – a 3,800-ton pressure chamber of the size of a 10-story building to be built by Larsen & Turbo is estimated worth more than $160 million [10].

Investments in manufacturing technology research will not only attract the existing nuclear industries but also attract new suppliers reluctant (due to complexity of manufacturing) to break into this sector.

ACKNOWLEDGMENT

Author is thankful to Dr. K. P. Singh (IPR) for suggestions during preparation of this paper.

KEYWORDS

- hot isostatic press
- electron beam welding
- tungsten inert gas

- oxide dispersion strengthened
- demonstration fusion power reactor
- test blanket module

REFERENCES

1. Tavassoli, A-A. F. (1998). Overview of advanced techniques for fabrication and testing of ITER multilayer plasma facing walls. *Fusion Engineering and Design 39–40*, 189–200.
2. Munoz, A., Monge, M. A., Pareja, R., Hernández, M. T., Jimenez-Rey, D., Román, R., González, M., García-Cortés, I., Perlado, M., & Ibarra, A. (2011). The materials production and processing facility at the Spanish National Centre for fusion technologies (TechnoFusión). *Fusion Engineering and Design 86*, 2538–2540.
3. Bo Huang, Yutao Zhai, Junyu Zhang, Chunjing Li, Qingsheng Wu, & Qunying Huang (2014). First wall fabrication of 1/3 scale china dual functional lithium lead blanket, *Fusion Engineering and Design 89*, 1181–1185.
4. Dong Won Lee, Suk Kwon Kim, Young Dug Bae, & Bong Geun Hong (2008). Fabrication and preparation of a high heat flux test with a mock-up for the first wall of the KO HCML TBM. Transactions of the Korean Nuclear Society Spring Meeting Gyeongju, Korea, May 29–30, 2008.
5. Jae Sung Yoon, Suk Kwon Kim, & Dong Won Lee (2011). Fabrication of three kinds of small mock-ups for an ITER TBM First Wall Transactions of the Korean Nuclear Society Spring Meeting Taebaek, Korea, May 26–27, 2011.
6. Qunying Huang, Qingsheng Wu Shaojun Liu, Chunjing Li, Bo Huang, Peng, L., Shuhui Zheng, Qian Han, Yican Wu, & FDS Team (2011). Latest progress on R&D of ITER DFLL-TBM in China, *Fusion Engineering and Design 86*, 2611–2615.
7. Shishir Despande & Predhiman Kaw (2013). Fusion research programme in India. Sadhana Vol. 38, Part 5, October 2013, pp. 839–848, available online at www.ias.ac.in/sadhana/Pdf2013Oct/5.pdf
8. Jayakumar, T., & Rajendra Kumar, E. (2014). Current status of technology development for fabrication of Indian Test Blanket Module (TBM) of ITER, *Fusion Engineering and Design 89*, 1562–1567.
9. Takanori Hirose Mikio Enoeda, Hiroyuki Ogiwara, Hiroyasu Tanigawa, & Masato Akiba (2008). Structural material properties and dimensional stability of components in first wall components of a breeding blanket module, *Fusion Engineering and Design 83*, 1176–1180.
10. Online reference, http://www.larouchepub.com/other/2014/4123india_frontier_fusion.html (accessed on 24 February 2015).
11. Online reference, http://www.gereports.com/post/102897646835/new-research-center-will-take-3d-printing-to-the. http://www.gefoundation.com/new-research-center-will-take-3d-printing-to-the-next-level/ (accessed on 11 April 2017).

12. Online reference, http://www.prweb.com/releases/2013/1/prweb10322359.htm (accessed on 24 February 2015).

13. Online reference, Presentation on ODS steel and their behavior under neutron irradiation by A. Moslang (accessed on 24 Feb 2015). Available online at https://static.iter.org/.../wp.../MIIFED-2013%20IrradiatedODS-SteelsMoslang5.pdf

14. Online reference, Presentation by Patrick Lorenzetto on EU procurement of in vessel components for ITER at ITER business Forum 2013, Toulon, 21–22 March 2013) (accessed on 24 Feb 2015). Available online at https://industryportal.f4e.europa.eu/IP_EXT_REFERENCE_DOCUMENTS/F4E%20-%20In-Vessel%20Components%20-%20Lorenzetto%20IBF-13.pdf

15. Report to the President Capturing Aa Domestic Competitive Advantage In Advanced Manufacturing: Report of the Advanced Manufacturing Partnership Steering Committee, Annex 2: Shared Infrastructure and Facilities Workstream Report. https://energy.gov/sites/prod/files/2013/11/f4/pcast_annex2_july2012.pdf

16. Online reference, http://epc.ac.uk/wp-content/uploads/2013/05/Dick-Elsy.pdf (accessed on 24 Feb 2015).

17. Online reference, http://www.amrc.co.uk/news/nuclear-amrc-launch/ (accessed on 24 February 2015).

18. Online reference, http://www.ornl.gov/science-discovery/advanced-materials/research-areas/materials-synthesis-from-atoms-to-systems/additive-manufacturing (accessed in August 2015).

19. Online reference, https://ec.europa.eu/growth/tools-databases/regional-innovation-monitor/sites/default/files/report/High%20Value%20Manufacturing%20Catapult_1.pdf (accessed on 7 August 2015).

CHAPTER 18

DAC-CONTROLLED VOLTAGE VARIABLE RF ATTENUATOR FOR GENERATING RF PULSES OF DIFFERENT SHAPES AND AMPLITUDES FOR ICRH SYSTEM

MANOJ SINGH, H. M. JADAV, RAMESH JOSHI, SUNIL KUMAR, Y. S. S. SRINIVAS, S. V. KULKARNI, and RF-ICRH GROUP

Institute for Plasma Research, Bhat, Gandhinagar – 382428, India, E-mail: parihar@ipr.res.in

CONTENTS

ABSTRACT

ICRH-DAC (Data Acquisition and Control) controls and monitors the RF power (1.5 MW, 20–40 MHz) to dummy load/Aditya/SST-1 tokamak [1]. ICRH system has to be used for interface and antenna conditioning, heating, pre-ionization, and current drive experiments in the tokamak machines. It is necessary that DAC system should provide flexibility to generate variable duty cycles and variable amplitude pulses upto maximum duration. Generation of this RF-shot pulse by DAC is very critical task as system operational performance monitoring and post-acquisition shot analysis is referenced from this pulse only.

During the high power testing of 1.5 MW tube, it was observed that HVPS overshoots as current drawn from power supply increased beyond 22 A, towards the end of pulse as this is unregulated supply and this limitations preventing further higher power commissioning work. To overcome this overshoot problem DAC hardware pulse generation facility is modified and new circuit is integrated with the DAC hardware and software. This report discusses functionality of in-house developed hardware and application software associated with pulse generation, integration with the system, and test results.

18.1   INTRODUCTION

ICRH-DAC of Aditya and SST Tokamak [1] system consists of VME-based real time system having Analog I/O, Digital I/O, timer modules. VME-based DAC system and its interfacing with subsystem is shown in the Figure 18.1.

All the modules are connected through VME processor by backplane connector. A 100 Mbps ethernet provides processor and PC connectivity.

Signals from different subsystem undergo signal conditioning in conditioning rack and are fed to the VME modules. ICRH system (hence DAC signals) is distributed physically in SST/Aditya hall and in RF-lab control room.

Analog input/output signals are used for time varying or trend monitoring and controls; whereas, digital input/output signals for status monitoring, controls, interlocks, and remote system protection.

RF shot duration, duty cycle, number of RF-shot pulses, online power control, multi-pulse operation, master-slave mode, and acquisition mode are various parameters controlled from the interactive GUI and compatible hardware support.

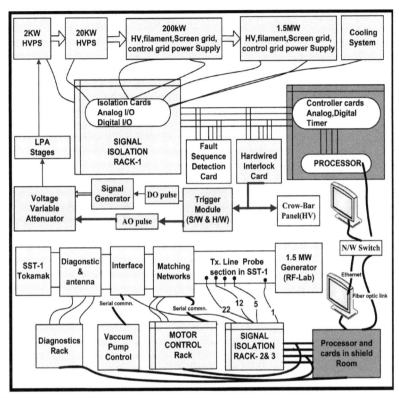

FIGURE 18.1   ICRH–DAC system for Aditya and SST.

The data acquired during the experiment from generator, high voltage and auxiliary power supplies, transmission line, matching networks, interface and diagnostic sections, etc., distributed at various locations is online monitored and also presented for post-shot analysis. For the post-shot analysis data acquired at the rate of 1 ms, for online monitoring data are acquired at 50 ms.

18.2   IMPORTANCE OF RF-SHOT PULSE

18.2.1   NEED OF RF-PULSE CONTROL IN ICRH SYSTEM

DAC system should provide flexibility to generate variable duty cycles and variable amplitude pulses. Generation of this RF-shot pulse is critical task as this pulse has been used for

- Define and control the duration of the RF power generated from the 1.5 MW cascaded amplifier.
- Generated pulse acts as a trigger pulse to the ICRH-DAC system situated at the SST-1 hall.
- Generated pulse acts as a trigger pulse to the various diagnostics placed and distributed at different locations.
- Pulse is used as a reference point for the acquisition and post-shot analysis of the system performance.
- Configured pulse is used to provide the time delay between two different system operations.
- To generate the pulse signal (TTL level) trigger the SG.
- To limit the over current of unregulated power supplies towards the end of RF shot.

18.2.2 CONTROL PARAMETERS OF RF-PULSE

$$\text{Total pulse time} = T_d + \{(T_{on} - T_{fall)} + T_{off}\} \times \text{No. of pulses}$$

Delay time, T_d: One time entering parameter when ICRH system run in slave mode and system is scheduled to launch power after some defined timed set by the master controller.

On time, T_{on}: Actual timeduration, during which the RF amplifier is activated to generate RF power of set parameters.

Off time, T_{off}: Off period between two pulses in the multi-pulse mode (in this period RF amplifier remains off).

Fall time, T_{fall}: Time to reach zero volts from the set voltages.

No of pulses: How many pulses of set parameter (on, off time, delay time, and amplitude) have to be launched in single shot.

Amplitude of pulses: This parameter controls the RF power by controlling the attenuation level of the RF attenuator.

Emergency pulse stop: RF pulse has to stop if system fault occurs during experiments; or RF pulse has to stop midway if situation demanded for that.

18.2.3 DIFFERENT MODES OF ICRH-OPERATION

18.2.3.1 Single Pulse Operation

In this mode after a pre-defined delay time, one pulse is generated of defined T_{on} time to trigger the RF-source.

18.2.3.2 Multi-Pulse Operation

This mode of ICRH operation is mostly used in the line conditioning, where a number of RF pulses of constant power are generated to condition the various component from transmission line to antenna section. In this multi-pulses of defined duty cycle, $\{T_{on}/(T_{on}+T_{off})\}$ is generated to trigger the RF-source.

18.2.3.3 Multi-Pulse Variable Duty Cycle

This mode of ICRH operation is mostly used in experiment, where it is needed to generate pulses of different duty cycles $(T_{on}/T_{on}+T_{off})$ but of same magnitude within the single shot.

18.2.3.4 Multi-Pulse Variable Amplitude

If the ICRH system is required for the breakdown as well as heating, this mode is needed. In this, first pulse's duty cycle and amplitude is set as per ICRH power needed to achieve the breakdown (keeping the control voltage low), and second pulses duty cycle and amplitude is set for the desired ICRH power requires for the heating (Figure 18.2).

18.3 LIMITATION OF EXISTING SYSTEM

18.3.1 HOW EXISTING SYSTEM WORKS?

Here we explain how ICRH system delivers RF power to dummy load/ Aditya/SST-1 Tokamak? Signal generator is set at operating frequency

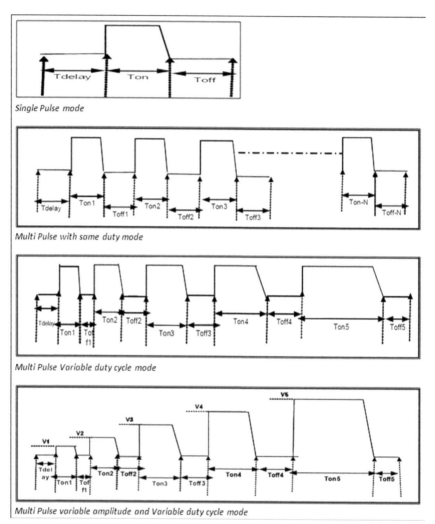

Single Pulse mode

Multi Pulse with same duty mode

Multi Pulse Variable duty cycle mode

Multi Pulse variable amplitude and Variable duty cycle mode

FIGURE 18.2   Various mode of ICRH operation.

and output of signal generatoris amplified by the cascading amplifiers as shown in Figure 18.3. This signal generator is remotely triggered from the DAC as per the operation requirement. Trigger signal is TTL (5V).

As shown in the Figure 18.4, a signal generator would provide the power of set dbm level at set frequency whenever it received the trigger pulse (TTL level) from the VME system. This TTL pulse is generated by

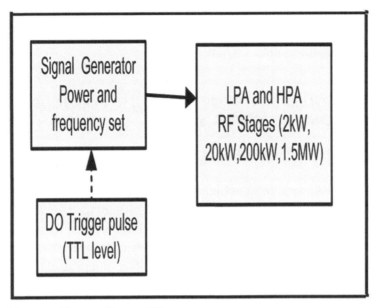

FIGURE 18.3   The cascading amplifiers.

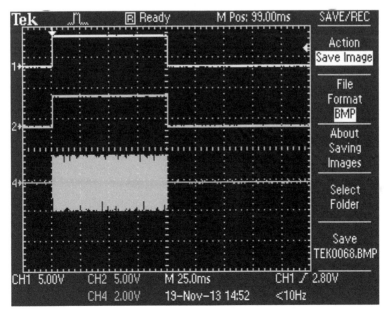

FIGURE 18.4   How the RF pulse triggers the signal generator and delivered power to LPA&HPA.

the host using timers available in the VME 480 module. This digital O/P pulse is used to define on time for signal generator used as a source for the cascaded amplifiers upto 1.5 MW stages.

18.3.2 REQUIREMENTS FOR NEW PULSE CONFIGURATION (LIMITATION OF EXISTING SYSTEM)

18.3.2.1 HVPS Requirement Connected with 1.5 MW Tube

During the RF experiment at higher power level, it is observed that if the power supplies are unregulated (or poorly regulated) then towards the last lag of on time (T_{on}) for given any duration HVPS tripped frequently. The reason is that if power supply is set at its maximum operating limit, then during the fall time of the pulse and due to sudden stoppage of the drawing current from supply, a high voltage overshoot occurs and complete system is shut down as this is critical signals connected with the hardwired protection cards of DAC. So experiment requirement is to control the fall time shape. This would be possible only through analog output signals.

It was observed in the experiment as shown in Figure 18.5, whenever plate current draws more than 24 A even at the 19 kV voltage setting

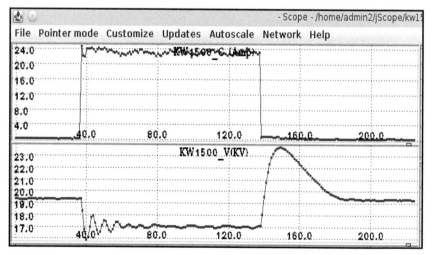

FIGURE 18.5   Experiment shot showing overvoltage signal at the end of RF pulse.

(operating parameter), towards the end of RF pulse voltage overshoot (upto 25 kV) signals frequently occurred and protection circuit sense this overshoot signal and trip the power supply (consequently other associated power supplies also tripped), and this way 1.5 MW amplifier can't generate higher power.

18.3.2.1.1 Solution

If the fall time of RF pulse is slowed down (ramp way), then current drawn from power supply is not immediately withdrawn, hence voltage overshoot of HVPS can be minimized as shown in the Figure 18.6.

18.3.2.2 Experiment Requirement At the Tokamak

Within the single pulse, RF power can't be varied as per present setup, because signal generator would trigger once it received TTL level and it would couple set dbm power (in signal generator) to LPA stages. So if ICRH experiment at tokamak machines requires two different power

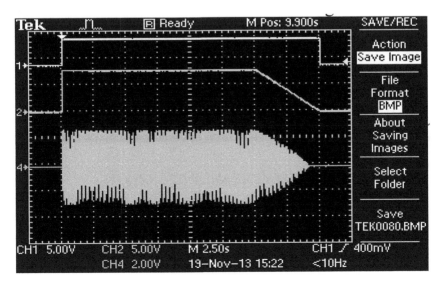

FIGURE 18.6 Slowing down the RF pulse to LPA by controlling fall time of pulse.

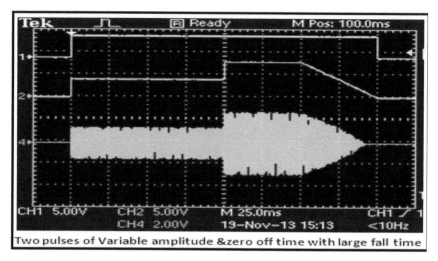

FIGURE 18.7 Two analog o/p pulse within the single DO pulses to control the RF power to LPA.

levels within same shot without off time between them (requires in preionization and RF heating experiment), as shown in the Figure 18.7, it can't be generated.

- CH1 shows TTL trigger pulse to signal generator.
- CH2 shows variable amplitude pulse within the single shot without off time between them.
- CH4 shows desired power variation within the single shot to perform different mode of ICRH experiments.

From the existing ICRH-DAC setup power profile control is not possible as shown in the Figure 18.7 at CH4. This needed for software and hardware modification in the pulse generation facilities.

18.4 SYSTEM REQUIREMENT ANDINTEGRATION

As mentioned in the Sections 18.3.2.1 and 18.3.2.2, it is preventing the higher power RF tube testing (1.5 MW commissioning work) as well some new experiments on the tokamak machine.

18.4.1 POWER CONTROL AND PULSE SHAPE CIRCUIT

To remove technical constraints in the system design, the following things has to be followed:

- need a circuit, which controls the input RF power to LPA.
- active RF attenuator controller needed variable control voltages for its proper functioning.
- control voltage within the single shot can be produced by VME programming only as it has to synchronously provide with the DO pulse (acts as trigger for signal generator).
- rise time and fall time of the controller's analog o/p pulse has to be controlled using software programming and these should be variable.

In order to generate the facility, system is modified with the addition of voltage variable attenuator in between signal generator and low power amplifiers as shown in Figure 18.8. Digital output pulse generated from the VME system acts as master trigger pulse to signal generator and analog output control pulse from VME would act as power control for LPA.

Figure 18.9 shows the basic flow to generate the multi-pulse operation of variable duty cycle and of variable RF power.

RF input signal from signal generator to LPA is controlled by the analog output pulse. It attenuates the RF power signal according to applied control pulse. By ramping down the falling edge of RF attenuator control pulse we have removed the overshoot in power supplies.

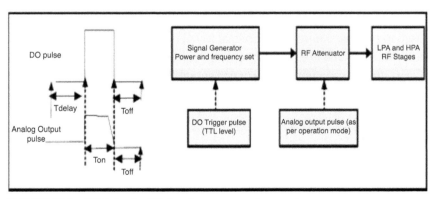

FIGURE 18.8   RF Pulse for RF signal generator and attenuator.

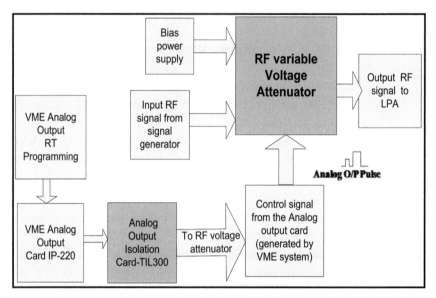

FIGURE 18.9   Block diagram showing multi-pulse and multi-voltage generation from the VME.

18.4.2   RF VARIABLE VOLTAGE ATTENUATOR

The function of RF voltage attenuator is to pass the RF signal present at its input section to output section as a function of control voltages present at its control pin. Magnitude and duty cycle of this control signal to RF attenuator is generated by the VME as per experiment requirement (Figure 18.10).

Voltage variable attenuator ZX73-2500 from the mini circuit has been used to provide controlled RF signal to LPA. Output RF signal depends on the control signal applied to attenuator as shown in the figure attenuation keeps down as magnitude of control signal increased (Figure 18.11).

18.4.3   VME-HARDWARE REQUIREMENT

To generate multiple pulse of variable, duty cycle timer card IP-480 is used; and for variable amplitude, analog output card IP-220 is used. From IP-220 card up to 12V of analog output signal can be programmed. Both IP modules are mounted on the VME-based IP carrier board.

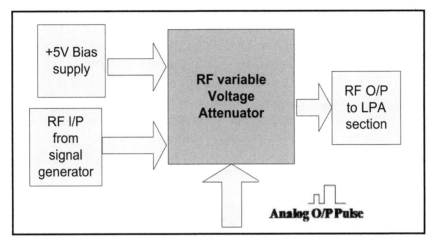

FIGURE 18.10 Voltage variable attenuator and control circuit.

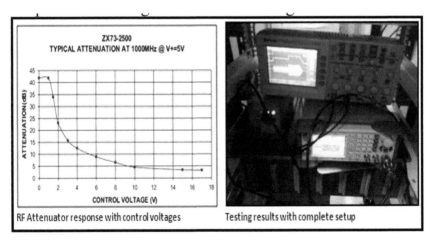

FIGURE 18.11 RF attenuator response with control voltages.

18.4.3.1 VME-based Timer Card

ICRH system requirement for safe operation is such that it needed external digital output signal to trigger the RF signal generator from the DAC and as long as this digital pulse remains high, signal generator keep providing RF signal of set operating frequency and amplitude to low power amplifier (LPA) stage.

18.4.3.2 VME-Analog Output Cards

Experiment requirement is to control the fall time shape to prevent HVPS overshoot during fall time of RF pulse. This was possible only through analog output signals. This analog output pulse is generated at the host using analog output modules available in the VME module IP 220. In this, falling edge of analog output AO pulse is ramped down in 5 ms.

18.4.4 ANALOG OUTPUT ISOLATION CARDS [6]

Generated analog control voltage from VMEIP220 card is connected to in-house developed TIL 300 based analog output cards to provide isolation needed between DAC and system side. This card can provide required attenuation and amplification. It shows highly linear response (Figure 18.12).

18.4.5 SYSTEM PROTECTION

With insertion of the new circuit, existing ICRH system and tube protection remains unaffected as even now RF signal generator gets trigged only when RF-pulse would apply from the VME (digital out) as per ICRH experiment requirement. In case of any fault inside the RF tube/HVPS/

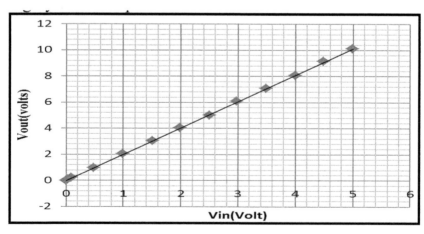

FIGURE 18.12 V_{in} vs. V_{out} for 0–10 V range.

auxiliary/cooling, etc., a TTL signal transition take placewithin micro-second in the hardwired interlock (from normally high to low transition). Output of the hardwired interlock card is logically AND with the DO trigger pulse (RF shot pulse).

18.5 TESTING RESULTS

18.5.1 TEST RESULTS WITH LOW POWER AMPLIFIER

The developed system is integrated with DAC and first tested with low power amplifier (50W) in all possible ICRH modes (Figure 18.14).

18.5.2 POWER CONTROL WITHIN THE SINGLE PULSE

In future ICRH experiment may requires, that within the single shot (without any off time) ICRH system has to do both pre-ionization as well

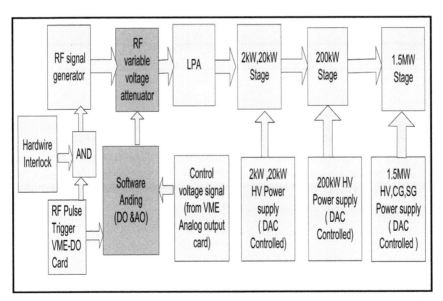

FIGURE 18.13 Block diagram of ICRH-DAC system for multiple pulses with varying RF power mode.

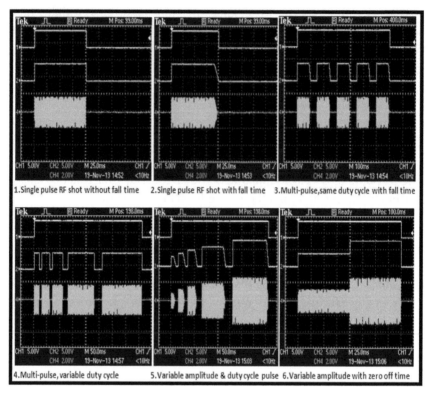

FIGURE 18.14   Testing results of control circuit in various mode of RF-pulse with LPA.

heating of the plasma. It means within the duration of DO pulse two different analog control signals of different duration have to provide to control the input RF power to LPA as shown in the Figure 18.15.

Two analog o/p signals of 5 V and 10 V, respectively, is applied to control pin of attenuator at fixed set value (dbm level and frequency) of signal generator and following power is noted in power meter as shown in the Figure 18.15.

18.5.3   POWER FROM 1.5 MW STAGE

The circuit and set up is integrated with the 1.5 MW amplifier chain, and it is tested with different power levels (Figure 18.16). Forward power signal (from direction copular connected in 1.5 MW transmission line) is

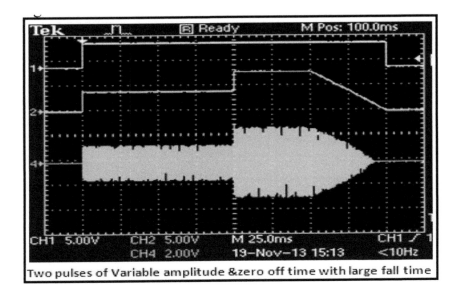

Two pulses of Variable amplitude &zero off time with large fall time

Signal Generator input to attenuator (in dbm)	Power(Watt) at Control voltage(5V)	Power(Watt) at Control voltage(10V)
-10	0.5	1.2
-8	1	2.5
-6	2	4
-4	2.5	7
-2	4	10
0	7	18
2	10	25
4	18	30
5	22	32

FIGURE 18.15   Results of power control within the single pulse.

observed and recorded. A comparison table is made to find out the effectiveness of new circuit is done as shown below. Three different test conditions is:

- Power from 1.5 MW without RF power controller set-up.
- Power from 1.5 MW with RF power controller set-up.
- Power from 1.5 MW with RF power controller and fine tuning of the system.

SG power level	Power from 1.5Mw without RF Power controller circuit			Power from 1.5MW stage with RF Power controller circuit			Power from 1.5MW stage after fine tuning of the system		
	Power in terms of Vpp(Volt)	Absolute power in Watt	Plate current in Amp	Power in terms of Vpp(Volt)	Absolute power in Watt	Plate current in Amp	Power in terms of Vpp(Volt)	Absolute power in Watt	Plate current in Amp
-6	4.96	39	10.2	2.72	9.7	5.5	9.12	131	19
-4	6.56	67.97	14.4	3.92	24.3	7.2	11.0	190	23
-2	8.4	111.5	18.4	4.9	37.9	14	11.7	216.1	24
-1	9.2	142.2	20.3	5.5	47.7	18	12.8	246.6	26.5
0	10	157.8	22.4	6.72	71.3	18.5	13.2	275.5	28
1	10.1	160.9	22.85	7.1	93.6	19.2	**14.1**	313.1	28.3
2	10.2	**164.1**	24Trip	8.72	120	20			
3				9.6	145.5	23			
4				10.8	184.5	25.5			
5				11	190.1	26.5			

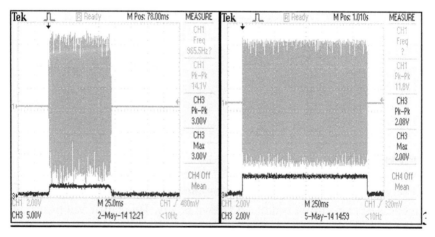

FIGURE 18.16(A)   313 kW power with 100 ms duration 220 kW power with 2 sec duration.

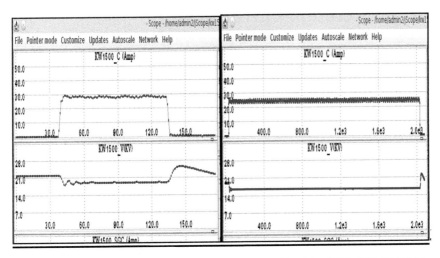

FIGURE 18.16(B)   313 kW (28.3 A, 21 kV, 100 ms)–Shot No. 2747; 220kW (25.5A, 19kV, 2sec)–Shot No. 2768.

18.6   CONCLUSIONS

From the table given in p. 302, the following conclusions are derived:

- Without controller HVPS trips continuously as current draws more than 24 A from HVPS connected with 1.5 MW tubes. Tube cannot be tested for higher power without new setup.
- With inserting the new control circuit, tube can be tested for more power (current can be continuously drawn more than 26 A without tripping).
- With inserting the new control circuit and fine tuning of the all stages continuously current drawn increases more than 28 A, without tripping.
- RF shot duration increases from 100 msec to 2 sec without tripping.
- Higher power achieved even with less drive (at 1 dbm SG setting power reached upto 313.1 kW), previously with this setting power delivered is around 160 kW.
- As the fall time duration increases in the RF pulse, the overshoot voltage of HVPS is reduced proportionally.

18.7 CONCLUSION AND FUTURE SCOPE

- Stability in ICRH-system is achieved as frequent tripping in HVPS is minimized.
- Improvements in the RF power generation and pulse length duration (upto 2 seconds).
- Pulse generation in different modes of operation.
- Synchronization between digital output (DO) and analog output (AO) pulse is achieved in every mode.
- Fall time adjustment at different rate is tested.
- Emergency shutdown feature tested to stop the DO and AO pulse midway during the shot.
- Acquisitions of AO and DO pulses is achieved for the post-shot analysis.
- Generated power can be varied with AO pulses as well by fall time adjustment.

18.8 FUTURE SCOPE

Voltage variable attenuator interfaced with the system having around –4 dbm loss and is analog controlled. In future, RF attenuator device can be explored, which have less attenuation level and digital control for more flexibility in the remote operation.

KEYWORDS

- **data acquisition and control**
- **front end rack**
- **ion cyclotron resonance heating**
- **radio frequency**
- **steady state tokamak-1**
- **transistor–transistor logic**
- **versa module euro cards**

REFERENCES

1. Bora, D. et al. (2005). Ion cyclotron resonance heating system on Aditya. *Sadhana* *30*(1), 21–46.
2. Bora, D. et al. (2006). Cyclotron resonance heating systems for SST-1. *Nuclear Fusion, 46*(3).
3. Sunil Kumar, Raj Singh, Kulkarni, S. V., & Bora, D., (1992). Layout of 200 kW RF Generator for Aditya ICRH System, IPR Technical Report, IPR/TR-47/192.
4. Manoj Singh Parihar, Kumar Rajnish, et al. (2007). "Testing and Optimization of Matching Response Time for the Real Time Feedback Controlled ICRH-Automatic Matching Network (AMN) System for SST-1," IPR Technical Report, IPR/TR-139/2007.
5. Kulkarni, S. V., Raj Singh, Sunil Kumar, & Bora, D., (1991). "RF Antenna for Heating Plasma near Ion Cyclotron Resonance Frequencies: Design Aspects," National Symposium on Science and Technology of Plasmas, PLASMA 91, Indore, December 17–21.
6. Manoj Singh, Jadav, H. M., Ramesh Joshi, Bhavesh, Kirit Parmar, Srinivias, Y. S. S., Kulkarni, S. V., & ICRH group, (2012). Design modification and testing of the Analog Isolation cards and its interfacing with ICRH system. IPR Technical Report, IPR/TR-207.

CHAPTER 19

INVESTIGATION OF THE EFFECT OF THERMAL CYCLE ON SS/CRZ BRAZED JOINT SAMPLE

K. P. SINGH, ALPESH PATEL, KEDAR BHOPE, S. BELSARE, NIKUNJ PATEL, PRAKASH MOKARIA, and S. S. KHIRWADKAR

Divertor and First Wall Technology Development Division, Institute For Plasma Research, Bhat, Gandhinagar – 382428, India, Tel.: +91-79-2396-2107; E-mail: kpsingh@ipr.res.in

CONTENTS

ABSTRACT

Stainless steel (SS 316L) is known to be structural material for plasma facing component (PFC) of ITER like tokamak. In the divertor PFC, heat sink material like CuCrZr slotted channel needs to be leaked proof attachment with the structural material. Joining of SS 316L with heat sink material (CuCrZr) has performed using vacuum brazing technique. The brazed joint specimen has undergone several numbers of thermal cycles (450°C, 5 sec, ON and cooled down to 300°C repeatedly) using thermo-mechanical simulator (Gleeble 3800) machine. The brazed joint of SS316L/CRZ is examined by non-destructive testing (NDT) using pulse echo immersion testing probe before and after the thermal cycles. The brazed samples were also undergone microstructural examination and micro hardness measurement. The detail of specimen preparation using vacuum brazing and the effect of thermal cycles on the brazed joint of SS/CRZ are studied and experimental results are presented in the chapter.

19.1   INTRODUCTION

Development of the joining technique for the plasma facing material with structural material is one of challenging area in fusion research [1]. In ITER like tokamak first wall and divertor plasma facing component (PFC) module, SS/CuCrZr (CRZ) joining has the mandatory requirement of good in thermal transfer and sound in structural joints [1].

In earlier design, the main purpose of SS/CRZ was only for good in thermal transfer. In the present, ITER design configuration, pressurized hot water could directly see the SS/CRZ interface, and hence leaked proof was also a mandatory. As improving in the design, the criticality of the manufacturing is also increased. Many researchers have attempted the joining of SS/CRZ using hot isostatic press (HIP) technique [2–4], explosive bonding [4], etc.

Each joining process has its own pros and cons but the baseline is to achieve the minimum ITER like tokamak acceptability requirement. HIP techniques are generally performed at very high temperature and hence inter-diffusion has made at the interface, there is worry of material deformation and also removal of casing is not easy.

Joining of CRZ on to SS material by casting technique was reported in elsewhere [3]. The CRZ material properties got degraded once it was processed at higher temperature. The material properties particularly CRZ material can be regained with subsequent heat treatment process after the joining procedure [4].

In this chapter, we present the details of specimen preparation using vacuum brazing technique and the effect of thermal cycles on SS/CRZ brazed joint specimen using Gleeble machine.

19.2 EXPERIMENTAL DETAILS

19.2.1 SAMPLE PREPARATION FOR VACUUM BRAZING

Two numbers of SS 316L tiles of dimension 10 mm (length) × 10 mm (width) × 10 mm (thickness) and two numbers of CRZ (C18150 Grade) copper alloy tiles having dimension 10 mm × 10 mm × 10 mm has been prepared. The surfaces have been polished and cleaned by using ultrasonic cleaning machine in acetone for 10 mins prior assembly for vacuum brazing.

Vacuum brazing of SS/CRZ has been performed @ 970°C using NiCuMn-37 filler material for 10 mins with the uniform load of 5 kPa under 10^{-6} mbar vacuum environment (see Figure 19.1). Prior to thermal cyclic test, the brazed joint of SS/CRZ was examined by NDT, followed by microstructural examination and micro hardness testing of the joint.

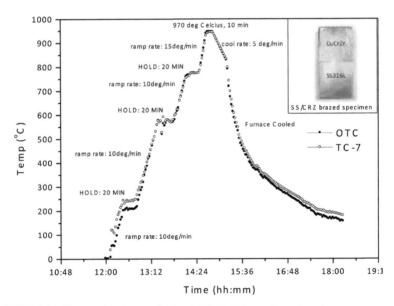

FIGURE 19.1   Vacuum brazing cycle, inset (SS/CRZ brazed specimen).

TABLE 19.1   Vacuum Brazing Experimental Parameters

Samples	Filler (Foil, 50 µm)	Brazing Temp, Holding Time	Environment	Uniform Loading (kPa)
SS316L/CuCrZr (10 × 10 × 20)	NiCuMn-37 9.5 Ni, 52.5 Cu, 38 Mn	970°C, 10 min	10^{-5} mbar Vacuum	5

19.2.2   SPECIMEN PREPARATION FOR THERMAL CYCLIC TEST IN GLEEBLE 3800

Sample preparation for the experiment in the Gleeble machine needs extra care particularly on thermocouple welding. Thermocouple welding is difficult on the surfaces particularly for a few materials such as tungsten, copper, molybdenum, etc. A small hole of 0.5 mm to 1 mm diameter with 2–3 mm depth is made for mechanically fixing the thermocouple wire. Generally, thermocouples are welded on the specimen by thermocouple spot welder. In the specimen, one leg of K-type thermocouple is welded on the SS316L material and other leg is attached on CuCrZr sample by making a hole of 1 mm diameter with 2 mm depth. Omega bond cement is used to apply to strengthen the attachment. Figure 19.2 shows the thermocouple attachment in the specimen in high temperature pocket jaw MCU.

The schematic diagram as shown in Figure 19.3 is the model of thermal cyclic experimental setup for SS/CRZ specimen in Gleeble machine. Graphite foil of ~1 mm thick was used to apply using nickel paste in between SS plungers and specimen on both the sides to avoid the edge localized heating due to current pass on the specimen.

FIGURE 19.2   Thermocouple attachment on the brazed SS316L/CuCrZr specimen.

MCU with square copper grip of 11 mm is used to hold the SS bar on both sides. An SS backing nut was used for applying a pressure of 3 kN load to the specimen during the experiment. Figure 19.4 shows the assembly of the specimen loading in the Gleeble machine before and after heating cycles.

SS plungers made of SS 304 square bar of dimension 11 mm are used as pins to apply the load from both the ends with 3 kN force to hold the specimen during the heating cycle. Figure 19.5 shows the zoom portion of thermal cyclic plots.

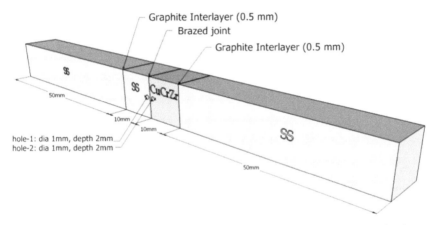

FIGURE 19.3   Schematic drawing of experimental setup high temperature pocket jaw.

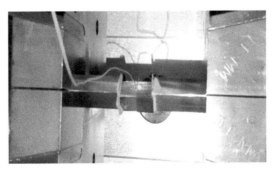

FIGURE 19.4   Specimen mounting in the high temperature pocket jaw MCU of Gleeble machine.

TABLE 19.2 Thermal Cyclic Experimental Parameters

Runs	Sample	No. of Cycle	Heating rate (°C/sec)	Heating temp (°C)	Holding time (sec)	Cooling rate (°C/sec)	Down temp (°C)	Total time/ cycle (sec)
1ˢᵗ run	SS/ NiCu	100	10	450	5	10	300	78
2ⁿᵈ run	Mn-37/ CuCrZr	100	10	450	5	10	300	78

Ultrasonic testing with the following test parameter has been performed on the SS/NiCuMn/CuCrZr specimen before and after the thermal cyclic test.

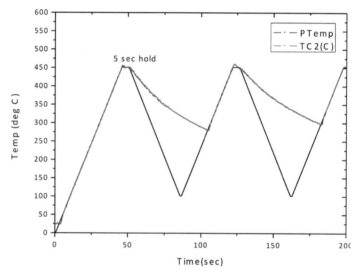

FIGURE 19.5 Thermal cyclic plot.

TABLE 19.3 Ultrasonic Testing (UT) Parameters

UT Technique	Probe	Gain	Specimen
Pulse-Echo Immersion	20 MHz, 6.35 dia, F = 50 mm	75 dB (starting), 74 dB (after 100 TC)	SS316L/CuCrZr

19.2.3 UT EXAMINATION BEFORE THERMAL CYCLIC TEST (TCT)

In the SS/NiCuMn/CuCrZr joint specimen, there is one interface where the debonding is expected. The detection of the de-bonding between SS/NiCuMn/CuCrZr was carried on using conventional immersion pulse-echo technique. However, Copper is acoustically soft and SS is acoustically harder than CuCrZr material. A 20 MHz frequency probe having 6.23 mm diameter and 50 mm focal length is used. Scan resolution are 0.1 mm in step and 0.1 mm in index direction are used to inspect the geometry. When seen from the CuCrZr (CRZ) side, a very small interface echo was observed due to acoustic difference between CuCrZr and SS material (Figure 19.6).

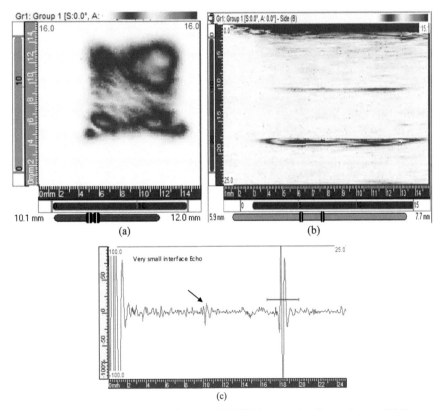

FIGURE 19.6 Ultrasonic testing image of SS/CRZ sample (a) C-scan image, (b) B-scan image, and (c) A-scan image.

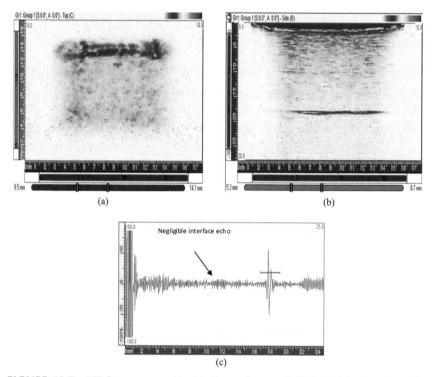

FIGURE 19.7 UT from copper side (a) C-scan image of CRZ_SS joint interface, (b) B-scan image, and (c) A-scan image.

It is observed from the UT images, there were no signatures of defects found at the interface region and the SS/CRZ joint detected to be good bonded.

19.2.3.1 UT Examination After 100 Units of TCT

UT examination was performed on the SS/CRZ specimen which was undergone initially 100 nos. of TCT. When seen from the copper side, negligible interface echo observed and which is merged with noise as shown in Figure 19.7(c) and it shows the increase of attenuation in copper and this may happened due to heating of sample. Gain is used to resolve is 74 dB.

From the UT results of C-scan, B-scan, and A-scan, it is depicted that no defect presents at the brazed joint area and joint seems to be good joint after 100 nos. of TCT. From the results, it is observed that the joint

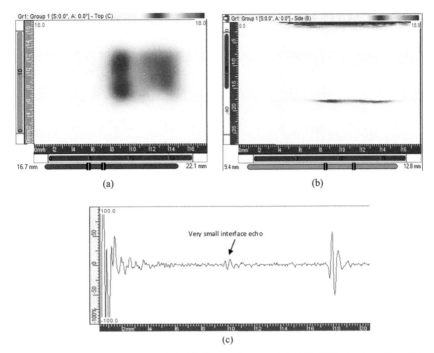

FIGURE 19.8 UT from copper side (a) C-scan image of CuCrZr/SS joint interface, (b) B-scan image, and (c) A-scan images.

interface is completely merged with the parent material. This is the indication of the good joint intact.

19.2.3.2 UT Examination After 200 Units of TCT

The specimen was further undergone 200 nos. of TCT. UT examination was performed on the SS/CRZ specimen which was undergone the 200 nos. of TCT with the same UT parameters. Here Gain used to resolve is 70 dB.

A very small interface echo observed and merged with noise as shown in A-scan. This indicates that the difference in impedance mismatch of CuCrZr/SS joint is negligible and which is represented as blue region in C-scan (Figure 19.8).

From the UT examination results on the 200 nos. of TCT sample, there was no signature of defect observed at the joint region of SS/CRZ and the joint was shown still good intact.

19.3   RESULTS AND OBSERVATION

The brazed joint of SS316L to CuCrZr sample has been ultrasonically examined at the in-house UT facility in the division. The conventional pulse-echo immersion technique has been employed to detect the de-bonding area at the interface region.

Ultrasonic testing has been performed on SS/CRZ brazed specimen before the thermal cyclic test (TCT) by using pulse-echo immersion technique from both side (copper side and SS side). A 20 MHz frequency, 6.23 mm diameter, and 50 mm focal length was used in the inspection. A-scan, B-scan, and C-scan images for both cases were shown that no defect was present at joint area and joint was observed to be good bonded. To check the joint integrity of the specimen which has undergone 100 nos. of thermal cyclic test @450C, ultrasonic testing has been carried out with same UT scanning parameters as mentioned above. From the results of the UT scans such as A-Scan, B-Scan and C-scan, it was depicted that no defect was detected at the joining interface and the joint was observed to be good bonded after 100 nos. of TCT.

Subsequently, the ultrasonic testing has been performed on SS/CuCrZr brazed specimen which has been undergone 200 nos. of thermal cycles @450°C. There was no indication of presence of any flaw or deboning at the brazed joint region. From the results of UT examination of A-scan, B-scan, and C-scan, it was depicted that the interface region was observed to be still intact and good bonded even after 200 nos. of TCT.

It was also observed that the attenuation in the CuCrZr material and SS316L materials increased non-uniformly due to repetitive heating (TCT) of the materials. As the attenuation increased in the materials, the gain value had been set to lower than the original gain value to minimize the noise level during the inspection.

It was observed that the SS/CRZ vacuum brazed specimen which was undergone 200 nos. of thermal cycles didn't detect any debonding signature after the test. Interestingly, the brazed joint was seen improving after the test. From this experimental study, it is derived that the SS/CRZ brazed joint can easily withstand more than 200 nos. of thermal cycles @ 450C, 5 sec, ON/OFF.

19.3.1 MICROHARDNESS MEASUREMENT

The Vickers microhardness measurement of the W/Cu/CRZ and SS/CRZ brazed joint samples have been carried out with the 100 g load, 10 sec hold at 14 different spots each across the polished brazed interface and calculated the average value of HV (see Figure 19.9).

The hardness value (HV) of the SS/CRZ interface was found increase after the thermal cyclic test. The HV value was 174 HV (after the thermal cyclic test) and 103 HV (before thermal cyclic test).

19.3.2 MICROSTRUCTURAL EXAMINATION

Microstructural examination has been conducted on the polished cross-section surface of the specimen using optical microscope with different magnification scale. Following images in Figure 19.10 is the optical images taken for SS/NiCuMn/CRZ brazed sample before the thermal cyclic test, after 100 nos. of TCT.

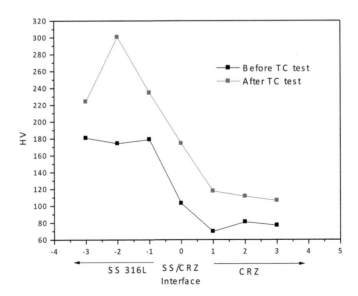

FIGURE 19.9   Vickers hardness (HV) of W/CRZ and SS/CRZ brazed samples.

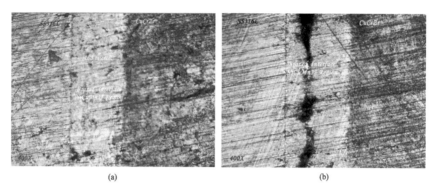

FIGURE 19.10   Optical images (a) before TCT and (b) after TCT. (a) SS/NiCuMn-37/ CRZ brazed showing interface thickness. (b) SS/NiCuMn-37/CRZ brazed interface having black burnout line mark after TCT.

From the optical image analysis on the polished cross-section of the SS/NiCuMn-37/CRZ specimen, it was observed that the homogenous distribution of the filler in between the parent materials, i.e., at CuCrZr side as well as copper side.

The thickness of the brazed joint was measured to be 163 micron. The blackening area was observed at some portion at the interface region after the 100 cycles of thermal cyclic test. It was expected the initiation of the oxidation at the interface region during the cyclic thermal load. The blackening area has not been seen after the polishing.

After completion of the 200 nos. of thermal cyclic tests on SS/NiCuMn/ CuCrZr brazed specimen, the surface was further polished with fine grit paper. There was no crack observed at the interface region and joint condition was observed to be good bond.

Some portion of the oxidized area was measured and the maximum width of the oxidized area was found to be approximately 80 micron.

19.4   CONCLUSION

The methodology for thermal cyclic test for SS/CRZ dissimilar materials brazed joint has been developed using Gleeble 3800 machine at IPR.

Brazing of SS/CuCrZr samples at 970°C for 10 min, 5 kPa load was performed using NiCuMn-37 filler material under 10^{-5} mbar vacuum environment.

Thermal cyclic test was done using thermo mechanical simulator (Gleeble 3800 system) for initially 100 cycles @ 450°C for 5 sec ON and cool down to 300°C repeatedly. The TCT has conducted up to 200 cycles without fail.

From the UT results, no debonding area observed at the brazed joint and the joint seems to be good bond after 100 cycles of thermal cycles. There is no debonding observed at the interface region after even 200 thermal cycles which depicted the sound brazed joint.

Microhardness testing was performed at the brazed joint of SS/CRZ and both the parent materials. The average HV value was found increase after the thermal cyclic test.

All interface joints of brazed sample were found homogeneous elemental distribution in both the samples (before the thermal cyclic test). In some portion at the interface region, blackening area is observed after the TC test which is likely due to oxidation occurred at the joint interface region.

KEYWORDS

- **NDT**
- **PFC**
- **thermal cyclic**
- **vacuum brazing**

REFERENCES

1. Barabash, V., Akiba, M., Mazul, I., Ulrickson, M., & Vieider, G., (1996). Selection, development and characterisation of plasma facing materials for ITER, *J. Nucl. Mater. 233–237,* 718–723.
2. Barabash, V., Akiba, M., Cardella, A., Mazul, I., Odegard, Jr. B., Plöchl, L., Tivey, R., & Vieider, G., (2000). Armor and heat sink materials joining technologies development for ITER plasma facing components, *J. Nucl. Mater. 283–287,* 1248–1252.
3. Gervash, A., Mazul, I., Yablokov, N., (2001). Study of alternative SS/Cu-alloy joining methods for ITER, *Fusion Eng. Des. 56–57,* 381–384.
4. Goods, S. H., & Puskar, J. D., (2011), Solid state bonding of CuCrZr to 316L stainless steel for ITER applications, *Fusion Eng. Des. 86,* 1634–1638.

CHAPTER 20

DESIGN AND TEST RESULTS OF A 200 kV, 15 mA HIGH VOLTAGE DC TEST GENERATOR

S. AMAL, URMIL M. THAKER, KUMAR SAURABH, and UJJWAL K. BARUAH

Institute for Plasma Research (IPR), Bhat, Gandhinagar, Gujarat – 382428, India, E-mail: amal@ipr.res.in

CONTENTS

ABSTRACT

A compact, low power, portable 200 kV high voltage DC power supply has been designed and developed at IPR, Gandhinagar. The design is based on symmetrical Cockroft-Walton voltage generator with a high frequency front-end converter. The high voltage is generated by a series fed seven-stage voltage multiplier circuit driven by a 15.8 kHz quasi sine wave inverter. A 17 kV–0–17 kV, 15.8 kHz ferrite core transformer interfaces the voltage multiplier circuit with IGBT based half-bridge inverter. The use of high frequency gives us advantages of less ripple, faster response, and low stored energy in the system. Additionally the scheme allows the use of smaller capacitor and magnetic parts thus minimizing the weight of components and improving portability of the system. With a brief introduction on design aspects of voltage multiplier, this chapter describes the design features and developmental aspects of various components for 200 kV power supply viz., high frequency inverter, high voltage high frequency transformer, high voltage feed through (HV bushings) and voltage multiplier circuit in detail. This power supply will be used as a DC source for HV testing of electrical installations in future.

20.1   INTRODUCTION

In this chapter, the main emphasis has been given on design, simulation and development of high voltage DC power supply. At the first stage of this work is to study voltage doublers circuit, Cockroft-Walton voltage multiplier circuits and to simulate the circuit for designed value of DC output voltage. And finally, assembly of components of high voltage test generator at the output DC voltage of 200 kV. Presently this power supply is used as a DC source for HV testing of electrical installations. This test set will be a friendly user in industries for field testing as well as in laboratory. The advantage of this set is low cost, high reliability, portability, and simple control. For safety considerations the multiplier unit is designed for minimum stored energy. Voltage ripple at output is kept low by selecting high frequency operation. Multiplier unit is fully contained in an oil filled tank, which provides excellent insulation.

The main application of the high voltage DC is to test cables having large capacitance, normally takes a very large current if it is tested with AC voltages [1, 8]. High voltages at relatively low current are generated for insulation testing of high voltage equipment under power frequency AC, DC, switching and lightning impulse voltages. High voltage DC also required for applications such as electron microscopes and x-ray units, electrostatic precipitators, particle accelerators in nuclear physics and so on [2, 8].

The main purpose of high voltage DC equipment is to study insulation behavior which the testing objects likely to encounter. Normally test voltages are higher than the normal working voltage by considering the safety margin. High voltages used in industry as well as research field can be divided into three classes: (a) alternating current voltages, (b) direct current voltages, and (c) transient voltages [3, 4].

Joseph [5] has presented his paper the basic operation of voltage multiplier circuits and discussed guidelines for electronic component selection for diode and capacitor. Tanaka et al. [6] has explained a new idea to develop 70 kV, 0.15-ampere DC power supply with a high frequency switching converter. Banwari et al. [7] has explained the design and developmental aspects of a −750 kV dc source for 750 keV industrial electron accelerator at CAT Indore. Mariun et al. [8] has explained design and construction of a 15 kV high voltage DC power supply using voltage multiplier circuits.

20.2   VOLTAGE MULTIPLIER CIRCUIT

One of the cheapest and popular ways of generating high voltages at relatively low currents is the classic multistage diode/capacitor voltage multiplier, known as Cockroft-Walton multiplier. The output voltage is several times the input voltage and is used where the load has high input impedance and is constant or where the input voltage stability is not the major issue. For increasing and decreasing voltage conventional transformer can be used which can step up or down the AC voltage and current which can be rectified using diodes or thyristors or other semiconductor devices and further can be filtered using capacitors. But here the problem is the weight of transformer, as the voltage which is to be stepped up increases the insulation level of the transformer increases and hence it becomes bulky. Apart from this the rectification process is carried out using rectifier grade diodes which are slow and filtering is done using high value capacitors which are heavy in weight.

So to mitigate all the above-mentioned problems the voltage multiplier circuits can be used where load current is less and voltage is high. Negative voltage can be achieved by inversing the diode direction. The following section describes the working principle, design, and test results of a 200 kV test generator.

20.3 COCKROFT-WALTON VOLTAGE MULTIPLIER CIRCUIT PRINCIPLE

Most voltage multiplier circuits, regardless of their topology, consist chiefly of rectifiers, and capacitors. The operating principle of half wave and full wave multiplier circuits is a essentially the same. Capacitors connected in series are charged and discharged on alternate half-cycles of the supply voltage. Rectifiers and additional capacitors are used to force equal voltage increments across each of these series capacitors.

The multiplier circuit's output voltage is simply the sum of these series capacitor voltages.

The most commonly used multiplier circuit is the half-wave series multiplier. All multiplier circuits can be derived from its basic operating principles. Thus, the half-wave series multiplier circuit is shown in Figure 20.1 to exemplify general multiplier operation. The example shown in Figure 20.1 assumes no losses and represents sequential reversals of transformer (T_S) polarity.

1. TS = negative peak: C1 charges through D1 to E_{pk}.
2. TS = positive peak: E_{pk} of TS adds arithmetically to existing potential C1, thus C2 charges to $2E_{pk}$ through D2.
3. TS = negative peak: C3 is charged to $2E_{pk}$ through D3.
4. TS = positive peak: C4 is charged to $2E_{pk}$ through D4.

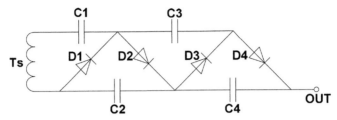

FIGURE 20.1   Half-wave voltage multiplier circuit.

Therefore, the total output voltage $= 2 \times E_{pk} \times N$ (where N = the number of stages). Thus, the multistage arranged in manner above enables to obtain very high voltage. The same operating principle extends to the full-wave voltage multiplier circuit. Although half-wave and full-wave multiplier circuits can provide equivalent output voltages, the output ripple frequency of the full-wave doubler is twice that of the input signal.

20.4   OVERALL SCHEME

This power supply is developed by using Cockroft-Walton voltage multiplier circuit at a frequency of 15.8 kHz. The block diagram of the power supply is shown in Figures 20.1 and 20.2 show the simplified circuit diagram. A 3-phase full wave uncontrolled converter with a power capacity of 17kVA feeds a fixed frequency quasi sine wave inverter which charges the voltage multiplier at 15.8 kHz frequency through a center tap ferrite core transformer. A single phase, 15.8 kHz, center tapped transformer is utilized to step up the output voltage of high frequency inverter. The terminal voltage of the multiplier is precisely controlled by controlling the output voltage of the half bridge inverter. A 10,000:1 voltage divider with RC compensation is used to measure the terminal voltage of the multiplier. All the electronic components of the voltage multiplier circuit and high frequency transformer are kept in oil filled tank and the high voltage terminals are brought out through HV bushings. Transformer oil is used here in the tank both for increasing the level of insulation as well as for providing a cooling medium.

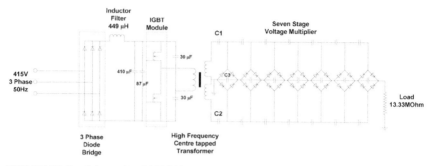

FIGURE 20.2   Schematic of 200 kV test generator.

20.5   THEORETICAL CONSIDERATIONS

The principle of Cockroft-Walton voltage multiplier has been known for many years (Figure 20.1 and Table 20.1). For the design of voltage multiplier circuit we consider only the most basic and approximate equations. The output voltage at no load is given by $V_{out} = 2 \times n \times E_{peak}$, where n is the number of stages, E_{peak} is the peak input voltage to the multiplier. The output voltage is reduced by the capacitive current and load current. The formulas for voltage drop and the ripple voltage caused by the load current are given in the Table 20.1.

It is evident from these formulas that there are three parameters which can be varied to maintain a specified voltage drop and ripple for a specified load current, namely; per stage capacitance, operating frequency and the number of stages. From the above equations we can find that increasing the frequency can dramatically reduce the ripple and the voltage drop

TABLE 20.1   Formulas for Voltage Drop and the Ripple Voltage

Parameter	Normal (Half wave) Multiplier	Full wave (Symmetric) Multiplier
ΔV	$\left[\dfrac{I}{(6fC)}\right] \times (4n^3 + 3n^2 - n)$	$\left[\dfrac{I}{(6fC)}\right] \times (n^3 + 2n)$
δV	$\left[\dfrac{I}{(2fC)}\right] \times n \times (n+1)$	$\left[\dfrac{I}{(2fC)}\right] \times n$

ΔV = voltage drop of the generator under load, δV = peak to peak ripple voltage, I = load current (Amps), C = perstage capacitance (Farads), f = operating frequency (Hz) and n = number of stages.

under load. Thus, a high frequency based symmetrical Cockroft-Walton voltage multiplier circuit was chosen for the generation of high voltage since this circuit effectively doubles the number of charging cycles per second, and thus cuts down the voltage drop and ripple factor when compared with normal half wave multiplier circuit. Other attractions which went in favor of such a circuit were: low stored energy and faster response.

20.6   DESIGN CONSIDERATIONS

For maintaining a low voltage drop an increase in supply frequency is in general more economic than increase of capacitance value. Small value of C also provides a dc power supply with limited stored energy which must be an essential design factor. However, practical limits of available components should always be observed while varying the above parameters during the design of high voltage generator.

20.6.1   OPERATING FREQUENCY (F)

For small voltage drop and less ripple, high frequency is desirable (Table 20.1). Very high frequencies put a demand for fast recovery diodes which are available in the market. From Figure 20.6 as we increases the frequency of multiplier circuit, the stage capacitance reduces to a value furthermore no drastic changes have been found as per the design. Hence, we fixed the frequency in a range of 15–25 kHz optimum for efficient working of this power supply. Considering the market availability of high frequency source and reverse recovery time of fast recovery diodes a moderate range of 15–20 kHz has been chosen as the operating frequency for the high voltage generator. The reverse recovery time of the fast recovery diodes chosen for the multiplier circuit are in hundreds of nano seconds (150 nS approx.).

20.6.2   CAPACITANCE (C)

Per stage value of the capacitance is limited by the stored energy of the voltage multiplier. Stored energy in excess of few kilojoules can cause considerable damage in the power supply itself in case of an accidental spark. Two numbers of 3.3 nF/50 kV ceramic disc capacitors with good dielectric strength are put in parallel to form a single unit of 6.6 nF/50 kV. When operated at 35 kV the effective capacitance of this unit remains about 50%, i.e., 3.3 nF when considering the de-rating factors. Thus, at 200 kV the stored energy in the generator is calculated to be less than 28 joules.

20.6.3 NUMBER OF STAGES (N)

The voltage drop is very much dependent on number of stages n, since it rises with the third power of n (Table 20.1). Decreasing the number of stages means increasing the voltage over the individual capacitors and diodes. This leads to nonlinear voltage distribution across these components, which in turn may lead to their voltage breakdown. Multiplication efficiency decreases as the number of stages is increased. In the present design the required number of stages is arrived mainly from the considerations on the market availability of component ratings. The 50 kV capacitor units are de-rated to a maximum operating voltage of 35 kV (70% of rated). Hence per stage voltage is taken as $2V_m = 35$ kV. The operating voltage of the power supply is 200 kV at 15 mA. The maximum design voltage at no load is taken as 220 kV to ensure stable operation. So the number of stages for the voltage multiplier comes out to be seven ($n = 7$). The voltage drop at a design current of 15 mA is calculated to be $V_{drop} = 18$ kV. Hence the no load terminal voltage in case of load rejection can reach up to 220 kV, which is in accordance with the designed no load voltage of the power supply. The maximum load induced ripple voltage is calculated to be 1.1 kV with a ripple factor of 0.53%.

20.7 SIMULATION RESULTS

The simulation work has been done by using PSIM, version 9.0 software (Figures 20.3–20.8). The Figure 20.3 represents simulated results of each stage and Figure 20.4 is the simulated result of output current.

20.8 VARIOUS SUBSYSTEMS OF 200 kV POWER SUPPLY

The design and constructional features of various subsystems of the high voltage generator are briefly described in following subsections.

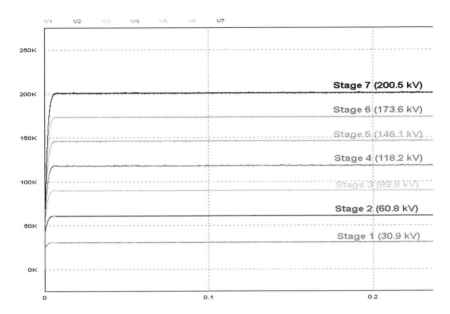

FIGURE 20.3   DC output voltages at different stages.

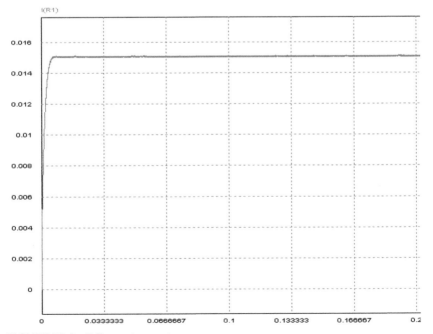

FIGURE 20.4   DC output current.

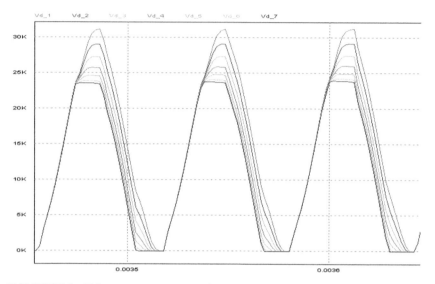

FIGURE 20.5   Voltage across stage capacitors.

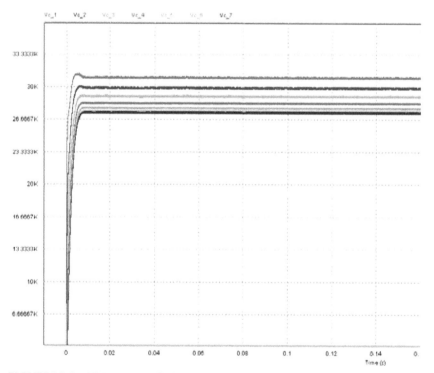

FIGURE 20.6   VRRM across diodes of successive stages.

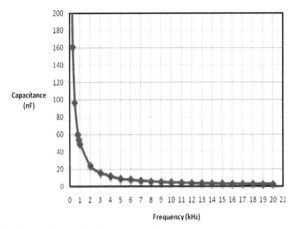

FIGURE 20.7 Capacitance with increase in frequency.

Parameter	Specification
Input Voltage	415VAC,50Hz, 3 phase
Input Current	Less than 22A per phase
Output Power	3kW maximum at full rated output voltage and current
Output Voltage	Maximum output voltage 200kV
Output Current	Up to 15mA Rx 200kV
Output Ripple voltage in kV	1.081
Output Ripple factor in %	0.13
Open circuit output voltage in kV	220
Voltage drop in multiplior in kV	20
Inverter frequency in kHx	16
Multiplier type	CW Full wave Type
Number of stages	7
Stage capacitance value in nF	3.3
Polarity	Positive output
Duty	Continuous
Dimension in m3	1.3x0.5x1

FIGURE 20.8 Technical specification of 200 kV test generator.

20.8.1 FRONT END CONVERTER

The three-phase AC mains is rectified and filtered by an uncontrolled bridge rectifier and resultant DC voltage is fed to the inverter. A half bridge topology is used for the inverter with a fixed switching frequency of 15.8 kHz. The inverter output is controlled by controlling its gate driver signal using PWM. The feedback signal for the inverter unit is derived from the terminal voltage of the HV generator through a RC compensated voltage divider (Figure 20.9).

20.8.2 CONTROL CIRCUIT MODIFICATION OF FRONT END CONVERTER

In the control circuit of the high frequency power source, a sample for feedback signal was output current, measured through Hall CT in the range of 0–3 V for 0–400 A. To use this HF power source for the development of 200 kV test generator, its feedback circuit has been modified. As sensing parameter is changed to output voltage from output current, sensing element is replaced by a voltage divider, and a signal conditioning circuit is added and calibrated in accordance with original ranges (0–3 V for 0–200 kV) which vary the output voltage of the inverter 0–100% duty cycle. The output voltage of a multiplier circuit only depends on the peak value of the input, i.e., here output voltage of the inverter. Since the charging time of the main capacitors is very less, the changes in the pulse width of the inverter output gives us a very narrow range of variation in the output voltage. To operate in the full range (0–200 kV) we decided to incorporate a three-phase variac at the input side of the HF power source and calibrated according for the range of 0–200 kV. Output voltage of 200 kV test generator is measured by 300 kV, 10,000:1 voltage divider.

20.8.3 HIGH FREQUENCY TRANSFORMER

A 290 V, 16 kV–0–16 kV center-tap transformer with VA rating of 7 kVA is an important link between the voltage multiplier circuit and the high frequency source and it plays a significant role on overall size, cost, performance, and

FIGURE 20.9 Front end converter.

efficiency of the system. The operating frequency of the high voltage generator is chosen to be 15.8 kHz. The high frequency helps in reducing the size
of the transformer. Ferrite cores are the ultimate choice at such frequencies
having features which include high electrical resistivity, low core losses, low
density suitable for light weight transformer, etc. Out of different shapes in
which ferrite cores are available U core (EPCOS make U 93 core) is considered where high voltage isolation is required (Figure 20.10).

20.8.4 VOLTAGE MULTIPLIER CIRCUIT

The seven-stage multiplier circuit is realized using 3.3 nF, 50 kV DC
ceramic disc capacitors and 0.6 A, 15 kV, fast recovery diodes. A high
frequency based symmetrical Cockroft-Walton voltage multiplier circuit
was chosen for generation of high voltage due to its design simplicity and
economical construction. Components of the multiplier circuits along with
its protective elements are distributed and arranged over Bakelite support
structure. There are seven stages in voltage multiplier circuit. Spacing
between the two successive stages has been maintained on the basis of
high voltage consideration and practices. The electronic components of
voltage multiplier circuit have been fitted on the reverse side of the top
cover of the HV tank (Figure 20.11).

20.8.5 HIGH VOLTAGE BUSHING

In order to connect the high voltage terminal of voltage multiplier, which
is inside the HV tank, to the load, flexible high voltage x-ray cable with

FIGURE 20.10 High frequency transformer.

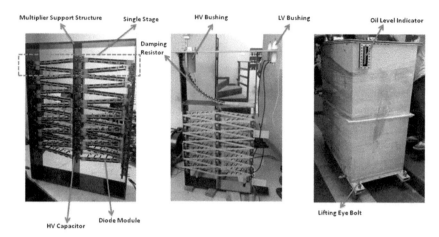

FIGURE 20.11   Voltage multiplier circuit.

specially designed connector housings is utilized. The connector housing consists of R28 (225 kV) generator receptacle fitted with aluminum flange and R28 (225 kV) straight connector fitted with C2236 (250 kV) high voltage x-ray cable.

20.8.6   TEST RESULTS

The test results are shown in Figures 20.12–20.15.

FIGURE 20.12   Testing of 200 kV test generator.

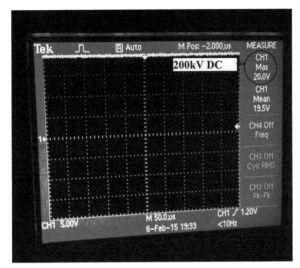

FIGURE 20.13   Oscillogram of 200 kV test generator.

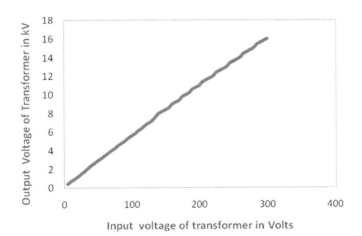

FIGURE 20.14   Test results of high frequency HV transformer.

20.9   CONCLUSION

The power supply described above is an attractive scheme for generation of high voltage, high power system especially in million volts level.
In comparison with other alternative schemes, it has better performance
with regards to voltage ripple and regulation. As there are less vulnerable

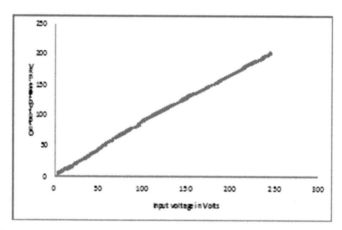

FIGURE 20.15   Test results of high voltage generator.

components and less stored energy in the system, it is also more reliable compared to other schemes. These systems are less bulky than conventional transformer rectifier sets. This type of power supply can be used only in special applications where the load is constant and has high impedance or where input voltage stability is not critical.

20.10   SUGGESTIONS FOR FUTURE WORK

The developed high voltage DC power supply based on Cockroft-Walton voltage multiplier circuit is a unique design and can be used for special applications like field testing of high voltage cables, for capacitor charging and as a power supply for special application like accelerators.

Inverter-based welding machines are available in the market at cheaper cost. So, it is used as high frequency source (15.8 kHz) for voltage multiplier circuit in 200 kV test generator. At present, the controlled entity of welding generator is output current. Closed loop control of output voltage in 200 kV test generator can be achieved by following ways.

1.   By modifying control circuit of the high frequency inverter.
2.   By designing a new converter with high frequency controllable output voltage.
3.   By introducing a servo controlled voltage stabilizer at the input side of the test generator and taking feedback signal from the HV divider.

ACKNOWLEDGMENT

We offer sincere thanks to Mr. Ujjwal K. Baruah for his technical guidance and constant motivation. We would also like to thank Animesh Bhat, Transformers and High Voltage Equipments Ltd. for his support during the development of project.

KEYWORDS

- DC power supply
- fast recovery diode
- frond end converter
- high frequency
- high voltage feed through
- high voltage multiplier
- high voltage transformer
- HV cable

REFERENCES

1. Kuffel, E., & Abdullah, M. (1984). High Voltage Engineering. Pergamon Press, Oxford.
2. Naidu, M. S., & Kamaraju, V. (2004). High Voltage Engineering. 3rd Edn., McGraw-Hill Company.
3. Khan, N., (2004). Lectures on Art and Science of High Voltage Engineering. Published in Pakistan.
4. Mazen, A. S., & Radwan, R. (2000). High Voltage Engineering Theory and Practice. 2nd Edn.
5. Joseph, M. B. (2008). Using rectifiers in voltage multipliers circuits. Vishay General Semiconductor.
6. Juichi, T. & Yuzurihara, I. (1988). The high frequency drive of a new multi-stage Rectifier Circuit. Kyosan Electric Mfg. Co. Ltd., 2–29, Heian-Cho, Tsurumi-Ku, Yokohama, 230, Japan, IEEE.
7. Banwari, R. et al. (2003). High Voltage Generator for 750 keV/20 kW DC Electron Accelerator.
8. Mariun, N. et al. (2006). Simulation, design, and construction of high voltage dc power supply at 15 kV output using voltage multiplier circuits. https://pdfs.semanticscholar.org/df09/bc67363ff785894d67e43efe53e81cdada3c.pdf

DESIGN, DEVELOPMENT, AND TESTING OF WATER-BASED CO-AXIAL BLUMLEIN PULSE GENERATOR

SANJAY SINGH, S. P. NAYAK, ASHUTOSH JAISWAR, T. C. KAUSHIK, and SATISH C. GUPTA

Applied Physics Division, Bhabha Atomic Research Centre, Trombay, Mumbai – 400085, India, E-mail:sanjay@barc.gov.in, spnayak@barc.gov.in, ashuj@barc.gov.in, tckk@barc.gov.in, satish@barc.gov.in

CONTENTS

ABSTRACT

In the present work, unbalanced water based co-axial Blumlein pulse generator has been designed and developed for generating fast 100 kV, 100 ns voltage pulses across a 25 Ω resistive load. Developed system is of overall length 1.7 m and diameter 450 mm. It has been tested upto 47 kV and characterized for change in pulse duration with change in dielectric temperature. A pulse of full width half maxima (FWHM) 133.33 ns is generated at 25°C dielectric temperature. Pulses of FWHM 120.00 ns and 115.50 ns are generated at dielectric temperature 38°C and 45°C, respectively. Pulse FWHM at different temperature is in agreement with decrease in the dielectric constant of de-mineralized (DM) water with increase in temperature. It is seen that generator length can be reduced for a pulse of given duration if dielectric (water) is used at low temperatures.

21.1 INTRODUCTION

High voltages are commonly generated by Marx generator in which number of capacitors are charged in parallel and then discharged in series. The voltage thus generated has relatively long rise time and pulse duration. So, output voltage generated by the Marx generator may have to be conditioned for applications requiring pulse with rise time of tens of ns. Inductive energy storage system offers a compact alternate method to generate desired voltage pulses. However, inductive storage systems need opening switches to generate voltage pulses. Opening switch technology is not well developed yet and thus designing it is relatively difficult. Whereas, fast rising ns pulses can be generated by transmission line based systems such as co-axial pulse forming line and Blumlein line. Blumlein line is selected for development in this work because output voltage of Blumlein line is equal to the charging voltage of the line whereas co-axial pulse forming line output voltage is half of the line charging voltage [8]. De-mineralized water has high dielectric constant which helps in generating a pulse of given duration with smaller generator length in comparison to other liquid dielectrics as shown in Table 21.1. However, energy stored in de-mineralized water gets discharged on its own within 7.16 μs [3] with water conductivity of 10^{-6} cm$^{-1}\Omega^{-1}$. Hence, care has to be taken while charging the line and it is charged in less than 2 μs. Other arrangements of Blumlein line can be used to generate the output voltage in multiple of line charging voltage.

TABLE 21.1   Line Length Comparison

Properties	Dielectric Constant (ε_r)	Line Length for 100 ns Pulse (m)
Liquid Dielectric:		
Transformer Oil	2.3	6
Costal Oil	4.7	4.2
DM Water	81	1.7

21.2   EXPERIMENTAL SETUP

Blumlein line has three coaxial conductors namely inner or center conductor, intermediate conductor and outer conductor. Inner and outer conductors are connected by a charging inductor. This inductor is selected such that it works as short circuit during charging of Blumlein line and as open circuit during discharging of the line [6]. Capacitor or transmission line formed between the inner and intermediate conductor (Line 1) and between intermediate and outer conductor (Line 2) as shown in Figure 21.1, are charged in parallel. During discharging, the transmission lines are discharged in series as charging inductor works as open circuit. Discharging of line is initiated by switching of Blumlein switch at desired voltage. Line is having outer diameter (OD) of inner, intermediate and outer conductors 10 mm,

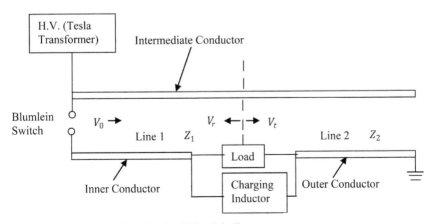

FIGURE 21.1   Equivalent circuit of Blumlein line.

82 mm and 450 mm, respectively. This gives 10.32 Ω charging impedance and 25 Ω discharging impedance. Unbalanced line has less volume hence is more compact than balanced line. However, in case of unbalanced line all the energy stored in the line is not discharged in the mainpulse, there are pulses after main pulse until whole energy is discharged. The length of the line is 1.7 m for generating 100 ns pulse across 25 Ω load with de-mineralized water as dielectric. Self-triggered pressurized spark gap is used as Blumlein switch. Dual resonance based Tesla transformer is used for stepping up of line charging voltage and parameters like inductance of primary and secondary windings of transformer are selected for negative charging of the line. Charging and output voltage of line are measured by resistive voltage dividers.

21.3   EXPERIMENTAL RESULTS AND DISCUSSION

The system developed by us is characterized for change in pulse duration with increase in dielectric temperature. Increasing the temperature of water breaks hydrogen bonds of water molecule and thus water becomes less polar due to which dielectric constant of water decreases with increasing temperature [2]. Due to decrease in dielectric constant of water, wave-velocity is increased resulting in decrease in pulse duration with increase in water temperature. The wave velocity (V) [5] is given as

$$V = \frac{c}{\sqrt{\mu_r \epsilon_r}} \, m/s \tag{1}$$

where c, ϵ_r, and μ_r are velocity of light in vacuum, relative permittivity and relative permeability of dielectric medium, respectively.

Similarly, characteristic impedance [5] of co-axial line increases as dielectric constant decreases with increase in temperature.

The characteristic impedance is given by,

$$Z_0 = 60 \sqrt{\frac{\mu_r}{\epsilon_r}} \, ln \left(\frac{r_2}{r_1}\right) \, \Omega \tag{2}$$

where r_1 and r_2 are the inner conductor OD and outer conductor ID, respectively.

The conductivity of water also increases with temperature due to which self discharge time of line will be reduced. This is described in Ref. [3]:

$$t_{sd} = \frac{\epsilon_r}{36 \times \pi \times 10^{11} \times \sigma} \ \text{s} \qquad (3)$$

where σ is conductivity of water in $cm^{-1}\Omega^{-1}$.

A typical output voltage pulse for 25 Ω load is shown in Figure 21.2 and it can be seen that there is pre-pulse as charging inductor does not work perfect short circuit during charging of the line and line is unbalanced hence there is post pulse as well. Effect of dielectric temperature on pulse duration and pulse amplitude is given in Tables 21.2 and 21.3, respectively.

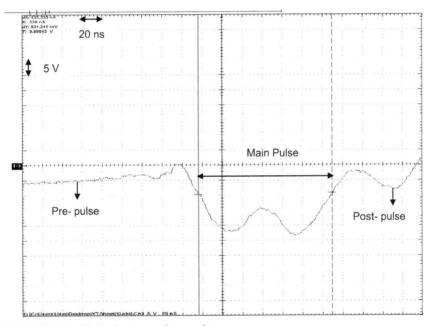

FIGURE 21.2   Typical output voltage pulse.

TABLE 21.2   Pulse Duration Variation with Dielectric Temperature

Sr. No.	Dielectric Temperature (°C)	Theoretical Pulse Duration (ns)	Experimental Pulse Duration (ns)
1	25	100.11	133.33
2	38	96.70	120.00
3	45	94.83	115.50

TABLE 21.3 Pulse Amplitude Variation with Dielectric Temperature

Sr. No.	Dielectric Temperature (°C)	Line Charging Voltage (kV)	Theoretical Pulse Amplitude (kV)	Experimental Pulse Amplitude (kV)
1	25	47.00	46.06	47.12
2	38	44.40	42.62	45.23
3	45	42.88	40.90	41.01

As shown in Table 21.2 there is difference in theoretical and experimental pulse duration. This appears to be arising from the capacitance and inductance of connecting wires, Blumlein switch, etc., while there is not much change in experimental and theoretical pulse amplitude due to change in the characteristic impedance of line because of increase in water temperature as shown in Table 21.3.

21.4 CONCLUSION

Co-axial Blumlein line has been designed and developed with de-mineralized water as dielectric medium. It is tested upto 47 kV and pulse of FWHM 133.33 ns is generated at room temperature, 25°C. It is observed from experimental data that increasing the water temperature reduces output pulse duration for a given line which is expected, while there is not much change in pulse amplitude due to change in the characteristic impedance of line because of increase in water temperature. Decreasing temperature will increase dielectric constant of water so wave velocity will decrease [2] that will give higher output duration pulse for a line of given length or else length of line can be reduced for a pulse of given duration by using colder water as dielectric medium.

KEYWORDS

- **Blumlein line**
- **de-mineralized water**
- **dielectric constant**

- **increase in water temperature**
- **pulse duration**
- **unbalanced line**
- **wave velocity**

REFERENCES

1. Hanjoachim, B., (2006). "Pulsed Power Systems: Principles and Applications," Springer-Verlag, Berlin-Heidelberg, Germany.
2. Cebedeau, (1965). Book of Water (Vols. 1 and 3) (Les constantesde l'eau), Cebedoc.
3. Mesyats, G. A., (2005). "Pulsed Power," Springer Inc., USA.
4. Naidu, M. S., & Kamaraju, V., (2011)."High Voltage Engineering," Tata McGraw-Hill, Delhi.
5. Pai, S. T., & Zhang, Q., (1995)."Introduction to High Pulse Technology," World Scientific Publishing Co. Pvt. Ltd., Singapore.
6. Pietruszka, B. R. (1989). "Operation and Characteristics of the Flash X-Ray Generator at the Naval Post Graduate School," Master of Sciencein Physics Thesis, Naval Postgraduate School Monterey, California.
7. Shotts, Z., Rose, F., Merryman, S., & Kirby, R., (2005). "Design Methodology for Dual Resonance Pulse Transformers," *IEEE Pulsed Power Conference,* 1117–1120.
8. Smith, P. W., (2002)."Transient Electronics: Pulsed Circuit Technology," John Wiley & Sons, Haboken, New Jersey.
9. Water Structure and Science. http://www.lsbu.ac.uk/water/explan3.html/. Accessed on 10.09.2012.

CHAPTER 22

OPTIMIZATION OF MgB$_2$-BRASS JOINT RESISTANCE FOR SST-1 SUPERCONDUCTING MAGNET CURRENT LEADS

U. PRASAD, V. L. TANNA, and S. PRADHAN

Institute for Plasma Research, Bhat, Gandhinagar, India,
E-mail: upendra@ipr.res.in

CONTENTS

ABSTRACT

MgB$_2$ superconducting strands characterization has been initiated for the suitability of SST-1 superconducting magnets current leads. The DC I-V characteristics of commercial strands are being studied in J$_c$ facility around 20 K at IPR. The conventional NbTi and copper based current leads will be replaced by MgB$_2$ and brass based current leads. The considerable amount of cryogenic cost saving is expected after adopting MgB$_2$ and brass based current leads because they will be operated ~20 K instead of 4.2 K. MgB$_2$ superconductor is used as a link between magnet leads and bottom of the brass wires. In order to optimize MgB$_2$-copper and MgB$_2$-brass joint resistance for the suitability for this type of current leads, various joint configurations has been fabricated and tested up to 4.2 K and encouraging results have been achieved. The joint configurations design, fabrication and validation are highlighted in this chapter.

22.1 INTRODUCTION

MgB$_2$ is an intermediate temperature superconductor (SC). Its critical temperature 39 K is higher than the previously known metallic superconductors. Because of the higher critical temperature it is considered as an attractive potential candidate for operation around 20 K which can be easily achieved using helium gas with ought liquid cryogen. The experimentally observed critical current of multi-core MgB$_2$ strands is ~200 A in the transverse magnetic field of 3–4 Tesla at operating temperature of ~ 20 K. Because of the higher critical current in low magnetic field at temperature higher than liquid helium it has potential future application in tokamaks for SC magnet current leads. The current leads are used in tokamak devices to deliver room-temperature electrical power to cryogenic SC magnets which are being operated at supercritical helium temperature (4.5 K) environment. The Joule heating produced at the interface of SC-metal joint by the transport electrical current due to the temperature difference between the two ends by means of conduction makes the current lead dominant heat source into the SC magnet system working at cryogenic temperature. In order to minimize the heat load at magnet system, the intermediate SC is preferred for the current leads.

The optimally designed current leads have temperature distribution right from supercritical helium to helium vapor which can conduct electrical current reliably as well as minimize the amount of heat load. The commercially manufactured MgB$_2$ strands [1, 2], are considered to be the alternative of NbTi superconductors for the current leads for fusion magnet applications, since the critical temperature of this superconductor is about 39 K, it can be operated either at helium gas ~ 22 K. Due to the large temperature margin, it is very much advantageous in reduction of helium consumption and remarkably low operating cost. The present work is focused on the preparation of low resistance joint of MgB$_2$ based superconducting wires with the brass wires for the SC magnets of SST-1 tokamak current leads [3] applications at the Institute for Plasma Research (IPR). Figure 22.1 shows a schematic view of the brass and MgB$_2$-based 10,000 A rated current leads for SST-1 tokamak. The joint configurations design, fabrication, and validation have been highlighted in this chapter.

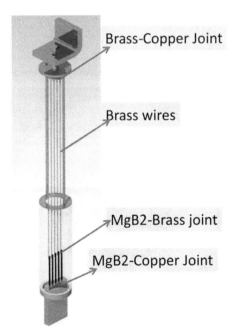

FIGURE 22.1 10,000 A rated brass and MgB$_2$-based current leads for SST-1 tokamak.

22.2   MGB2 STRANDS DETAILS

Recently, we procured MgB_2 strands from M/s. Hyper Tech Research Inc., with the following technical specifications as summarized in Table 22.1. Figure 22.2 shows the mono filament architecture of MgB_2 wire with sheath material as Monel and niobium as barrier material. Recent advances in MgB_2 conductors are leading to a new level of performance.

TABLE 22.1   Technical Details of MgB_2 Wires from M/s. Hyper Tech Research, Inc.

Parameters	Specifications
Strand Type	Monofilament (MgB_2)
Wire diameter (in mm)	0.83
Wire cross-sectional area (in mm²)	0.54
Sheath material	Monel
Thickness of Monel layer (µm)	~109
Barrier material	Nb
Thickness of Nb layer (in µm)	60
MgB_2 core diameter (in µm)	496
Superconducting fraction (in %)	26
Ic of the strand at self-field 4.2 K (in A)	270

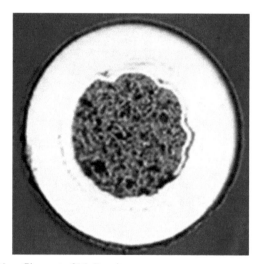

FIGURE 22.2   Monofilament of MgB_2 wire.

Based on the use of proper powders, proper chemistry, and an architecture which incorporates internal Mg diffusion (IMD), a dense MgB$_2$ structure with not only a high critical current density (*Jc*), but also a high engineering critical current density (*Je*), can be obtained. The best layer *Jc* for the sample is reported as 1.07×10^5 A/cm^2 at 10 T, 4.2 K, and the best *Je* is seen to be 1.67×10^4 A/cm^2 at 10 T, 4.2 K.

22.3   BRASS WIRE DETAILS

The technical specification of brass wire has been shown in Table 22.2. The brass wires bundle used for joint making has been shown in Figure 22.3. One end of this bundle has been soldered with MgB$_2$ wires and another end with copper lug.

TABLE 22.2   Technical Details of Brass Wires from M/s. Senor Metals Pvt. Ltd.

Parameters	Specifications
Strand type	Brass
Alloy	Cu:Pb:Fe:Zn
Chemical composition (in %)	78.90:0.002:0.003:Reminder
Wire diameter (in mm)	1.0
Wire cross-sectional area (in mm^2)	0.785
Brass RRR (measured)	1.995
Engineering current density (in A/mm^2)	2.5
Density (in kg/m^3)	8664
Average specific heat [300–600 K] (in J/kg-K)	570
Average resistivity [300–600 K] (in Ohm-m)	72×10^{-9}

FIGURE 22.3   Bunch of brass (80% Cu: 20% Zn) wires.

22.4   COPPER AND SOLDER DETAILS

Copper of residual resistance ratio (RRR) around 10 has been used for sleeve and silver based solder has been used for joint making. The technical specification of copper and solder has been shown in Tables 22.3 and 22.4.

22.5   JOINT SAMPLE MAKING

Copper sleeve was fabricated and cleaned with nitric acid and ringed with soap water. The sample was finally cleaned ultrasonically before assembly with MgB_2 wire. The solder filled assembly of brass, copper and MgB_2 was heated around 250°C for solder melting. Joint sample was acetone cleaned after cooling at room temperature. The joint assembly has been shown in Figure 22.7.

22.6   RRR AND JOINT RESISTANCE MEASUREMENT

RRR and joint resistances have been measured with standard four probe technique using 100 mA direct current (DC) current source and precision

TABLE 22.3   Technical Details of Copper Sleeve Material

Sr. No.	Constituents	Composition
1	Cu	99.90%
2	Bi	~ 5 PPM
3	O	400 PPM
4	Pb	50 PPM
5	Others	~300 PPM

TABLE 22.4   Technical Details of Solder Material

Sr. No.	Particular	Observed (%)
1	Copper (Cu)	Nil
2	Lead (Pb)	43.35
3	Tin (Sn)	56.50
4	Antimony (Sb)	0.100
5	Total impurities	0.05

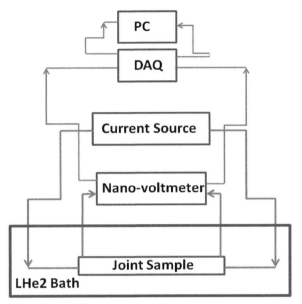

FIGURE 22.4 Schematic of residual resistivity ratio (RRR) and joint resistance measurement.

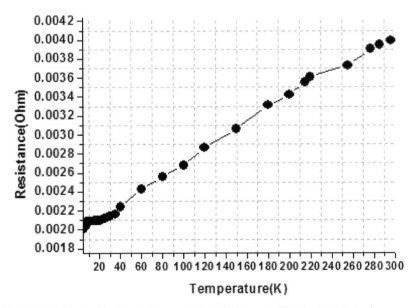

FIGURE 22.5 Residual resistivity ratio (RRR) of brass (80% Cu: 20% Zn) wires.

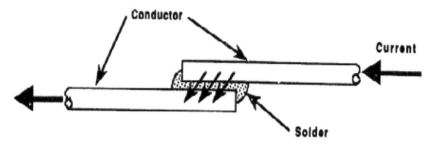

FIGURE 22.6   Typical joint concept for strand-to-strand joint.

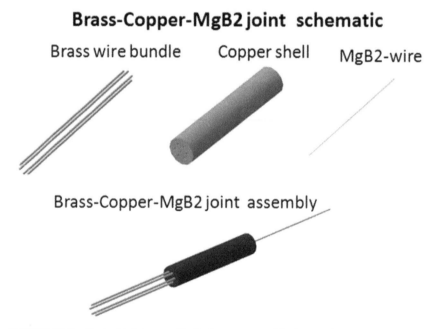

FIGURE 22.7   Typical joint assembly for brass-copper-MgB$_2$ strands.

Nano voltmeter in the temperature range of 4.2 K–25 K. The schematic of measurement technique has been shown in Figure 22.4, which includes liquid helium (LHe) bath, sample, current taps, current flow, voltage taps for voltage measurement, data acquisition (DAQ) system, and computer (PC).

22.7   JOINT CONFIGURATION USING BRASS – MGB$_2$

The first successful splice, reported in 2005, involves splicing NbTi wire and MgB$_2$ tape for an MgB$_2$ coil operated at 4.2 K in persistent mode [4]. Another MgB$_2$ – MgB$_2$ joint result, based on a field-decay measurement of MgB$_2$ loop carrying 30 to 40 A at 20 K in zero background field, was reported in 2006, having a joint resistance of <10^{-14} Ω [5]. We have tried three different geometries for manufacturing the joint under optimized joint resistance measurements, out of which the best geometry was selected as MgB$_2$ and brass wires in copper sleeve with an experimental RRR measurements reveal that RRR of brass and MgB$_2$ strand: 1.99 and 39.6, respectively.

22.8   PERFORMANCE VALIDATION TEST RESULTS OF JOINT RESISTANCE

RRR measurements of brass wire and MgB$_2$ wire has been shown in Figures 22.1 and 22.8. The brass-MgB$_2$ joint, MgB$_2$-copper joint and

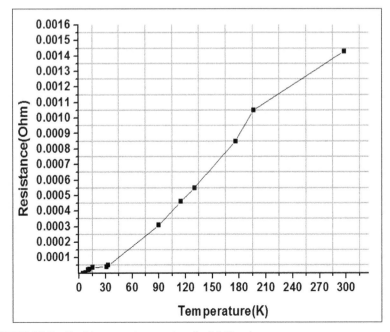

FIGURE 22.8   Resistance vs. temperature for MgB$_2$ wires.

brass-copper-MgB$_2$ joint resistances as a function of temperature have been shown in Figures 22.9–22.11, respectively. The measurement results have been summarized in Table 22.5.

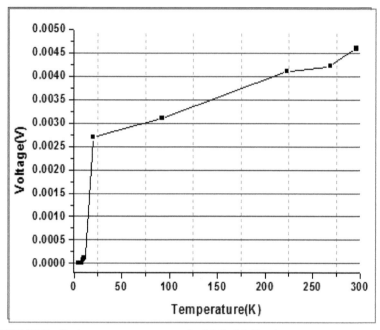

FIGURE 22.9   Resistance vs. temperature for MgB$_2$-MgB$_2$ wires.

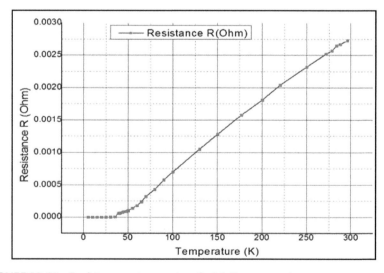

FIGURE 22.10   Resistance vs. temperature for MgB$_2$-copper wires.

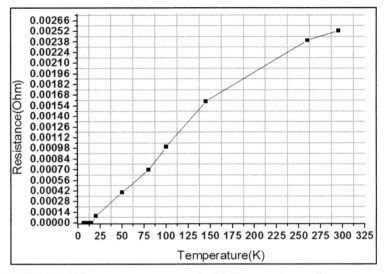

FIGURE 22.11   Voltage–temperature curve for MgB$_2$-copper-brass wires.

TABLE 22.5   Joint Resistance Measurement Summary

Parameters	Description/typical value
Suitable joint geometry	MgB$_2$ and Brass wires in copper sleeve
RRR of brass	1.99
RRR of brass of MgB$_2$	39.69
MgB$_2$-brass joint resistance @ 5 K, 20 K and 25 K	$7.2 \times 10^{-7}\,\Omega$, $2.48 \times 10^{-6}\,\Omega$, $2.64 \times 10^{-6}\,\Omega$
MgB$_2$-copper joint resistance @ 5 K, and 25.27 K	$1 \times 10^{-7}\,\Omega$, $2.7 \times 10^{-6}\,\Omega$
Brass-copper-MgB$_2$-joint resistance @ 5 K	$4.7 \times 10^{-7}\,\Omega$

22.9   CONCLUSIONS

In this chapter, we tried to summarize the selection of proper strands wires of brass and MgB$_2$ with suitable electrical properties for the higher current carrying applications, where there is a great importance of optimization of joint resistance and establishing a procedure and protocols for the process of manufacturing brass-copper and MgB$_2$ wires with measured MgB$_2$-brass joint resistance at different operation temperature to save the precious cryogenic cold capacity as $7.2 \times 10^{-7}\,\Omega$ at 4.2 K, $2.48 \times 10^{-6}\,\Omega$ at

20 K, and $2.64 \times 10^{-6}\,\Omega$ at 25 K, respectively. The reported value of MgB_2-copper joint resistance is $1.0 \times 10^{-7}\,\Omega$ at 4.2 K, $2.70 \times 10^{-6}\,\Omega$ at 25 K, and finally using optimized method of joint, the measured MgB_2-copper-brass strands joint resistance was found to be $4.7 \times 10^{-7}\,\Omega$ at 4.2 K.

KEYWORDS

- **current leads**
- **low resistance joint**
- **MgB_2 strand**
- **RRR**
- **SC magnets**
- **tokamak**

REFERENCES

1. Collings, E. W., & Lee, E. (2003). Continuous and batch-processed MgB_2/Fe strands transport and magnetic properties, *Physica C, 386*, 555–559.
2. Iwasa, Y. (1976). Superconducting joint between multi-filamentary wires (Part II)—Joint evaluation technique. *Cryogenics, 16*, 217.
3. Tanna, V. L., Sarkar, B. et al. (2004). Superconducting current feeder system with associated test results for SST-1 Tokamak. *IEEE Trans. Appl. Supercond. 2*(2).
4. Takahashi, M., & Tanaka, K. et al. (2006). Relaxation of trapped high magnetic field in 100 m-long class MgB_2 solenoid coil in persistent current mode operation. *IEEE Trans. Appl. Superconductivity, 16*, 1431.
5. Penco, R., & Grasso, G. (2007). Recent development of MgB_2-based large scale applications. *IEEE Trans. Appl. Supercond. 17*(2), 2291.

CHAPTER 23

STUDIES ON EFFECT OF GASEOUS QUENCHING MEDIA ON PERFORMANCE OF ELECTRICALLY EXPLODED FOILS

S. P. NAYAK, A. CHOWDHURY, M. D. KALE, T. C. KAUSHIK, and S. C. GUPTA

Applied Physics Division, Bhabha Atomic Research Centre, Trombay, Mumbai – 400085, India, E-mail: spnayak@barc.gov.in, ankurc@barc.gov.in, kalemd@barc.gov.in, tckk@barc.gov.in, satish@barc.gov.in

CONTENTS

ABSTRACT

Electrically exploded foil (EEF) has been previously used to sharpen current pulses generated by relatively slow discharging energy storage capacitors. They are operated as opening switches to generate fast rising voltage/current pulses in loads connected across them. Magnitude of voltage/current pulse in load and its rise-time is greatly influenced by the performance of EEF acting as opening switches. Parameters such as cross-sectional area of EEF, its constituent material and quenching medium surrounding it play a major role in deciding its performance. In the present work, EEF has been placed in two gaseous quenching medium, i.e., dry air and N_2. Keeping circuit parameters fixed geometrical parameters of EEF as also the gaseous quenching medium and the pressure at which gas is filled in experimental chamber has been varied. Effects on parameters such as peak load current, time of burst, switch current and peak switch voltage of switch have been studied and several inferences have been derived.

23.1 INTRODUCTION

Fast rise-time high pulsed high currents have been used for many interesting applications such as generation of high temperature plasma for fusion and X-rays [4, 5, 14], shock less compression of materials up to high pressure [1, 6, 15], electric guns [7, 13], etc. All such applications require a sub-microsecond rise-time, high current pulsed current to drive its load. Fast capacitor banks technology has been vastly used for generation of high current pulses [1, 6, 14]. Earliest technique used large numbers of capacitors charged and discharged in parallel. Such capacitors are carefully designed to store kJs of energy, capable of being charged to high voltages and provide equivalent series inductance (ESL) of the order of few tens of nano-henry. Such capacitors have been successfully used to generate current rise rate of the order of 10^{12} A/s. However, capacitors have smaller ESL with inherent high cost and thus fast current pulses become relatively difficult to generate. Other methods developed for such applications are Marx generators coupled with pulse forming lines and magnetically insulated transmission lines (MITL) [11] and linear transformer drivers (LTD) [8, 9]. However, such systems become bulky and occupy relatively large space if used to generating several hundreds of kilo-amperes to mega-ampere range current pulses. Another technique which has been demonstrated using capacitor bank and electrically exploded conductor (EEC)

[10] also generates fast rise-time high current pulses. Inductive energy storage systems (IESS) utilize this technique. In IESS, energy is stored as magnetic energy in inductance. As a result, it has higher energy density than electrostatic energy storage systems.

Janes and Koritz [12] first demonstrated voltage pulse sharpening using electrically exploding wire based IESS. Early and Martin [3] reported generation of fast rising high pulsed current system and were able to achieve a current rate of 1.5×10^{13} A/s. Maisonnier et al. [10] studied the effect of opening switch parameters on time of burst and gave a detailed empirical theory on estimating geometrical parameters for desired time of burst. However, effects of changing opening switch parameters on parameters such as opening time, final resistance of EEC, load current, etc., have not been studied in details. Di Marco et al. [2] studied the effects of changing switch parameters on parameters such as voltage developed across EEC and its opening time. However, a study on its effect on load current has not yet been reported.

In the present work, a detailed study has been done on systems consisting of capacitor banks having rise-time of the order of a few microseconds and EEC. In this study, parameters of capacitor bank such as its overall capacitance and inductance has been kept constant. Thickness of EEC has also been kept constant as per as availability. Width, length, and quenching medium have been varied systematically so as to understand its effect on parameters such as opening time, peak resistance achieved by EEF upon opening, energy deposited in EEF and energy transferred into load. In next section, description of overall system has been presented. Following section contains experimental results obtained. In the final section, conclusions and inferences have been noted based on the results.

23.2   SYSTEM DESCRIPTION

Figure 23.1 shows the equivalent circuit diagram of entire setup, respectively. The input circuit consists of an energy storage capacitor, stray inductance of the circuit, and EEC-based opening switch. Capacitor is charged to desired voltage and discharged into opening switch through a triggered spark gap. EEC explodes and interrupts the current thereby generating a pulsed high voltage across it. A combination of peaking gap and low inductance load is connected in parallel with the EEC. As a result,

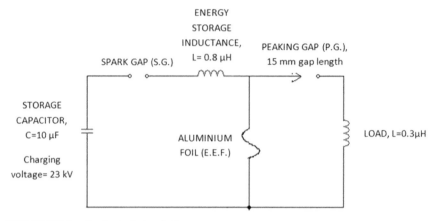

FIGURE 23.1   Equivalent circuit diagram.

same voltage gets applied across the peaking gap and low inductance load. This leads to sudden flow of current in the load and transfer of energy from capacitor to stray circuit inductor and finally into load. In order to prevent electrical breakdown across EEC at the time of opening, it is surrounded by a quenching medium. 99.5% pure aluminum foils have been chosen in these experiments as EEC. In our study, gaseous quenching medium dry air and N2 have been chosen and filled around EEC. Gases have been filled at 2 and 3 bars to look at the effect of pressure on parameters under study. A stainless steel container has been designed to contain shock waves generated due to EEC explosion. Thickness of the foil has been kept constant at 25 microns as per availability whereas width and length have been varied. Charging voltage, stray inductance of the circuit, and capacitance has been kept constant throughout the experiments in order to ascertain the effect of geometrical variations of EEC. Width of the foil and length have been changed and placed under gaseous quenching medium at different pressures.Current and voltage measurements have been done using Rogowski coils and aqueous $CuSO_4$ resistive dividers. Current discharged by the capacitor, current flowing through EEC, and current flowing through load has been measured. Voltage generated across EEC and load has been recorded. Differentiating type Rogowski coils have been used and their output has been passively integrated using RC integrators. Rogowski coils have been calibrated against a standard Pearson make CT and voltage dividers have been calibrated against a standard North

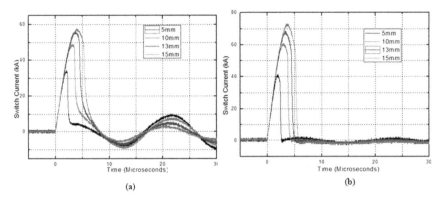

FIGURE 23.2   Switch current for varying cross-section within dry air quench medium having length. (a) 150 mm and (b) 250 mm.

Star make voltage probe. Signals have been recorded using digital storage oscilloscopes placed inside a faraday cage.

23.3   OBSERVATIONS AND RESULTS

The entire experimental plan was divided into three stages. Firstly, the quenching medium is taken as dry air at 2 bar (gauge) pressure. Experimental shots had been taken in four sets, where in each set, the length of the EEF is kept constant and the width has been taken as 5 mm, 10 mm, 13 mm, and 15 mm. In the next set, the length is increased and the same procedure is followed. Each shot (corresponding to a definite length and width) has been repeated to ensure consistency of the results. The length is also varied four times as 100 mm, 150 mm, 200 mm, and 250 mm. Then, the same procedure is repeated for dry air at 3 bar gauge pressure. Finally, the entire procedure has been carried out for N_2 at 2 bar gauge pressure. This way, the effect of change in foil parameters, change in quenching medium and also, change in pressure of quenching medium has been observed. Variation of time of burst for a 150 mm and 250 mm long EEC, dry air as quenching medium filled at a 2 bar gauge pressure and varying cross-section area of EEC have been shown in Figure 23.2.

From Figure 23.2, it can be seen that with increasing cross-section (width) of EEC, time of burst also increases. By varying length and keeping cross-section constant, time of burst almost remains same. A detailed

TABLE 23.1   Time of Burst

Quenching medium	Length/ Width	100 mm	150 mm	200mm	250 mm
Dry air; 2 bar (g)	5 mm	2.2	2.25	2.5	2.5
Dry air; 3 bar (g)		2.25	2.5	2.5	2.5
Nitrogen; 2 bar (g)		—	2.5	3.0	2.5
Dry air; 2 bar (g)	10 mm	3.8	3.5	4.0	5.5
Dry air; 3 bar (g)		4.0	3.5	4.0	3.75
Nitrogen; 2 bar (g)		—	3.75	3.75	3.75
Dry air; 2 bar (g)	13 mm	4.4	4.5	4.75	5.0
Dry air; 3 bar (g)		4.25	4.5	5.0	5.0
Nitrogen; 2 bar (g)		—	4.75	4.5	4.75
Dry air; 2 bar (g)	15 mm	5.0	5.5	4.5	5.25
Dry air; 3 bar (g)		5.0	5.0	5.5	5.5
Nitrogen; 2 bar (g)		—	5.25	5.5	5.75

TABLE 23.2   Peak Switch Voltage (kV)

Quenching medium	Length/ Width	100 mm	150 mm	200mm	250 mm
Dry air; 2 bar (g)	5 mm	65	96	62	58
Dry air; 3 bar (g)		42	50	58	24
Nitrogen; 2 bar (g)		—	40	50	35
Dry air; 2 bar (g)	10 mm	32	76	—	50
Dry air; 3 bar (g)		36	40	56	30
Nitrogen; 2 bar (g)		—	30	26	34
Dry air; 2 bar (g)	13 mm	32.5	70	92	48
Dry air; 3 bar (g)		28	32	29	44
Nitrogen; 2 bar (g)		—	34	36	33
Dry air; 2 bar (g)	15 mm	31.5	45	98	38
Dry air; 3 bar (g)		30	26	34	34
Nitrogen; 2 bar (g)		—	42	30	36

variation of time of burst with change in cross-section, length, quenching medium and its fill pressure is presented in Table 23.1. These observations are consistent with results already reported in literature [2, 10]. It can also be observed that time of burst is independent of quenching gas and its fill pressure. Similarly, Figure 23.3 shows variation in switch voltage for 150 mm and 250 mm long EEC, dry air quenched filled at 2 bar gauge pressure up on varying cross-section.

Table 23.2 shows the different peak switch voltages obtained for varying width and length of EEC, gaseous quenching medium and its fill pressure. Here, it can be seen that switch voltage peaks with 150 mm or 200 mm length and is lesser for lower and higher lengths. Possible explanation for such an observation may be that final resistance is optimally high and rate of rise of switch resistance is higher in comparison to other cases. It can also be observed from Table 23.2 that highest switch voltage is achieved when dry air was filled at 2 bar gauge pressure. It could be understood on a way that higher fill pressure provides opposition to optimum rate of expansion of vapors of EEC which ultimately leads to lower final resistance and lower rate change of resistance of switch. Figure 23.4 shows load current for 150 mm and 250 mm long EEC, dry air as quenching medium filled at 2 bar gauge pressure. Table 23.3 shows peak load current generated and Table 23.4 shows time taken for load current to reach the peak. It has been observed that peak load current also reduces as cross-section of EEC increases which can be attributed to lower energy transfer to load in such cases. Upon increasing gas fill pressure, peak load current increases. This may be because lower final resistance is achieved and lesser energy is dissipated in switch after burst. Hence, greater energy is transferred into load and higher load current is generated in load.

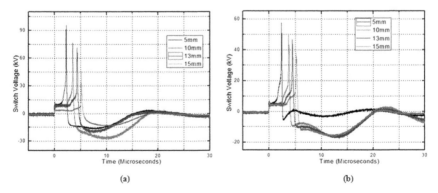

FIGURE 23.3 Switch voltage for varying cross-section within dry air quench medium having length. (a) 150 mm and (b) 250 mm.

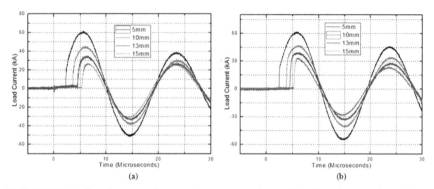

FIGURE 23.4 Load current for varying cross-section within dry air quench medium having length. (a) 150 mm and (b) 250 mm.

TABLE 23.3 Peak Load Current (kA)

Quenching medium	Length/Width	100 mm	150 mm	200mm	250 mm
Dry air; 2 bar (g)	5 mm	15	20	30	25
Dry air; 3 bar (g)		15	25	30	35
Nitrogen; 2 bar (g)		—	20	20	35
Dry air; 2 bar (g)	10 mm	12.5	18	25	33
Dry air; 3 bar (g)		15	20	30	40
Nitrogen; 2 bar (g)		—	18	30	30
Dry air; 2 bar (g)	13 mm	10	15	20	30
Dry air; 3 bar (g)		10	20	28	32
Nitrogen; 2 bar (g)		—	12	25	35
Dry air; 2 bar (g)	15 mm	10	10	—	30
Dry air; 3 bar (g)		7.5	20	25	30
Nitrogen; 2 bar (g)		—	12	28	30

Change in gaseous quenching medium and increase in gas fill pressure results in slight change in time to reach peak load current which shows that effects of these are negligible. Experiments with higher pressure have not been conducted yet and are being planned.

TABLE 23.4 Time to Peak Load Current

Rise Time of Load Current (µs) (0–100%)	Quenching medium	100 mm	150 mm	200mm	250 mm
5 mm	Dry air; 2 bar (g)	3.496	3.42	—	3.352
	Dry air; 3 bar (g)	3.28	3.352	3.218	3.13
	Nitrogen; 2 bar (g)	—	3.216	3.46	3.148
10 mm	Dry air; 2 bar (g)	2.112	2.528	—	2.266
	Dry air; 3 bar (g)	2.328	2.12	2.116	2.258
	Nitrogen; 2 bar (g)	—	2.548	2.396	2.3
13 mm	Dry air; 2 bar (g)	1.632	1.79	—	1.364
	Dry air; 3 bar (g)	1.9	1.712	1.232	0.92
	Nitrogen; 2 bar (g)	—	1.656	1.568	1.648
15 mm	Dry air; 2 bar (g)	1.304	1.1	—	0.688
	Dry air; 3 bar (g)	1.49	1.3	1.028	0.6
	Nitrogen; 2 bar (g)	—	1.432	1.122	0.668

23.4 CONCLUSION

In the present experiments, EEF have been placed in two gaseous quenching medium, i.e., dry air and N_2. With fixed cross-section and quenching medium, peak voltage generated across EEF increases up to a particular length and then decreases. However, upon increasing fill pressure, voltage decreases. Higher peak voltage across switch is obtained when dry air is used instead of N_2. Peak load current and load power increases up to a particular length and then decrease at constant cross-section and quenching medium. Upon increasing gas fill pressure, both peak load current and load power increases. It is again observed that with dry air as quenching medium, higher peak load current has been demonstrated. Quenching medium and its fill pressure have been observed not to play a significant role in current rise rates. However, scope of a greater detailed study still exists and needs to carried out in order properly understand effect on these parameters. In present study, dry air filled at lower pressure has been seen to perform better than N_2 or dry air filled at higher pressure.

KEYWORDS

- electrically exploded conductor
- high current interruption
- inductive energy storage system
- opening switches
- pulsed currents/voltages
- pulsed power system

REFERENCES

1. Asay, J. R. (2001). 12th APS Topical Conference on Shock Compression of Condensed Matter, 849.
2. Di Marco, J. N., & Burkhardt, L. C. (1970). Characteristics of a Magnetic Energy Storage System using Exploding Foils. *Journal of Applied Physics, 41*(9), 3894–3899.
3. Early, H. C., &Martin, F. J., (1965). Method of Producing a Fast Current Rise from Energy Storage Capacitors. *Review of Scientific Instruments, 36*(7), 1000–1002.
4. Garanin, S. G., Ivanovsky, A.V., & Mkhitariyan, L. S., (2011). An ICF system based on Z-Pinch Radiation Produced by an Explosive Magnetic Generator. *Nuclear Fusion 51*(103010), 1–15.
5. Haines, M. G. (2011). A Review of the Dense Z-Pinch. *Plasma Physics and Controlled Fusion, 53*(093001), 1–168.
6. Hall, C. A., Asay, J. R., Knudson, M. D., Stygar, W. A., Spielman, R. B., Pointon, T. D., Reisman, D. B., Toor, A., & Cauble, R. C. (2001). Experimental Configuration for isentropic compression of solids using Pulsed Magnetic Loading. *Review of Scientific Instruments, 72*, 3587.
7. Kaushik, T. C., Kulkarni, L. V., Auluck, S. K. H. (2002). Feasibility Studies on Performance Enhancement in Electrically Exploding Foils Accelerators. *IEEE Transactions on Plasma Science, 30*(6), 2133–2138.
8. Kim, A. A., Mazarakis, M. G., Sinebryukhov, V. A., Kovalchuk, B. M., Visir, V. A., Volkov, S. N., Bayol, F., Bastrikov, A. N., Durakov, V. G., Frolov, S. V., Alexeenko, V. M., McDaniel, D. H., Fowler, W. E., LeChien, K., Olson, C., Stygar, W. A., Struve, K. W., Porter, J., & Gilgenbach, R. M. (2009). Development and Test of a fast 1MA linear transformer driver stages. *Physical Review Special Topics–Accelerator and Beams, 12*(050402), 1–10.
9. Kovalchuk, B. M., Vizir, V. A., Kim, A. A., Kumpjak, E. V., Loginov, S. V., Bastrikov, A. N., Chervjakov, V. V., Tsou, N. V., Monjaux, P., & Kh'yui, D. D. (1997). *Russian Physics Journal 40*, 1142.

10. Maisonnier, C. H., Linhart, J.G., & Gourlan, C. (1966). Rapid Transfer of Magnetic Energy by Means of Exploding Foils. *Review of Scientific Instruments, 37*(10), 1380–1384.

11. McDaniel, D. H., Mazarakis, M. G., Bliss, D. E., Elizondo, J. M., Harjes, H. C., Ives, H. C., Kitterman, D. L., Maenchen, J. E., Pointon, T. D., Rosenthal, S. E., Smith, D. L., Struve, K. W., Stygar, W. A., Weinbrecht, E. A., Johnson, D. L., & Corley, J. P. (2002). Conference Record of the Twenty-Fifth International Power Modulator Symposium, 2002 and 2002 High-Voltage Workshop, 252.

12. Sargent Janes, G., & Kortiz, H. (1959). High Power Pulse Steepening by Means of Exploding Wires. *Review of Scientific Instruments, 30*(11), 1032–1037.

13. Saxena, A. K., Kaushik, T. C., & Gupta, S. C. (2010). Shock Experiments and Numerical Simulations on Low Energy Portable Electrically Exploding Foil Accelerators. *Review of Scientific Instruments, 81*(3), 033508.

14. Spielman, R. B., Deeney, C., Douglas, M. R., Chandler, G. A., Cuneo, M. E., Nash, T. J., Porter, J. L., Ruggles, L. E., Sanford, T. W. L., Stygar, W. A., Struve, K. W., Matzen, M. K., McDaniel, D. H., Peterson, D.L., & Hammer, J. H. (2000). Wire Array *z* Pinches as Intense X-Ray Sources for Inertial Confinement Fusion. *Plasma Physics and Controlled Fusion, 42*, B157–B164.

15. Tasker, D. G., Goforth, J. H., & Oona, H. (2007). Design Improvements to High-Explosive Pulsed-Power Isentropic-Compression Experiments. *IEEE Transaction on Plasma Science, 38*(8), 1828–1834.

PART IV

LASER PLASMA AND INDUSTRIAL APPLICATION OF PLASMA

CHAPTER 24

PARTICLE ACCELERATION BY WHISTLER PULSE IN HIGH DENSITY PLASMA

PUNIT KUMAR, ABHISEK KUMAR SINGH, and SHIV SINGH

Department of Physics, University of Lucknow, Lucknow – 226007, India, E-mail: punitkumar@hotmail.com

CONTENTS

ABSTRACT

The acceleration of electron by the ponderomotive force of a Gaussian whistler pulse is a magnetized high density quantum plasma obeying Fermi-Dirac distribution using the recently developed quantum hydrodynamic (QHD) model. Effective acceleration takes place when the peak whistler amplitude exceeds a threshold value and the whistler frequency is greater than cyclotron frequency. The threshold amplitude decrease with ratio of plasma frequency to electron cyclotron frequency. The electron gain at velocities about twice the group velocity of the whistler.

24.1 INTRODUCTION

Whistler waves are important not only in space plasma due to wave particle interactions, but also in laboratory plasmas as helicons for efficient plasma production as well as acceleration [15, 15, 18, 21–23, 34]. Large amplitude whistlers propagating in a magnetized plasma can initiate a great variety of non-linear effects, modulation instability [34], and the subsequent soliton formation [15, 18, 21]. Whistlers also contribute to fast magnetic reconnection and plasma dynamics in two beam laser–solid density plasma interaction experiment. Whistler can make electrons go faster to high energies. Chen et al. [3] showed clearly that particles may go faster by plasma magnetowaves, and plentiful in astrophysical settings. When PIC simulations become active they have shown good results for celestial acceleration [4]. Lower hybrid waves have been used to maintain runaway electrons in a tokamak to give the toroidal current. Recent experiments have observed MeV electrons during lower hybrid heating/current drive in the tokamak.

Later, a significant progress has been made in the field of plasma-based accelerators for attaining high electron energies [1, 5, 12, 20, 27, 30]. The advent of high-power lasers have led to scheme that can produce large phase velocity plasma waves and it can accelerate electron to hundreds of MeV energy. Work on acceleration by ponderomotive force of the laser pulse have been pursued vigorously and electron acceleration approaching GeV energy is being achieved [14, 17, 24, 31, 33, 35]. The electron energies are far in excess of ponderomotive potential energy and the acceleration is a consequence of direct exchange of energy between electron and laser via betatron resonance. Gahn et al. [7] reported generation of multi-MeV electrons by direct laser acceleration in high-density plasma channels. Tsakiris et al. [32] developed a theoretical laser model of direct acceleration. Microwave and radio waves provide another frequency regime for particle acceleration.

The low temperature dense electron plasmas are degenerate and obey Fermi-Dirac statistics. The high-density and low-temperature quantum Fermi plasma is significantly different from the low-density and high-temperature "classical plasma" obeying Maxwell-Boltzmann distribution. The study of highly dense quantum plasma becomes important, when the de-Broglie thermal wavelength associated with the charged particle i.e., $\lambda_B = \hbar/(2\pi m k_B T)$ approaches the electron Fermi wavelength λ_{Fe} and exceeds the electron Debye radius λ_{De} (*viz.*, $\lambda_B \sim \lambda_{Fe} \sim \lambda_{De}$). Furthermore, the quantum effects associated with the strong density correlation start playing a significant role when λ_B becomes of the same order or larger than the average inter-particle distance ($\sim n_0^{-1/3}$), i.e., $n_0 \lambda_B^3 \geq 1$ hold in degenerate plasma. However, the other condition for degeneracy is that the Fermi temperature (T_F), which is related to the equilibrium density (n_0) of the charged particles must be greater than the thermal temperature (T) of the system [9, 10, 25, 26].

In the present work, we focus on the recently developed quantum hydrodynamic model (QHD) [8]. The QHD model consists of a set of equations describing the transport of charge density, momentum (including the Bohm potential) and energy in a charge particle system interacting through a self consistent electrostatic potential [26]. The validity of QHD is a macroscopic model and is limited to those systems that are large compared to Fermi length of the species in the system. The advantages of the QHD model over kinetic descriptions are its numerical efficiency, the direct use of the macroscopic variables of interest such as momentum and energy and the easy way the boundary conditions are implemented.

In this chapter, we analytically examine electron acceleration by the ponderomotive force associated with a right circularly polarized Gaussian whistler pulse in a magnetized quantum plasma. Initially, the electrons have speeds lower than the group velocity of the whistler pulse and the ponderomotive force pushes electron ahead of the pulse. The group velocity of the whistler pulse is significantly less than the speed of light and hence electron can resonantly interact with the pulse. One expects the saturation to occur when electron velocity exceeds the pulse group velocity and the electron outruns the whistler pulse. The theoretical formalism has been explained in Section 24.2, and the conclusions are given in Section 24.3.

24.2 PONDEROMOTIVE ACCELERATION

We consider the propagation of a right circularly polarized whistler pulse in high density quantum plasma in the direction of static magnetic field $B_s\hat{z}$. The electric and magnetic fields of the whistler are [29]

$$\vec{E} = (\hat{x} + i\hat{y})A_o(z - v_g t)\exp[-i(\omega t - kz)] \tag{1}$$

$$\vec{B} = -\left[\frac{1}{\omega}\left(1 - \frac{v_g k}{\omega}\right)\frac{\partial A_o}{\partial z} + \frac{ik}{\omega}A_0\right](\hat{x} + i\hat{y}) \tag{2}$$

where A_o for a Gaussian pulse is given by

$$A_o^2 = A_{oo}^2 \exp\left[-(t - z/v_g)^2/\tau^2\right] \tag{3}$$

where, $k = (\omega/c)[1 - \omega_p^2/\omega(\omega - \omega_c)]^{1/2}$, $v_g = \partial\omega/\partial k$ is the pulse group velocity, τ is pulse duration, $\omega_p^2 = (4\pi n_o e^2/m)$ is the electron plasma frequency, $\omega_c = eB_s/mc$ is the electron cyclotron frequency, m is the rest mass of electron, c is the velocity of light.

The QHD equations governing the motion of electron in the presence of laser field and the static magnetic field is given by [13]

$$\frac{d\vec{p}}{dt} = -e\vec{E} - \frac{e}{c}(\vec{v} \times \vec{B}) - \frac{v_F^2}{3n_o^2}\frac{\vec{\nabla}n^3}{n} + \frac{\hbar^2}{2m}\vec{\nabla}\left(\frac{1}{\sqrt{n}}\vec{\nabla}^2\sqrt{n}\right) \tag{4}$$

$$\frac{\partial n}{\partial t} + \vec{\nabla}.(n\vec{v}) = 0 \tag{5}$$

where, $n(= n_0 + n^{(1)})$ is the electron density and $\hbar$ is the Planck's constant divided by 2π, $v_F(= (\hbar/m)(3\pi^2 n)^{1/3})$ is the Fermi velocity. The third term on the right–hand side of Eq. (4) denotes the Fermi electron pressure $(= mv_F^2 n^3/3n_0^3)$. The fourth term is the quantum Bohm force and is due to the quantum corrections in the density fluctuation. The classical equation may be recovered in the limit of $\hbar=0$. The ponderomotive force of the high-frequency laser pulse drives longitudinal waves with a frequency much smaller than ω, but fast enough for the dynamics to take place on the electron time scale. The ions form a neutralizing background in dense plasma. A correct relativistic treatment of quantum effects should rely on the moments of a relativistic wigner function. Simultaneous solution of Eqs. (4) and (5) give the electron velocity and electron density as

$$\vec{v} = \frac{e\omega\vec{E}}{mi\left[\left(v_F^2 + \frac{\hbar^2}{4m^2}k^2\right)k^2 - \omega(\omega - \omega_c)\right]} \tag{6}$$

$$n = i\frac{en_0 k\vec{E}}{\left[\left(v_F^2 + \frac{\hbar^2}{4m^2}k^2\right)n_0 k^2 - m\omega(\omega - \omega_c)\right]} \tag{7}$$

Equation (6) represents the oscillatory motion imparted by whistler pulse to the plasma electron.

To evaluate the ponderomotive force in the next order, one must remember that $\vec{v}$ and $\vec{B}$ in our representations are complex; however, actual $\vec{v}$ and $\vec{B}$, and their real parts are implied in $\vec{F}_p$. We may use the complex number identity $\operatorname{Re}\vec{A} \times \operatorname{Re}\vec{B} = (1/2)\operatorname{Re}(\vec{A} \times \vec{B} + \vec{A} \times \vec{B}^*)$ to simplify it, where * refers to complex conjugate. The ponderomotive force on electron due to the whistler pulse is thus

$$\vec{F}_P = -\frac{m}{2}\left(\vec{v}.\vec{\nabla}\right)\vec{v}^* - \frac{e}{2}\left(\vec{v} \times \vec{B}^*\right) \tag{8}$$

Assuming the magnetic field to vary as $\vec{B} = be^{-i(\omega t - kz)}$, we get

$$\frac{\partial\vec{B}}{\partial t} = \left(\frac{\partial\vec{b}}{\partial t} - i\omega\vec{b}\right)e^{-i(\omega t - kz)} \tag{9}$$

and

$$\vec{b} = -\frac{1}{\omega}\left(\frac{\partial A_0}{\partial z} + ikA_0\right)(\hat{x} + i\hat{y}) - \frac{k}{\omega^2}\frac{\partial A_0}{\partial t}(\hat{x} + i\hat{y}) \tag{10}$$

For $\partial/\partial t = -(\partial/\partial z)v_g$, the above equation reduces to

$$\vec{b} = -\frac{1}{\omega}\left(\frac{\partial A_0}{\partial z} + ikA_0\right)(\hat{x} + i\hat{y}) - \frac{k}{\omega^2}\frac{\partial A_0}{\partial t}(\hat{x} + i\hat{y}) \qquad (11)$$

where,

$$\left(1 - \frac{v_g k}{\omega}\right) = \frac{\omega_P^2(2\omega - \omega_c)}{2\omega(\omega - \omega_c)^2\left(1 + \dfrac{\omega_P^2\omega_c/\omega}{2(\omega - \omega_c)^2}\right)} \qquad (12)$$

and

$$\frac{v_g}{c} = 1 - \frac{\omega_P^2/\omega^2(2 - \omega_c/\omega)}{2(1 - \omega_c/\omega)^2\left(1 + \dfrac{\omega_P^2/\omega^2(\omega_c/\omega)}{2(1 - \omega_c/\omega)^2}\right)} \qquad (13)$$

The above equation shows that there occurs a cyclotron resonance at $\omega = \omega_c$ and the group velocity of the whistler pulse vanishes at the cyclotron resonance. Substituting Eq. (11) in Eq. (9), we get

$$\vec{B} = -\left[\frac{1}{\omega}\left(1 - \frac{v_g k}{\omega}\right)\frac{\partial A_0}{\partial z} + \frac{ik}{\omega}A_0\right](\hat{x} + i\hat{y}) \qquad (14)$$

Substitution of relevant quantities in Eq. (8) gives the parallel component of the ponderomotive force (with respect to static magnetic field)

$$F_{pz} = -\frac{e^2}{2m\left[\left(v_F^2 + \dfrac{\hbar^2}{4m^2}k^2\right)k^2 - \omega(\omega - \omega_c)\right]}\left(1 - \frac{v_g k}{\omega}\right)\frac{\partial A_0^2}{\partial z} \qquad (15)$$

which may be further solved to yield

$$F_{pz} = -\frac{e^2}{2m\left[\left(v_F^2 + \dfrac{\hbar^2}{4m^2}k^2\right)k^2 - \omega(\omega - \omega_c)\right]}\left(1 - \frac{v_g k}{\omega}\right)\frac{\partial A_0^2}{\partial z} \qquad (16)$$

From the above equation it is evident that the ponderomotive force F_{pz} reverses its direction at $\omega = \omega_c/2$. For $\omega > \omega_c/2$, the front of the whistler pulse exerts a longitudinal force on the electron and there is a net energy gain by the electrons. The equation of motion for an electron in the presence of the ponderomotive force is

$$\frac{d^2z}{dt^2} = -\frac{e^2}{2m^2\left[\left(v_F^2 + \frac{\hbar^2}{4m^2}k^2\right)k^2 - \omega(\omega - \omega_c)\right]2\omega(\omega - \omega_c)^2\left(1 + \frac{\omega_p^2\omega_c/\omega}{2(\omega - \omega_c)^2}\right)} \cdot \frac{\omega_p^2(2\omega - \omega_c)}{1}$$

$$\times \frac{2A_0^2}{v_g\tau^2}\left(t - \frac{z}{v_g}\right)\exp\left[-(t - z/v_g)^2/\tau^2\right].$$

(17)

Defining $\dfrac{z - v_g t}{v_g t} = \xi$; $\dfrac{dz}{dt} = v_g\tau\dfrac{d\xi}{dt} + $; Eq. (17) can be expressed as

$$\frac{d^2\xi}{dt^2} = \Omega\xi e^{-\xi^2}$$

(18)

Multiplying both sides of Eq. (18) by $2\dfrac{d\xi}{dt}$ and integrating we get,

$$\left(\frac{d\xi}{dt}\right)^2 = -\Omega e^{-\xi^2} + c_1$$

where c_1 is the constant of integration. Using Eq. (18), we get

$$(v_z - v_g)^2 = (v_g - v_0)^2$$

(19)

At $t = -\infty$, $z = 0$ and $v_z = v_0$, So we get $c_1 = (v_z - v_0)^2$.
At $t = \infty$, implying

$$(v_z - v_g)^2 = (v_g - v_0)^2$$

here $v_z = v_g \pm \sqrt{(v_g - v_0)^2} = 2v_g - v_0$, when the wave amplitude is large;
$v_z = v_g \pm \sqrt{(v_g - v_0)^2} = v_0$ when the wave amplitude is small.

It is obvious that, the electron with small v_0 (small positive initial velocity) gains less velocity (energy). Counter moving electron will gain larger velocity.

Now, we calculate the normalized threshold whistler amplitude for which $v_z = v_0$ at $\zeta = 0$; i.e., electron velocity is equal to the group velocity of the whistler pulse. Under these conditions Eq. (19) reduces to

$$\left(v_z - v_0\right)^2 - \Omega = 0 \tag{20}$$

giving,

$$a_0^2 = \left(v_g - v_0\right)^2 \left(\left(v_F^2 + \frac{\hbar^2}{4m^2}k^2\right)k^2 - \omega\left(\omega - \omega_c\right)\right) \frac{2\left(\omega - \omega_c\right)\left(1 + \dfrac{\omega_p^2 \omega_c/\omega}{2\left(\omega - \omega_c\right)^2}\right)}{\omega_p^2 \left(2\omega - \omega_c\right)} \tag{21}$$

Figure 24.1 shows, the variation of normalised whistler amplitude a_0^2 with ω_c/ω for $\omega_p^2/\omega^2 = 0.3$ and $v_0 = v_g/10$. It is observed that the normalised whistler amplitude decrease with ω_c/ω. In Figure 24.2, the variation of normalised whistler amplitude a_0^2 with has ω_p^2/ω^2 been shown for $\omega_c/\omega = 1.31$ and $v_0 = v_g/10$. The variation shows a nearly parabolic pattern.

24.3 SUMMARY AND DISCUSSION

The ponderomotive force associated with a 1D whistler pulse is axial. For $\omega > \omega_c/2$, the ponderomotive is directed away from the intensity maximum while for $\omega < \omega_c/2$, the ponderomotive force reverses to direction and tends to pull electron towards the intensity peak. For acceleration of electrons, one requires a whistler pulse of $\omega > \omega_c/2$. Maximum ponderomotive force is obtained near resonance $\omega > \omega_c/2$. A pulse of ω close to ω_c exerts much stronger axial ponderomotive force. In 1D, the electron gains energy during the rising front of the pulse, the rear acts as a backward ponderomotive force and slows down the electron to its original energy. Hence, to achieve energy gain the value of peak pulse amplitude should be above the threshold value. The threshold value of the whistler amplitude depends on the quantum plasma density as well as on ω_c. The threshold corresponds to $v_z = v_g$ at the

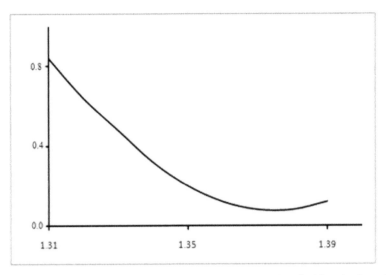

FIGURE 24.1   The variation of normalized whistler amplitude a_0^2 with ω_c/ω for ω_p^2/ω^2 = 0.2 and $v_0 = v_0/10$.

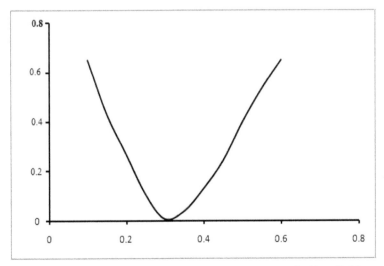

FIGURE 24.2   The variation of normalized whistler amplitude a_0^2 with ω_p^2/ω^2 for ω_c/ω = 1.3 and $v_0 = v_0/10$.

pulse peak, i.e., the electron velocity is equal to the group velocity of the whistler pulse as the electron reaches the pulse peak. The threshold value of amplitude (a_0^2) for $\omega_c/\omega = 1.03$ and $\omega_p^2/\omega^2 = 0.3$ is 0.82. This threshold value is much greater than that obtained for classical plasma and hence a higher intensity laser pulse is required for quantum plasma for effective acceleration. This is due to the diffraction effects induced by the quantum diffraction terms. The threshold amplitude decreases with the ratio of plasma frequency to electron cyclotron frequency, ω_p/ω_c. However, above the threshold amplitude, the acceleration energy decreases with ω_p/ω_c. The electrons gain velocities about twice the group velocity of the whistler. The threshold amplitude approches zero as $\omega_c/\omega \to 1$. However, the group velocity of the whistler pulse vanishes at the cyclotron resonsnce and there is no net energy gain.

ACKNOWLEDGMENT

This work was performed under the financial assistance from the University Grants Commission (UGC) India, under project no. 39: 458/2010(SR).

KEYWORDS

- electron gain
- particle acceleration
- ponderomotive force
- QHD model
- quantum plasma
- whistler pulse

REFERENCES

1. Brodin, G., & Lundberg, J. (1998). Excitation of electromagnetic wake fields in a magnetized plasma. *Phys. Rev. E, 57,* 704130.
2. Bohm, D. (1952). A suggested interpretation of the quantum theory in terms of "Hidden" variables. *Phys. Rev., 8,* 5166.
3. Chen, P., Tajima, T., & Takahashi, Y. (2002). Plasma Wakefield Acceleration for Ultrahigh-Energy Cosmic Rays. *Phys. Rev. Lett., 89,* 161101.
4. Chen, P. et al. (2009). A new type of plasma wakefield accelerator driven by magnetowaves. *Plasma Phys. Controlled Fusion, 51,* 024012.

5. Caldwell, A., Lotov, K., Pukho, A., & Simon, F. (2009). Proton-driven plasma-wakefield acceleration. *Nat. Phys, 5,* 363.

6. Cao, J., & Ren, H. (2008). Quantum effects on Rayleigh-Taylor instability in magnetized plasma. *Phys. Plasmas, 15,* 012110.

7. Gahn, C. et al. (1999). Multi-MeV Electron Beam Generation by Direct Laser Acceleration in High-Density Plasma Channels. *Phys. Rev. Lett., 83,* 4772.

8. Gardner, C. L., & Ringhofer, C. (1996). Smooth quantum potential for the hydrodynamic model. *Phys. Rev. E, 53,* 157.

9. Haque, Q., Mahmood, S., & Mushtaq, A. (2008). Nonlinear electrostatic drift waves in dense electron-positron-ion plasmas. *Phys. Plasmas, 15,* 082315.

10. Haas, F., Garcia, L. G., Goedert, J., & Manfredi, G. (2003). Quantum ion-acoustic waves. *Phys. Plasmas, 10,* 3858.

11. Hass, F. et al. (2002). A multistream model for quantum plasmas. *Phys. Rev. E, 62,* 2763.

12. Joshi, C. (2007). The development of laser- and beam-driven plasma accelerators as an experimental field. *Phys. Plasmas, 14,* 055501.

13. Kumar, P., & Tewari, C. (2012). Electric, magnetic Wakefields, and electron acceleration in quantum plasma. *Laser and Particle Beams, 30,* 267–273.

14. Kalmykov, S., Yi, S. A., & Shvets, G. (2009). All-optical control of nonlinear focusing of laser beams in plasma beat wave accelerator. *Plasma Phys. Control. Fusion, 51,* 024011.

15. Karpman, V. I., & Washimi, H. (1977). Two-dimensional self-modulation of a whistler wave propagating along the magnetic field in a plasma. *J. Plasma Phys., 18,* 173.

16. Karpman, V. I., Hansen, F. R., Huld, T., Lynov, J. P., Pecseli, H. L., & Rasmussen, J. J. *Phys. Rev. Lett. 64,* 890.

17. Liu, C. S., & Tripathi, V. K. (2005). Ponderomotive effect on electron acceleration by plasma wave and betatron resonance in short pulse laser. *Phys. Plasmas, 12,* 043103.

18. Rao, N. N. (1988). Theory of near-sonic envelope electromagnetic waves in magnetized plasmas. *Phys. Rev. A, 37,* 4846.

19. Shukla, P. K. (2009). Generation of wakefields by electromagnetic waves in a magnetized electron-positron-ion plasma. *Plasma Phys. Controlled Fusion, 51,* 024013.

20. Schlenvoigt, H. P. et al. (2008). A compact synchrotron radiation source driven by a laser-plasma wakefield accelerator. *Nat. Phys., 4,* 130.

21. Shukla, P. K., & Stenflo, L. (1984). Nonlinear propagation of electromagnetic waves in magnetized plasmas. *Phys. Rev. A, 30,* 2110.

22. Shukla, P. K. (1978). Relation between monthly variations of global ozone and solar activity. *Nature (London), 274,* 874.

23. Spatschek, K. H., Shukla, P. K., Yu, M. Y., & Karpman, V. I. (1979). *Phys. Fluids, 22,* 576.

24. Sazegari, V., Mirzaie, M., & Shokri, B. (2006). Ponderomotive acceleration of injected electrons in tenuous plasmas by intense laser pulses. *Phys. Plasmas, 13,* 033102.

25. Shukla, P. K., & Akbari-Moghananghi, M. (2012). Comment on "quantum plasma: A peal for a common sense." *EPL, 99,* 65001.

26. Shukla, P. K., & Eliasson, B. (2010). Nonlinear aspects of quantum plasma physics. *Physics-Uspekhi, 53*(1), 51.

27. Shukla, P. K., Brodin, G., Marklund, M., & Stenflo, L. (2009). Excitation of multiple wakefields by a short laser pulses in quantum plasmas. *Phys. Lett. A, 373,* 3165.
28. Shukla, P. K., & Eliasson, B. (2006). Formation and Dynamics of Dark Solitons and Vortices in Quantum Electron Plasmas. *Phys. Rev. Lett., 96,* 245001.
29. Singh, R., & Sharma, A. K. (2010). Ponderomotive acceleration of electrons by a whistler pulse. *Appl. Phys. B, 100,* 535–538.
30. Tajima, T., & Dawson, J. M. (1979). Laser Electron Accelerator. *Phys. Rev. Lett., 43,* 267.
31. Tochitsky, S. Y. et al. (2004). Enhanced Acceleration of Injected Electrons in a Laser-Beat-Wave Induced Plasma Channel. *Phys. Rev. Lett., 92,* 095004.
32. Tsakiris, G. D., Gahn, C., & Tripathi, V. K. (2000). *Phys. Plasmas, 7,* 3017.
33. Zhidkov, J., Fujii, T., & Nemoto, K. (2008). Electron self-injection during interaction of tightly focused few-cycle laser pulses with under dense plasma. *Phys. Rev. E, 78,* 036406.
34. Zhao, J. S., Lu, J. Y., & Wu, D. J. (2010). Observation of anisotropic scaling of solar wind turbulence. *Astrophys. J., 714,* 138.

CHAPTER 25

GENERATION OF TERAHERTZ RADIATIONS BY FLAT TOP LASER PULSES IN MODULATED DENSITY PLASMAS

DIVYA SINGH[1,2] and HITENDRA K. MALIK[1]

[1]*Department of Physics, PWAPA Laboratory, Indian Institute of Technology Delhi, New Delhi – 110016, India,*
E-mail: divyasingh1984@gmail.com

[2]*Department of Physics & Electronics, Rajdhani College, University of Delhi, New Delhi – 110015, India*

CONTENTS

ABSTRACT

In this chapter, we present a mechanism of THz radiation generation by excitation of nonlinear currents due to two flat top laser pulses of same intensities but different frequencies and wavenumbers in a plasma of modulated density. In this process, the ponderomotive force is developed, which leads to nonlinear current density that resonantly excites the radiation in the THz frequency range. The effect of beam width and density ripples is discussed for these currents and THz generation and comparative studies are made for flat top beams with Gaussian one.

25.1 INTRODUCTION

EM radiations have been a part of all walks of life, based on their application. Of the EM spectra the range 0.1–10.0 THz rapidly had become an important area of research since last decade due to its diverse applications in material characterization, medical imaging, topography, remote sensing [1], chemical and security identification [2, 3], etc. In order to develop high-power and efficient sources, several schemes [4–25] have been proposed, such as tunable THz generation by superluminous laser pulse interaction with large band gap semiconductors and electro-optic crystals viz. ZnSe, GaP, and LiNbO$_3$ [4–7]. THz radiation generation also has been reported by the nonlinear interaction of an intense short laser pulse with a semiconductor and dielectric [8–14]. Due to the material damage and low conversion efficiency, it is difficult to obtain powerful THz emission from the THz emitters, such as electro-optic crystals, semiconductors, synchrotrons, etc. Therefore, many experiments use plasma as a nonlinear medium for THz generation using subpicosecond laser pulses [15], as plasma has advantage of supporting very high fields and shows very strong nonlinear effects [16]. Malik et al. [17] have analytically investigated the THz generation by tunnel ionization of a gas jet with superposed femtosecond laser pulses impinging onto it after passing through an axicon. Yoshii et al. [18] theoretically and Yugami et al. [19, 20] experimentally have demonstrated the THz radiation generation when the Cerenkov wake is excited by a short laser pulse in a perpendicularly magnetized plasma. Cook and Hochstrasser [21] considered the THz generation when the fundamental and second harmonic lasers are simultaneously focused into the air. Sheng et al. [22] proposed a scheme in which a short laser pulse excites a large amplitude plasma wake field, which, in the presence of an axial density gradient, produces radiation at the plasma frequency (ω_p) via mode

conversion. Antonsen et al. [23] have proposed the employment of a cor-
rugated plasma channel for phase matched THz radiation generation by
the ponderomotive force of a laser pulse. So far many schemes have been
discussed employing plasma as a medium for THz generation but none of
them is for flat top lasers while shape of the lasers pulse is an important
ingredient to cause effective THz radiation generation.

25.2 MECHANISM AND NONLINEAR CURRENT

In the present mechanism, two flat top laser pulses are taken to propagate
along z direction in the presence of a plasma having periodic density modulation of wavenumber α such as $N = N_0 + N_\alpha e^{i\alpha z}$. Due to the spatial intensity
variation of laser fields in the transverse direction, i.e., along y axis, force
known as ponderomotive force arises in the direction of propagation of
lasers and perpendicular to it. This force gives rise to current that oscillates
at the beating frequency of the lasers. At resonance, when wavenumber and
frequency match, maximum amount of nonlinear currents are generated
and these currents are responsible for the generation of THz radiation.

We consider two ultra-short fs (femtosecond) flat top lasers (shown
later in Figure 25.4) of frequencies ω_1 and ω_2 and wavenumbers k_1 and k_2,
respectively, co-propagating in the z direction and polarized along the y
direction. We also consider laser-produced plasma of modulated density
ripples (Figure 25.1). As the laser fields have space variation along the
y axis, a transverse component of the nonlinear ponderomotive force is
realized in the y direction at frequency $\omega = \omega_1 - \omega_2$ and wavenumber $k = k_1 - k_2$.

The laser fields are chosen as

$$\vec{E}_1 = E_0 \exp\left[-\left(\frac{y}{b_w}\right)^8\right] e^{i(k_1 z - \omega_1 t)} \hat{y} \,, \vec{E}_2 = E_0 \exp\left[-\left(\frac{y}{b_w}\right)^8\right] e^{i(k_2 z - \omega_2 t)} \hat{y} \quad (1)$$

In the above expressions of flat top lasers, b_w is the beamwidth of laser
pulse. The nonlinear ponderomotive force imparted by lasers is given by

$$\vec{F}_p^{NL} = -\frac{m}{2e} \vec{\upsilon}_1 \cdot \vec{\upsilon}_2^* = -\frac{e^2 E_0^2}{2m\omega_1\omega_2} \exp\left[-2\left(\frac{y}{b_w}\right)^8\right]\left[\frac{2p}{b_w}\left(\frac{y}{b_w}\right)^7 \hat{y} - ik\hat{z}\right] e^{i(kz - \omega t)} \quad (2)$$

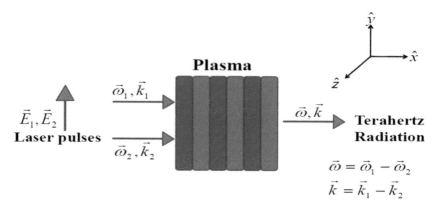

FIGURE 25.1   Schematic of generation of terahertz radiations by flat top laser pulses in modulated density plasmas.

Under the effect of this ponderomotive force, electron oscillations are modified and some nonlinear perturbations in plasma density are created those are evaluated using equation of continuity as $N^{NL} = -\frac{n_0}{m\omega^2}\vec{\nabla}.\vec{F}_p^{\;NL}$. Due to this nonlinear perturbation in electron density, some local electrostatic fields is φ are developed which further introduces some kind of linear density perturbations given as $N^L = -\frac{N_0 e\vec{\nabla}.\vec{\nabla}\phi}{m\omega^2}$. We use Poisson's equation under combined effect of densities perturbations $\vec{\nabla}.\vec{\nabla}\phi = 4\pi e\left(N^L + N^{NL}\right)$ to find out resultant electrostatic field which is used further to evaluate resultant nonlinear velocities of plasma electrons. Hence, the nonlinear oscillatory current is obtained as

$$\vec{j}^{NL} = -\frac{1}{2}Ne\vec{\upsilon}^{NL} = -\frac{1}{2}N_\alpha e\frac{i\omega\vec{F}_p^{\;NL}}{m[\omega^2 - \omega_p^{\;2}]} \tag{3}$$

Transverse component of nonlinear oscillatory current is

$$\vec{j}_y^{NL} = -\frac{1}{4}\frac{i\omega N_0 e^3 E_0^{\;2}}{m^2\omega_1\omega_2[\omega^2 - \omega_p^{\;2}]}e^{\left[-2\left(\frac{y}{b_w}\right)^8\right]}\frac{2p}{b_w}\left(\frac{y}{b_w}\right)^7 e^{i\{kz - \omega t\}}$$

where $k = k_1 - k_2 + \alpha, \omega = \omega_1 - \omega_2$. This condition should be met for generation of THz radiations.

25.2.1 ELECTRIC FIELD OF EMITTED THz RADIATION

Wave equation governing emission and propagation of THz radiations within plasma is as follows

$$-\nabla^2 \vec{E} + \vec{\nabla}(\vec{\nabla}.\vec{E}) = -\frac{4\pi i\omega}{c^2} \vec{J}^{NL} + \frac{\omega^2}{c^2} \varepsilon\vec{E} \tag{4}$$

Here E represents the THz field, wherein we symbolize as E_{THz}. Thus the magnitude of emitted THz is calculated by taking the divergence of the above equation,

$$E_{THz} = \frac{pN_\alpha eE_0^2\omega^2\omega_p^2 e^{-2\left(\frac{y}{b_w}\right)^8}}{2N_0 mb_w\omega_1\omega_2\left(\omega^2 - \omega_p^2\right)^2} \left(\frac{y}{b_w}\right)^7 \tag{5}$$

The given amplitude of THz is obtained only if the following phase matching condition is satisfied

$$\frac{c\alpha}{\omega_p} = \frac{\omega}{\omega_p}\left[\left(1 - \frac{\omega_p^2}{\omega^2}\right)^{1/2} - 1\right] \tag{6}$$

Since quality of emitted THz radiations is determined by their amplitude, band width and efficiency, these are the key factors those are required to be achieved. Therefore, following analytical studies are made to study the nature of emitted THz radiations while flat top laser pulses are co-propagating in a modulated density plasma and further adequate optimization of laser and plasma parameters are done.

25.3 RESULTS AND DISCUSSION

We examine the profile of the field amplitude of emitted THz radiation through Figure 25.2 for Gaussian laser (GL) and flat top laser (FTL) for the parameters $\omega/\omega_p = 1.05$; $\omega_1 = 2.4\times10^{14}$ rad/sec and $\omega_2 = 2.0\times10^{13}$ rad/sec. From the following Figure 25.2 it is evident that the THz field amplitude increases and acquires a maximum value for a particular value of y/b_w, i.e., critical transverse distance (y_o). For FTL pulse $y_o = 0.8$ and for GL pulse $y_o = 0.5$. The reason for the highest magnitude of THz field for a specific value of y/b_w can be understood based on the magnitude of y component of the ponderomotive force F^{NL}_{py}. Actually, for a small but particular value of y/b_w, the component F^{NL}_{py} acquires maximum magnitude and, hence, generates the highest nonlinear current J^{NL}, This in turn, results in the emission of the highest amplitude THz radiation.

25.3.1 EFFECT OF BEAMWIDTH (B_w)

From Figure 25.3 the amplitude of THz is found to fall with increasing beam width but both Figures 25.2 and 25.3 shows that for $y_o = 0.8$, FTL pulses gives better amplitude of THz fields than GL pulses. From Figure 25.2 it is also very obvious that for modulated density plasma the profile of normalised amplitude of emitted THz when GL pulses used is

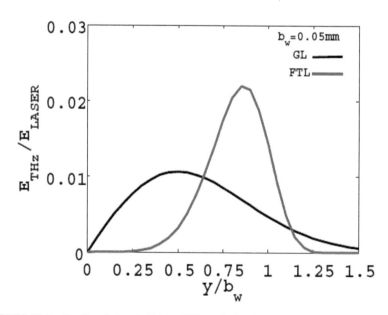

FIGURE 25.2 Profile of electric field of THz radiation flat top laser (FTL) and Gaussian (GL) laser.

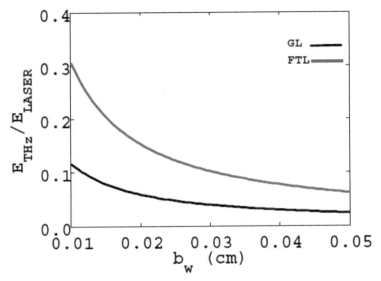

FIGURE 25.3 Variation in the normalized THz field with beam width of FTL and GL profiles.

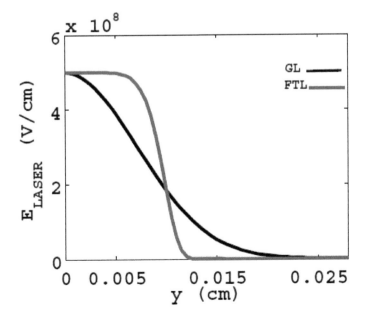

FIGURE 25.4 Profiles of flat top and Gaussian laser.

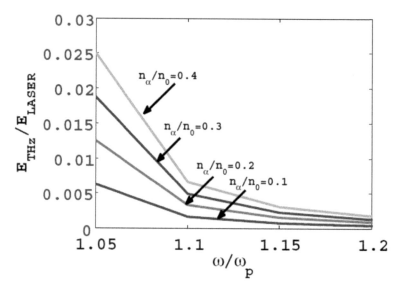

FIGURE 25.5 Variation in the normalized THz field with normalized beat frequency of different density ripples.

unsymmetrical about y/b_w whereas for FTL pulses we get more symmetric profiles, i.e., more synchronised highly focussed emitted THz radiations. Thus, small, beam width flat top laser pulse provides highly focussed and collimated THz radiations than Gaussian pulses.

The explanation why flat top lasers are better than the Gaussian lasers for THz generation can be understood with the help of Figure 25.4 that clearly depicts the half beam profile of Gaussian and flat top lasers. The latter has nearly rectangular shape of beam having high value of intensity gradient than Gaussian lasers, therefore FTL gives large ponderomotive force and nonlinear currents. Hence this leads to higher amplitudes of THz radiation along with symmetric profile.

25.3.2 EFFECT OF DENSITY RIPPLES ON PHASE MATCHING

Figures 25.5 and 25.6 explain that the density ripples help in achieving high amplitudes of THz radiation, resonance condition is also shown to be close to $\omega = \omega_p$. As beating frequency varies, the resonance condition departs leading to a massive reduction in the amplitude of THz fields.

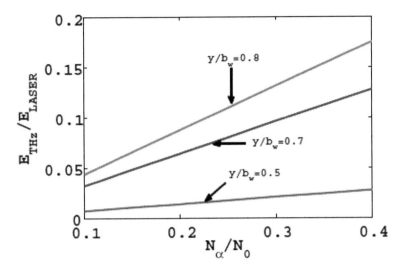

FIGURE 25.6 Variation in the normalized THz field with amplitude of density ripples.

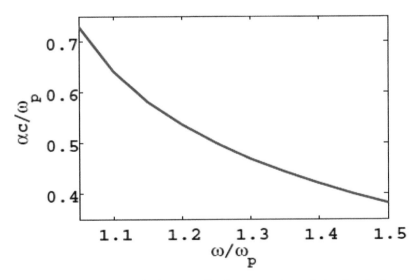

FIGURE 25.7 Variation of the normalized wavenumber of density ripples with normalized beat wave frequency.

The resonance condition can be achieved by proper phase matching of wavenumbers using density ripples in plasma. The large number of density ripples attributes to the high contribution of electrons in the generation of nonlinear currents. Figure 25.7 clearly explains that ripples must be formed closer for achieving best condition of phase matching, which leads to resonance and maximum transfer of energy to get enhanced amplitude of emitted THz radiation.

KEYWORDS

- **density ripples**
- **flat top laser**
- **laser plasma interaction**
- **ponderomotive force**
- **resonance**
- **terahertz**

REFERENCES

1. Ferguson, B., & Zhang, X. C., (2002). *Nat. Mater. 1*, 26.
2. Shen, Y. C., Today, T., W. P. F., Cole, B. E., Tribe, W. R., & Kemp, M. C., (2005). *Appl. Phys. Lett. 86*, 241116.
3. Zheng, H., Redo-Sanchez, A., & Zhang, X. C., (2006). *Opt. Express 14*, 9130.
4. Faure, J., Tilborg, J. V., Kaindl, R. A., & Leemans, W. P., (2004). *Opt. Quantum Electron. 36*, 681.
5. Shi, W., Ding, Y. J., Fernelius, N., & Vodopyanov, K., (2002). *Opt. Lett. 27*, 1454.
6. Zhao, P., Ragam, S., Ding, Y. J., & Zotova, I. B., (2010). *Opt. Lett. 35*, 3979.
7. Jiang, Y., Li, D., Ding, Y. J., & Zotova, I. B., (2011). *Opt. Lett. 36*, 1608.
8. Chen, M., Pukhov, A., X.-Peng, Y., & Willi, O., (2008). *Phys. Rev. E 78*, 046406.
9. Penano, J., Sprangle, P., Hafizi, B., Gordon, D., & Serafim, P., (2010). *Phys. Rev. E 81*, 026407.
10. Melrose, D. B., & Mushtaq, A., (2010). *Phys. Rev. E 82*, 056402.
11. Hashimshony, D., Zigler, A., & Papadopoulos, K., (1999). *Phys. Rev. Lett. 86*, 2806. (2001); *Appl. Phys. Lett. 74*, 1669.
12. Holzman, J. F., & Elezzabi, A. Y., (2003). *Appl. Phys. Lett. 83*, 2967.
13. Ma, G. H., Tang, S. H., Kitaeva, G. K., & Naumova, I. I., (2006). *J. Opt. Soc. Am. B 23*, 81.
14. Budiarto, E., Margolies, J., Jeong, S., Son, J., & Bokor, J., (1996). *IEEE J. Quantum Electron. 32*, 1839.

15. Kim, K. Y., Taylor, A. J., Glownia, T. H., & Rodriguez, G., (2008). *Nat. Photonics 153*, 1.
16. Rothwell, E. J., & Cloud, M. J., *Electromagnetics*, 2nd ed. (CRC/Taylor & Francis, London, 2009), p. 211. 016401-9.
17. Leemans, W. P., J. van Tilborg, Faure, J., Geddes, C. G. R., Toth, C., Schroeder, C. B., Esarey, E., Fubioni, G., & Dugan, G., (2004). *Phys. Plasmas 11*, 2899.
18. Pukhov, A., (2003). *Rep. Prog. Phys. 66*, 47.
19. Malik, A. K., Malik, H. K., & Kawata, S., (2010). *J. Appl. Phys. 107*, 113105.
20. Cook, D. J., & Hochstrasser, R. M., (2000). *Opt. Lett. 25*, 16.
21. Z.-Sheng, M., Mima, K., Zhang, J., & Sanuki, H., (2005). *Phys. Rev. Lett. 94*, 095003. (2005); *Phys. Plasmas 12*, 123103.
22. T. M. Antonsen Jr., Palastra, J., & Milchberg, H. M., (2007). *Phys. Plasmas 14*, 033107.
23. Bartel, T., Reimann, K., Woerner, M., & Elsacsser, T., (2005). *Opt. Lett. 30*, 2805.
24. You, D., Jones, R. R., Bucksbaum, P. H., & Dykaar, D. R., (1993). *Opt. Lett. 18*, 290.
25. Wu, H. C., Z.-Sheng, M., & Zhang, J., (2008). *Phys. Rev. E 77*, 046405.

CHAPTER 26

SURFACE MODIFICATION OF HIGH DENSITY POLYETHYLENE AND POLYCARBONATE BY ATMOSPHERIC PRESSURE COLD ARGON/OXYGEN PLASMA JET

R. SHRESTHA,[1,2] J. P. GURUNG,[1] A. SHRESTHA,[1] and D. P. SUBEDI[1]

[1]Department of Natural Science, Kathmandu University, Dhulikhel, Nepal

[2]Department of Physics, Basu H.S.S./Basu College, Kalighat, Byasi, Bhaktapur, Nepal, E-mail: rajendra.ts2002@gmail.com

CONTENTS

ABSTRACT

In this chapter, atmospheric pressure plasma jet (APPJ) and its application for polymer surface modification is reported. Atmospheric pressure plasma jet sustained in argon/oxygen mixture has been used to modify the surface properties of high density polyethylene (HDPE) and polycarbonate (PC). The surface properties of the untreated and plasma treated HDPE and PC samples were characterized by contact angle measurement with water and glycerol. The contact angles were used to determine the surface energy and its polar and dispersion component. The effects of treatment time, frequency of the applied voltage and distance of the sample from the nozzle on the wettability of the sample were studied. The water contact angle of untreated PC and HDPE samples were 89° and 94.6°, respectively. After treatment for 2 minutes, the contact angles were found to be less than 40° for both the samples. Moreover, it was found that, the best plasma treatment can be obtained with frequency 27 kHz and a distance of 3.5 cm between surface of samples (HDPE and PC) and plasma jet's nozzle. Our result showed that atmospheric pressure nonthermal plasma can be effectively used to enhance the surface wettability of HDPE and PC. Argon/oxygen plasma jet is more efficient on treatment of polymers than pure argon jet.

26.1 INTRODUCTION

Polymers surfaces generally have low surface energy and high chemical inertness, and so they usually have poor wetting and adhesion properties [1]. Hence, surface modification of polymeric materials plays an important role to improve surface properties such as wetting and adhesion for coatings, inking and printing processes, biomaterials, and certain types of composites materials. Therefore, various modification techniques have been used to overcome such type of problems. This includes the use of the UV irradiation, laser, chemical treatment flame, grafting, as well as plasma treatment [2]. One of the most interesting methods used to increase their surface energy and improve adhesive properties is plasma activation of their surfaces. Plasma treatment for activating polymers provides uniformity on the surface of the polymer. Plasma used for these applications, although not fully ionized, are composed of ions, free electrons, photons, neutral atoms, and molecules in ground and excited electronic states. Each of these components has the potentials of interaction with surfaces with which they come in contact. The plasma process results in a physical and/ or chemical modification of the first few molecular layers of the surface, while maintaining the properties of the bulk phase. The depth of modification with plasma treatments is generally less than 10 nm [3].

Polycarbonate (PC) has wide industrial applications because of its excellent transparency, high impact resistance and high temperature resistance. But it has low surface free energy. Thus, surface treatment is required to achieve satisfactory adhesion for printing and metallization, etc. Subedi et al. [4, 5] studied the influence of low pressure RF plasma on surface modification of PC and plasma treatment at low pressure for the enhancement of wettability of polycarbonate. Shrestha et al. [6] have reported the surface modification of polycarbonate for improvement of wettability using mesh electrode at atmospheric pressure discharge at 50 Hz.

Polyethylene is one of the world's most used thermoplastics. This is due to its good physical properties, chemical, and mechanical resistance, in addition to the economic competitiveness of its industrial production. Polyethylene, due to its apolar nature, has a low wettability, which induces poor adhesive bonding. Aragão et al. [7] fabricated atmospheric pressure microplasma jet (APMJ) and used it for polyethylene (PE) surface modification.

In this work, APPJ device was easily fabricated in a pencil-type to produce stable cold plasma suitable for applications and research in a low manufacturing cost. High density polyethylene (HDPE) and PC substrates have been used to investigate the surface modification applying APPJ. The modified surface of polymers (PC and HDPE) was characterized by contact angle analysis.

26.2 EXPERIMENTAL PART

The schematics diagram and Photograph of the experimental setup are show in Figure 26.1(a-b). The setup consists of a hollow cylindrical tube of inner diameter about 3 mm in which two copper electrodes is attached as show in figure. A high voltage AC source (0–20) kV and frequency (0–30) kHz was applied to electrodes through a current limiting resister of 10 kΩ.

Argon gas is fed as a working gas from upper end of the tube as in Figure 26.1(a). The applied voltage and current were measured by high voltage probe (Tektronics). Optical emission spectra were recorded by using optical emission spectrometer (Linear Array Spectrometer 190 nm–900 nm).

The jet length was varied upto 6.5 cm by adjusting power supply and flow rate of argon. The PC and HDPE samples are cleaned using ultrasonic cleaner for 10 minutes and subjected for treatment under plasma jet at distance of 3.5 cm from the nozzle, which is generated at applied frequency 27 kHz and 5 kV for different time period.

26.3   RESULTS AND DISCUSSION

Figure 26.1(b) shows image of the jet argon flow rate 2 L/m at fixed discharge voltage 5 kV and discharge frequency 27 kHz. It appears uniform to the human eye and can reach a length of 6.5 cm out side of the nozzle. The typical waveform of discharge voltage and discharge current at fixed frequency 27 kHz are plotted in Figure 26.2. In the figure, the positive current pulse and a negative current pulse per cycle of applied voltage is the typical characteristics of atmospheric pressure glow-like discharge. Figure

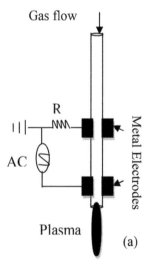

FIGURE 26.1   Schematic diagram and photograph of plasma jet.

26.3 is the optical emission spectra of APPJ at a distance of 3.5 cm from the orifice. From the analysis of spectra electron density (n_e) and electron temperature (T_e) were found to be $3.36 \times 1016 \text{ cm}^{-3}$ and 0.786 eV, respectively, at the jet length 3.5 cm downstream of the jet.

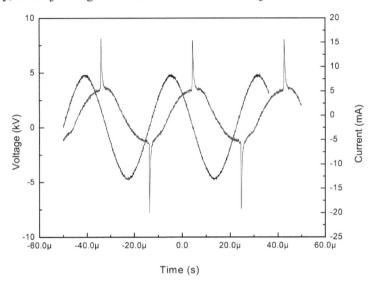

FIGURE 26.2   Current and voltage waveform of APAPJ.

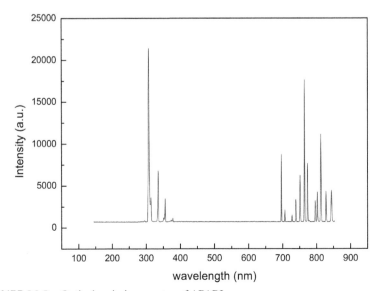

FIGURE 26.3   Optical emission spectra of APAPJ.

26.3.1 CONTACT ANGLE MEASUREMENTS AND SURFACE ENERGIES ESTIMATION

Contact angles after treatment were measured immediately on polymers modified with APAPJ exposure time 10–120 seconds by goniometry with static water drop method. The measurements of the contact angle were performed using distilled water and glycerin on different positions of the treated polymers. Surface energies were calculated using the Owens-Wendt Kaelble method. In this method, it is possible to determine the solid surface energy (γ) as the sum of polar (γ^p) and dispersive (γ^d) contribution using at least two different test liquids [8].

$$\gamma_l\left(1+\cos\theta\right)=2\left[\gamma_l^d\gamma_s^d\right]^{\frac{1}{2}}+2\left[\gamma_l^p\gamma_s^p\right]^{\frac{1}{2}} \tag{1}$$

The dependence of contact angle and surface energy on exposure time for the polymers is shown in Figures 26.4 and 26.5. It is evident from the figure that the value of contact angle decreases as the time of treatment increases and consequently the surface energy increases. The increase in surface free energy is attributed to the functionalization of the polymer surface with hydrophilic groups.

26.4 CONCLUSION

Atmospheric pressure argon plasma jet of jet length 6.5 cm has been produced and characterized by optical and electrical methods. Electron density (n_e) and electron temperature (T_e) were found to be 3.36×10^{16} cm^{-3} and 0.786 eV, respectively, at the jet length 3.5 cm downstream of the jet. Electron density and electron temperature depends on the distance from the nozzle of the jet

APPJ treatment of PC and HDPE surface resulted an improvement on hydrophilicity. It is mainly due to the increase in the polar component of the surface free energy after plasma treatment which suggests the formation of polar functional groups on the surface. The treatment leads to improve the adhesion to PC and HDPE via surface cleaning, cross-linking, or formation of chemical bonds. The improvement of wettability of PC and HDPE strongly depends on the treatment time. This system can be very useful for the treatment of thermally sensitive materials and also for the treatment of biological samples.

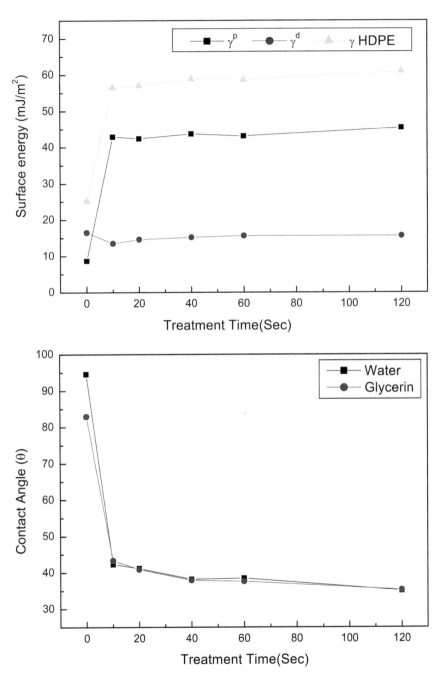

FIGURE 26.4 Contact angle and surface energy as function of treatment time for HDPE.

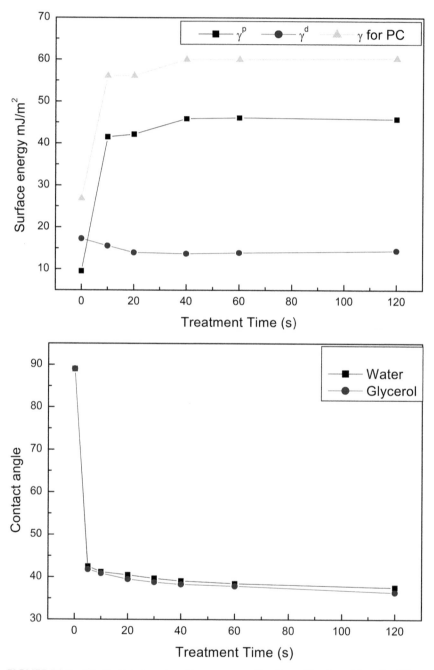

FIGURE 26.5   Contact angle and surface energy as function of treatment time for PC.

KEYWORDS

- atmospheric pressure plasma jet
- contact angle
- polymer surface modification
- wettability

REFERENCES

1. Shenton, M. J., Lovell-Hoare, M. C., & Stevens G. C. (2001). Adhesion enhancement of polymer surfaces by atmospheric plasma treatment. *Journal of Applied Physics, 34*, 2754–2760.
2. Noeske, M., Degenhardt, J., Strudthoff, S., & Lommatzsch, U. (2004). Plasma treatment of five polymers at atmospheric pressure surface modifications and the relevance for adhesion. *Int. J. Adhesion and Adhesives, 24*, 171–177.
3. Choi, Y. H., Kim, J. H., & Paek, K. H. (2005). Characteristics of atmospheric pressure N_2 cold plasma torch using 60-hz ac power and its application to polymer surface modification. *Surface & Coatings Technology, 193* (1–3), 319–324.
4. Subedi, D. P., Zajickova, L., Bursikova, V., & Janca, J. (2003). Surface modification of the polycarbonate by low pressure RF plasma. *Himalayan Journal of Science 1*(1), 115–118.
5. Subedi D. P., Madhup D. K., Adhikari, K., & Joshi, U. M. (2008) Plasma treatment at low pressure for the enhancement of wettability of polycarbonate. *Indian Journal of Pure and Applied Physics. 46,* 540–544.
6. Shrestha, A. K., Dotel, B. S., Dhungana, S., Shrestha, R., Baniya, H. B., Joshi, U. M., & Subedi, D. P., (2014). Surface modification of polycarbonate for improvement of wettability using mesh electrode at atmospheric pressure discharge at 50 Hz. *Proceedings of National Symposium on Emerging Plasma Techniques for Material Processing and Industrial Applications (NSEPMI)*, India.
7. Aragão, E. C. B. B., Nascimento, J. C., Fernandes, A. D., Barbosa, F. T. F., Sousa, D. C., Oliveira, C., Abreu, G. J. P., Ribas, V. W., & Sismanoglu, B. N. (2014). Low temperature microplasma jet at atmospheric pressure for inducing surface modification on polyethylene substrates. *American Journal of Condensed Matter Physics, 4*(3A), 1–7.
8. Park, S. J., & Lee, H. Y. (2005). Effect of Atmospheric-Pressure Plasma on Adhesion Characteristics of Polyimide Film. *Journal of Colloid and Interface Science 285,* 267–272.
9. Enkiewicz, M. (2007). Methods for the calculation of surface free energy of solids *Journal of Achievements in Materials and Manufacturing Engineering, 24*, 13–145.

CHAPTER 27

OPTICAL IMAGING OF SST-1 PLASMA

MANOJ KUMAR, VISHNU CHAUDHARY, CHESTA PARMAR,
AJAI KUMAR, and SST-1 TEAM

*Institute for Plasma Research, Gandhinagar – 382428, India,
E-mail: mkg.ipr@gmail.com*

CONTENTS

ABSTRACT

With the development of heterogeneous camera networks working at different wavelengths and frame rates, and covering a large surface of vacuum vessel, the visual observation of a large variety of plasma and thermal phenomena (e.g., hotspots, ELMs, MARFE, arcs, dusts, etc.) becomes possible. In the domain of machine protection, a phenomenological diagnostics with a key-element towards plasma/thermal event proves a dangerousness assessment during real time operation. It is also of primary importance to automate the extraction and the storage of phenomena information for further off-line event retrieval and analysis, thus leading to a better use of massive image databases for plasma physics studies. To this end, efforts have been devoted to the development of image processing algorithms dedicated to the recognition of specific events. But a need arises now for the integration of techniques developed so far in both hardware and software directions. We present in this chapter our latest results in the field of real time phenomena recognition and management through our image understanding software platform.

27.1 INTRODUCTION

The interest in direct imaging diagnostics in nuclear fusion devices has gained a lot of attention during the last decades [1–3]. Initially dedicated to specific physics analysis such as thermal studies from targeted in-vessel components monitoring (e.g., divertor tiles), imaging diagnostics are now considered as fully required for both plasma control systems and safety plasma operations [4]. One of the diagnostics tools that can be used to study plasma is examination of shape and behavior of the plasma emission region [5]. Plasma mainly emits radiation in x-ray and vacuum ultra violet spectral regions but near material surfaces the temperature is cool enough for plasma to emit in visible region. This allows analysis using simple diode arrays and CCD detectors. Using filtered view it is possible to determine structure of various line emitters present in the plasma. This information can then be used to study edge phenomenon such as impurity recycling, plasma detachment, MARFE, etc. [6].

The recent years, the authors have witnessed a huge development and widespread use of video and computer technology [7]. Digital video cameras have become a tool of convenience, being used in a variety of devices to record information. Multimedia software and hardware are nowadays standard tools [8] for handling images and videos, allowing information sharing, and with a clear impact on society. Science follows naturally this evolution, image processing becoming a standard scientific tool for various scientific applications, and large experiments. In particular, CCD cameras have become very popular in magnetic confinement nuclear fusion devices. Various types of cameras are routinely used for imaging in Tokamaks, in the IR and visible wavelength regions, for various purposes, from surveying and control to scientific investigations. SST-1 tokamak is also equipped with number of imaging diagnostics. In this chapter the initial results observed during different phases of experimental campaigns are reported. In this chapter only the results from optical imaging diagnostics are discussed.

27.2 EXPERIMENTAL SETUP AND DATA ACQUISITION

The SST-1 (steady state superconducting tokamak) is a mid size tokamak with major radius 1.1 meters and minor radius 0.2 meters. For the initial phase of operation of SST-1 optical imaging diagnostics is installed which give information on plasma size, shape, movement, etc. This diagnostics consists of three CCD cameras. Two cameras have wide angle view which is placed diametrically opposite to cover almost 90% of the vacuum vessel

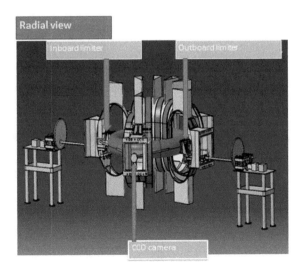

FIGURE 27.1 Radial view of plasma (70° view).

(Figure 27.1). Each camera has a viewing angle of around 70°. The third camera is so position to get the tangential view of the plasma. It gives information about emission distribution in poloidal cross-section (Figure 27.2). Figures 27.1 and 27.2 shows the arrangement of the radial and tangential view respectively on the machine.

27.3   CAMERA SYSTEM

Two micro cameras (Toshiba: IK-UM51H) are placed at two different ports of tokamak vessel for tangential viewing of plasma inner wall. The camera has charge couple device (CCD) sensor with 752 X 582 pixels (4.89 mm X 3.64 mm). The camera system is divided in two parts; (i) camera head, and (ii) camera control unit (CCU). The camera head selected is miniature in size (length = 40 mm, diameter = 12 mm) for better viewing of plasma without any obstruction. The camera control unit (Toshiba: IK-CU51) is connected to camera head using 5 meters data cables. The CCU supports interlace scanning PAL (25 fps) and NTSC (30 fps) camera interfacing with external trigger for synchronization purpose. The CCU gives facility to program the exposure time (100 µsec to 20 msec) using computer via

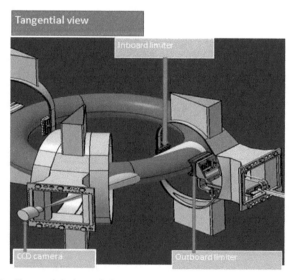

FIGURE 27.2   Tangential view of plasma.

RS232 interfacing option. The cameras are operating in harsh environment; the CCU uses shielded cables for data connectivity to camera for signal integrity and higher signal to noise ratio (46 dB). The video output from CCU is connected to video to fiber optic media converter (SI Tech, 2809). To eliminate noisy ground loops and radio frequency noise of tokamak machine; the electrical signal is converted to optical signal for better signal integrity. At host side the optical signal is converted back to video signal using fiber optic to video converter (SI Tech, 2810). Figure 27.3 shows schematic of visible imaging diagnostic; the power supply and media converter are installed near to tokamak machine and image acquisition system is installed at diagnostic laboratory.

27.4 IMAGE ACQUISITION SYSTEM

The PCI-based image acquisition hardware (NI, PCI 1410) is opted for real time image acquisition to computer memory. The connector block

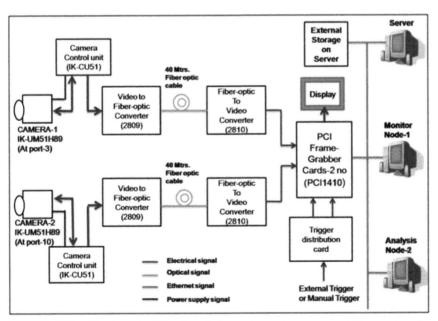

FIGURE 27.3 The layout of developed electronics and image acquisition system for visible imaging diagnostic.

(NI, IMAQ 6822) is used to couple trigger signal and video signal to PCI 1410 card. Two image acquisition cards are used to acquire videos from two cameras located at different viewing port of tokamak. The PCI 1410 cards digitize the video signal to 10-bits resolution with signal to noise ratio of 56 dB. Two image acquisition cards are used to avoid losing frame rates when two or more cameras are acquired simultaneously. The timing diagram for imaging diagnostic is shown in Figure 27.4.

The image acquisition setup follows two modes: (i) continuous mode, and (ii) burst mode. The continuous mode is used for camera alignment purpose and in glow discharging phase of machine. In camera alignment procedure; the standard light source illuminates the inner area of vessel and the camera is projecting those areas. The output of alignment procedure is stored in computer database as reference images. In glow discharge operation, the cameras and image acquisition system is tested for their readiness for actual plasma operation. The hardwired trigger signal (start acquisition command) is not required in continuous mode. The burst mode is used for actual plasma monitoring operation. The trigger signal (–200 msec before plasma breakdown signal) from SST1 timing electronics starts the burst mode operation of visible imaging diagnostic for entered numbers of frames. The graphical user interface (GUI) screen is developed using LabVIEW (National instrument) software. The GUI gives on-line monitoring facility and control parameter input editing screen to change number of frames parameter and status output criteria. The status output facility is developed to give interlock signal in faulty condition. The PCI

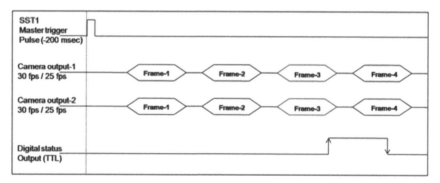

FIGURE 27.4   Timing diagram for visible imaging diagnostic; Red: frame with saturated pixel gives digital output.

1410 board and GUI are programmed in a way that gives digital signal status output when the input image is having saturated pixels. The status output is useful to give feedback to other experiment and for setting exposure time of CCU. The background frames are acquired to subtract environment light from actual plasma information. The offline image analysis is carried out using LabVIEW with image subtraction and edge detection. The offline image analysis is useful for image annotation purpose where the observed feature of plasma is added to image as footnote.

The acquired images are stored in local computer and main server in raw data format using Ethernet connectivity. The MATLAB software is used for extracting plasma parameter from acquired images. The analyzed images are stored in server for uses to different diagnostic as a feedback reference. Figure 27.5 shows the complete process from image detection to image processing.

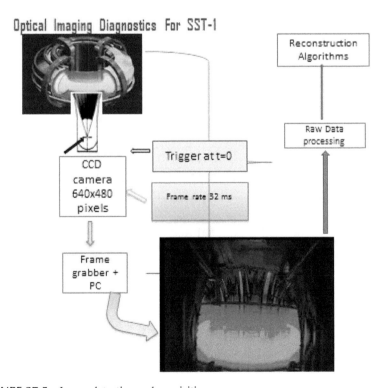

FIGURE 27.5   Image detection and acquisition.

27.5 RESULTS

For acquisition of plasma images during experimental campaigns the GUI runs for full day operation in automatic way (without need of operator). Figure 27.6 shows radial view of the images acquired during experimental campaign of SST-1 tokamak. Images are acquired at frame rate of 40 fps. In this particular shot electron cyclotron resonance heating (ECRH pulse) is of 120 ms. It is observed that during the ECRH pulse duration the vacuum vessel is filled with plasma because the ECRH pulse is launched vertically. After the termination of the ECRH pulse, i.e., 120 ms in this case, the plasma column is formed which is around 50 cm in diameter. This diameter is measured with respect to known coordinates from the machine with the help of optical imaging diagnostics.

Since these were the initial experimental campaigns of SST-1 so it played an important role to diagnose the plasma both from operation point of view and for interpretation of various phenomena. In Figure 27.7 the images from two diametrically opposite cameras are shown. These images

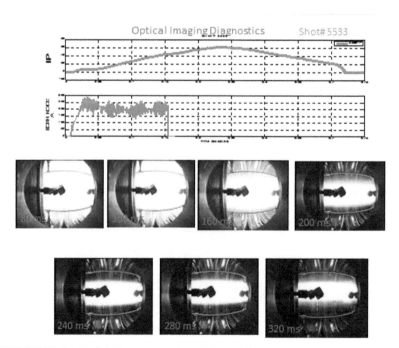

FIGURE 27.6 Radial view-temporal evolution of plasma.

Toroidally symmetric plasma shots

Cam 2   **Cam 1**

FIGURE 27.7   Images from 180° opposite views.

show that plasma in SST-1 is toroidally symmetric. The shape, vertical position, and plasma diameter observed with the two cameras are same.

Figure 27.8 shows the different frames of plasma image from the tangential viewing camera. The tangential viewing camera, if it is H-alpha (656.4 nm) filter, interfere to see distribution in the poloidal cross-section of the plasma. Because of the temperature gradient in the tokamak plasma, the characteristics of the radiation from core to edges changes. Since core is the hottest part, it emits radiation in the soft x-ray region. In the edge the temperature is low enough to radiate in the visible region. In between edge and core UV and VUV radiation are emitted. This distribution gives information on the emission loss from the boundary of the plasma. In first frame the ECRH power was launched so emission is filled vertically in the vacuum vessel and in subsequent frames the plasma current is driven by available loop voltage. The central dark part in the poloidal cross-section is the hot core of the plasma. Since the hot plasma core emits radiation only in soft x-ray region, it is not detected in the visible camera. The edge

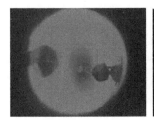 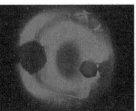 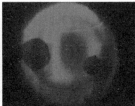

FIGURE 27.8   Tangential view-poloidal cross-section of plasma.

region shows the distribution of H-alpha radiation. This distribution of H-alpha can give information on plasma wall interaction.

27.6 CONCLUSIONS

The optical imaging diagnostics was successfully installed and operated on SST-1 machine. Both the radial and tangential views of the optical imaging diagnostics provided useful information on SST-1 plasma during experimental campaign. Toroidal symmetry could be established with the help of two diametrically opposite cameras. Plasma diameter could be estimated from the radial view of imaging. H-alpha (656.4 nm) distribution in the poloidal cross-section of the plasma was observed.

KEYWORDS

- **magnetically confined plasma**
- **optical imaging**
- **plasma**
- **plasma diagnostics**
- **plasma parameters**
- **tangential view**
- **Tokamaks**

REFERENCES

1. Reichle, R., Andrew, P., Counsell, G., Drevon, J.-M., Encheva, A., Janeschitz, G., Johnson, D., Kusama, Y., Levesy, B., Martin, A., Pitcher, C., Pitts, R., Thomas, D., Vayakis, G., & Walsh, M. (2010). "Defining the infrared systems for ITER," *Review of Scientific Instruments, 81*(10), 130.
2. Reichle, R., Andrew, P., Counsell, R. G., Drevon, J.-M., Encheva, A., Janeschitz, G., Johnson, D., Kusama, Y., Levesy, B., Martin, A., Pitcher, C. S., Pitts, R., Thomas, D., Vayakis, G., & Walsh, M. (2010). "Defining the infrared systems for ITER," *Review of Scientific Instruments, 81*(10), 135.

3. Arena, P., Basile, A., De Angelis, R., Fortuna, L., Mazzitelli, G., Migliori, S., Vagliasindi, G., & Zammataro, M. (2005). "Real time monitoring of radiation instabilities in Tokamak machines via CNNs," Plasma Science, *IEEE Transactions, 33* (3), 1106–1114.

4. Ghendrih, P., Sarazin, Y., Becoulet, M., Huysmans, G., Benkadda, S., Beyer, P., Figarella, C., Garbet, X., & Monier-Garbet, P. (2003). "Patterns of ELM impacts on the JET wall components," *Journal of Nuclear Materials, 313–316,* 914–918.

5. Alonso, J. A., Andrew, P., Neto, A., de Pablos, J. L., de la Cal, E., Fernandes, A., Fundamenski, W., Hidalgo, C., Kocsis, G., Murari, A., Petravich, G., Pitts, R. A., Rios, L., & Silva, C. (2009). "Fast visible imaging of ELM-wall interactions on JET," *Journal of Nuclear Materials, 390*(91), 797–800.

6. Murari, A., Camplani, M., Cannas, B., Mazon, D., Delaunay, F., Usai, P., & Delmond, J. (2010). "Algorithms for the automatic identification of MARFEs and UFOs in JET database of visible camera videos," *IEEE Trans. On Plasma Science, 38*(12), 3409–3418.

7. Maqueda, R., Maingi, R., Tritz, K., Lee, K., Bush, C., Fredrickson, E., Menard, J., Roquemore, A., Sabbagh, S., & Zweben, S. (2007). "Structure of MARFEs and ELMs in NSTX," *Journal of Nuclear Materials, 363–365,* 1000–1005.

8. Love, N. S., & Kamath, C. (2007). "Image analysis for the identification ofcoherent structures in plasma," *Proc. SPIE 6696, 66960D.*

CHAPTER 28

PULSED ELECTRICAL EXPLODING WIRE FOR PRODUCTION OF NANOPOWDERS

S. BORTHAKUR, N. TALUKDAR, N. K. NEOG, and
T. K. BORTHAKUR

Centre of Plasma Physics, Institute for Plasma Research, Sonapur, Kamrup, Assam – 782402, India, E-mail: tkborthakur@yahoo.co.uk

CONTENTS

ABSTRACT

Pulsed electrical exploding wire (PEEW) for production of various kinds of nanopowders is a well-known method. In this chapter, development of a PEEW system for production of copper nano powder is discussed.

The pulsed power system that is used in this experiment consist of a High voltage (HV) energy storage capacitor of rating 10 µF, 20 kV; an ignitron switch with driver circuit, controls and related HV charger. The Capacitor is charged up to 8 kV and it is discharged through the ignitron switch into the copper wire. The voltage and the current are measured by HV probe and a current transformer respectively. Copper wires of which, the two ends are connected to two metallic rods to the pulsed power input are placed inside an evacuated chamber. After discharge, the wires are exploded, melted, and condensed in collision with ambient air forming tiny particles of copper wire. Collection of nano particles produced is thereby done at the bottom of the chamber. A special arrangement is done to fix multiple copper wires inside the chamber and discharge it one by one without breaking the vacuum. The collected copper particles are observed in SEM, which shows formation of nano-sized particles.

28.1 INTRODUCTION

Nanopowders possess interesting physical and chemical properties of a material that are not found in their bulk form. Because of these properties, nanosized powders have a very wide application in magnetic media, catalysts, gas sensors and other numerous industrial fields. There are various physical methods for the synthesis of nano powder in addition to chemical methods. The physical methods, such as the laser ablation method [1] and thermal plasma method [2], pulsed electrical exploding wire method [3–6] have been developed to produce high-purity nanopowders. Among these, pulsed electrical exploding wire (PEEW) method is a well known, highly efficient, low-cost and high-purity nanosized powder production technique. This is basically a physicochemical technique which uses high-density metal vapor plasma to obtain fine particles in the micron, sub-micron and nano size range. In this process, nano particles are generated by the rapid heating of a metal wire by a large pulsed current which is usually produced by the discharge of a capacitor bank. When a capacitor bank is charged to a specific voltage, energy is stored inside it in the form of electrical energy. This energy is released on the thin metal from the capacitor in the form of a pulse with an instantaneous high density energy and current. As soon as this high-density current pulse passes through the thin metal wire, the density of energy in the wire considerably exceeds the binding energy because of the high rate of energy injection and expansion lag upon the thin wire. Henceforth, the material melts up and evaporates giving a bright light flash, and then a mixture of superheated vapor along with boiling droplets of the exploding wire material with a shockwave, scatter to the ambient atmosphere. This high temperature superheated vapor is the plasma of the material which gradually cools due to the interaction with background gas molecules resulting in a vapor of the wire material that finally condenses uniformly to form nano sized particles. The production of ultrafine powders by the exploding wire technique is not a new technique. It was apparently begun by Abrams in 1946, who studied radioactive Al, U, and Pu aerosols. Since then, the wire explosion method is used by researchers for the synthesizing of nanoparticles of different metals, metal oxides, metal nitrides and alloys [7–10]. Several studies [11–14] have been reported to study the effect of different experimental parameters on the characteristics of the nanoparticles produced by the wire explosion process. Among these parameters are the ambient gas species and pressure, amount of deposited energy and the initial crystalline structure of the wire material.

In this chapter, we have described our experiment setup of pulsed electrical wire discharge and discussed a demonstrative experiment to produce copper powders by this method. In our experiment, we have placed some glass slides near the wire, so that the metal vapor deposits over its surface during condensation. This process involves melting, boiling, evaporation and condensation of the wire one after the other, ultimately yielding nano particles. The amount of energy applied for explosion is set depending on the physical and chemical properties of the material selected which is described here. This method has the advantage of high energy conversion efficiency for the generation of metal vapor plasma and high production rate for the preparation of nanoparticles. In addition, in this technique we have the flexibility of controlling the size of the prepared nano particles by controlling the pressure of the ambient gas and input energy.

28.2 EXPERIMENTAL SETUP

The basic circuit diagram of the PEEW system is shown in Figure 28.1. It consists of a capacitor bank, HV DC charger, a chamber, a rotary pump and a HV switch. The capacitor bank is charged from the HV DC charger and discharged through the metal wire (to form powder) via the HV switch. The explosion of the wire takes place when the capacitor energy is higher than the evaporation energy of the material. The total energy required to evaporate the metal wire is the sum of the melting energy, latent heat of fusion, evaporation energy and latent heat of evaporation. In our experiment, we have used copper wires of diameter 0.3 mm and length 40 mm. The calculated total energy required to evaporate the wire is found to be around 151 Joule. In practice, the required energy for explosion of the wire and to form plasma is significantly greater than the calculated value. In our experiment, we have applied energy twice more than the vaporization energy of the copper wire. To achieve this criterion, we have used a capacitor of 10 μF which is charged to 8 kV so that the discharged energy of the capacitor becomes 320 Joule.

Essentially, the wire is placed inside a vacuum chamber of diameter 150 mm and length 210 mm. The chamber is evacuated by a rotary pump. A special arrangement is made inside the vacuum chamber to fit 8 numbers of copper wires by connecting it's both ends to two metallic rods, to the pulsed power input and exploded one by one without breaking vacuum inside the chamber. The photograph of the experimental setup and the schematic diagram of the wire holder are shown in Figures 28.2(a) and 28.2(b),

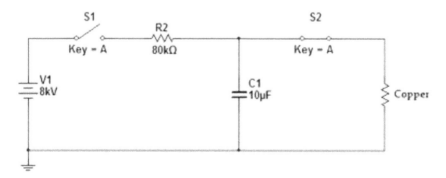

FIGURE 28.1   Basic circuit diagram of the PEEW system.

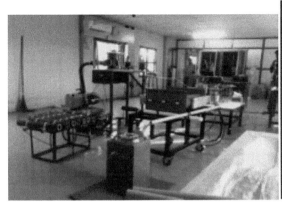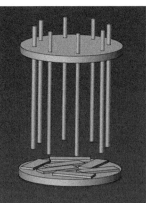

FIGURE 28.2   (a) Photograph of the experimental setup. (b) Schematic diagram of the wire holder.

respectively. To discharge the capacitor to the load wire, a HV switch called as ignitron is used which is placed inside a squirrel cage holder.

28.3   RESULTS AND DISCUSSIONS

As stated earlier, experiments were done to explode electrical grade copper wires of length 40 mm and diameter 0.3 mm. The discharge was made inside the vacuum chamber at 100 mbar of pressure in air. The capacitor is discharged through the HV ignitron switch to the load (copper wire). Four numbers of URM 67 cables, each 3 m long along with tin coated copper braids were attached to the discharge path to shape the output current pulse. The peak discharge current was found to be around 10 kA. A typical time evolution of discharge current and voltage are shown in Figures 28.3(a) and 28.3(b). The small hump in the voltage wave form and the small dip in the discharge current wave form at around 11.5 µs indicate the explosion of the wire due to joule heating. At that instant, the energy deposition exceeds the total energy required for evaporation of the wire resulting breakup of the wire. Due to breakup of the wire, a sudden discharge current drop (current interruption) occurs resulting in a dip formation in the discharge current signal and a sudden rise in the voltage. The energy deposited during the heating of the wire is calculated from the discharge voltage and the current as shown in Figure 28.3(c). The curve shows that the deposited energy at the time explosion of the wire was around 207 joule which is higher than the required evaporation energy (151 joule) of the wire and is around 65% of applied energy of the capacitor. Just after the moment of explosion, the formation of arc plasma takes place that acts as the conducting medium and completes the discharge further. During this period, the metal plasma starts to interacts with the background gas medium and condenses to form tiny particles. For collection of these particles we have placed some small glass chips of dimension 1 sq. cm near the explosion area so that the fine particles form a coat over it. These glass samples were examined under scanning electron microscope (SEM). A SEM micrograph of samples is shown in Figure 28.4. The SEM analysis shows that the particles are in the micron, sub-micron, and nano-scale size. Typically size varied from 500 nm to 50 nm. Moreover, the particles were found to be of similar shape when observed locally. At present there is no valid argument supporting that the particles are purely of

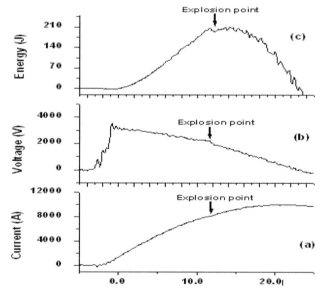

FIGURE 28.3   (a) Discharge current (b) discharge voltage, and (c) energy deposited.

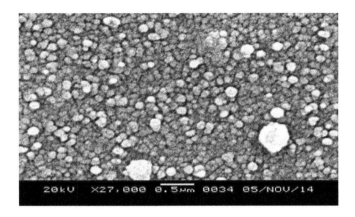

FIGURE 28.4   SEM micrograph of samples.

copper only and not any copper compound. To estimate this we have to do the XRD analysis of these samples. Experiments are being carried out to collect sufficient amount of specimen samples for XRD analysis. For this, the present setup has to be modified considerably to increase the number of copper wire for explosion. Also, the nano particles that are expected to form is difficult to collect. This is due to the fact that the nano particles are accelerated away just after explosion and gets scattered in distant atmosphere.

28.4 CONCLUSION

A PEEW system for production of copper nano powder was successfully established for the production of metal powder. Copper wire of length 40 mm and diameter 0.3 mm was exploded by discharging a capacitor at 8 kV, giving a peak discharge current of 10 kA. The energy deposited during the discharge was calculated from discharge signals and it is found to be around 207 joule during the explosion, which is 65% of the input energy. The tiny particles generated during the process were collected for SEM analysis. The SEM micrograph shows formation of nano particles embedded amongst larger particles. Further analysis of XRD will give us the information of the nature of the material.

ACKNOWLEDGMENT

We acknowledge the Centre Director of Centre of Plasma Physics, Institute for Plasma Research, Assam for providing the necessary infrastructure and financial assistance for successfully carrying out the research work. The assistance of Mr. N. Kathar is highly appreciated.

KEYWORDS

- energy storage capacitor
- nano particles
- pulsed electrical exploding of wire (PEEW)
- pulsed plasma devices
- pulsed power
- scanning electron microscope (SEM)

REFERENCES

1. Maser, W. K., Muñoz, E., Benito, A. M., Martínez, M. T., de la Fuente, G. F., Maniette, Y., Anglaret, E., & Sauvajol, J. L. (1998). Production of high-density single-walled nanotube material by a simple laser-ablation method. *Chemical Physics Letters, 292*(4–6), 587–593.
2. Bora, B., Aomoa, N., Bordoloi, R. K., Srivastava, D. N., Bhuyan, H., Das, A. K., & Kakati, M. (2012). Free-flowing, transparent g-alumina nanoparticles synthesized by a supersonic thermal plasma expansion process. *Current Applied Physics, 12*, 880-884.
3. Kotov, Y. A. (2003). Electric explosion of wires as a method for preparation of nanopowders. *Journal of Nanoparticle Research, 5*, 539–550.
4. Sindhu, T., Sarathi, R., & Chakravarty, S. (2007). Generation and characterization of nano aluminum powder obtained through wire explosion process. *Bulletin of Materials Science, 30*, 187–195.
5. Channarong, S., Yoshiaki, K., Tsuneo, S., Weihua, J., & Kiyoshi, Y. (2001). Synthesis of Nanosize Powders of Aluminum Nitride by Pulsed Wire Discharge. *Japanese Journal of Applied Physics 40*, 1070–1072.
6. Yilmaz, F., Lee, D., Song, J., Hong, H., Son, H., Yoon, J., & Hong, S. (2013). Fabrication of cobalt nano-particles by pulsed wire evaporation method in nitrogen atmosphere. *Powder Technology, 235*, 1047–1052.

7. Lee, Y. S., Bora, B., Yap, S. L., Wong, C. S., Bhuyan, H., & Favre, M. (2012). Investigation on effect of ambient pressure in wire explosion process for synthesis of copper nanoparticles by optical emission spectroscopy. *Powder Technology, 222*, 95–100.

8. An, V., Bozheyev, F., Richecoeur, F., & Irtegov, Y. (2011). Synthesis and characterization of nanolamellar tungsten and molybdenum disulfides. *Materials Letters 65*, 2381–2383.

9. Satoru, I., Hisayuki, S., Tadachika, N., Tsuneo, S., & Koichi, N. (2011). Nano-sized particles formed by pulsed discharge of powders. *Materials Letters, 67*, 289–292.

10. Lee, Y. S., Bora, B., Yap, S. L., & Wong, C. S., (2012). Effect of ambient air pressure on synthesis of copper and copper oxide nanoparticles by wire explosion process. *Current Applied Physics, 12*, 199–203.

11. Kwon, Y. S., Ilyin, A. P., Tikhonov, D. V., Yablunovsky, G. V., & An, V. V., (2008). Characteristics of nanopowders produced by wire electrical explosion of tinned copper conductor in argon. *Materials Letters, 62*, 3143–3145.

12. Mao, Z., Zou, X., Wang, X., Liu, X., & Jiang, W. (2009). Circuit simulation of the behavior of exploding wires for nano-powder production. *Laser and Particle Beams, 27*, 49–55.

13. Sedoi, V. S., & Ivanov, Y. F., (2008). Particles and crystallites under electrical explosion of wires. *Nanotechnology, 19*, 145710.

14. Sarathi, R., Sindhu, T. K., & Chakravarthy, S. R., (2007). Influence of binary gas on particle formation by wire explosion process. *Materials Letters, 61*, 1823–1826.

INDEX

T - #0805 - 101024 - C468 - 229/152/21 - PB - 9781774630433 - Gloss Lamination